The Great Dinosaur Controversy

The Great Dinosaur Controversy

A Guide to the Debates

Keith M. Parsons

A B C ⬤ C L I O

Santa Barbara, California
Denver, Colorado
Oxford, England

Library of Congress Cataloging-in-Publication Data
Parsons, Keith M., 1952–
    The great dinosaur controversy : a guide to the debates / Keith M. Parsons.
        p.   cm.——(Controversies in science)
    Includes bibliographical references and index.
    ISBN 1-57607-922-8 (hardcover: alk. paper) ISBN 1-57607-923-6 (e-book)
    1. Dinosaurs. I. Title. II. Series.
QE861.4.P38   2003
567.9——dc22   2003017991

09   08   07   06   05   04   10 9 8 7 6 5 4 3 2 1

ABC-CLIO, Inc.
140 Cremona Drive, P.O. Box 1911
Santa Barbara, California 93116-1911

This book is printed on acid-free paper ∞.
Manufactured in the United States of America.

Contents

Preface

A vertebrate paleontologist I knew at the Carnegie Museum of Natural History in Pittsburgh used to wear a T-shirt imprinted with the motto "I hate dinosaurs." Many paleontologists feel this way—or at least they feel considerable ambivalence. On the one hand, dinosaurs are great public relations for paleontology. People will flock to see dinosaurs when they would not cross the street to see the greatest trilobite collection in the world. Dinosaur merchandise flies off the shelves in museum gift shops, and nothing garners media exposure like a new dinosaur display. On the other hand, it has to be galling to see 98 percent of the public's attention going to a group of creatures that most paleontologists do not regard as all that important. As one paleontologist explained it to me, to really learn about evolution you need to look at rodent teeth. Rodents outnumbered dinosaurs by many orders of magnitude, so their teeth are much more abundantly preserved than dinosaur bones. I sympathized with her argument, but, alas, like the vast majority of nonpaleontologists, I just cannot find rodent teeth as exciting as dinosaurs.

Dinosaurs elicit deep passion, and it is a mystery just why they do so. One reason, of course, is their stupendous size. When standing beside a skeleton as long as two city buses parked end to end, it is natural to feel a sense of awe. Not long ago I read a book by a philosopher who said that sheer size did not awe him; after all, a fat man is not necessarily more impressive than a thin one. Maybe not, but if the Grand Canyon were only twenty feet deep, it just would not be grand. More importantly than sheer size, dinosaurs and the other enormous Mesozoic creatures such as pterodactyls and mosasaurs strike a deep mythical resonance within us. They are storybook creatures that really lived—gargoyles, dragons, griffins, and sea serpents that once stalked the land, sea, and air of our planet.

Some of the historical conflicts between dinosaur specialists have taken on a mythical or at least legendary air. The nineteenth-century "bone war" between O. C. Marsh and E. D. Cope is the most famous example. This book focuses on the scientific debates over dinosaurs, but the Cope-Marsh feud was hardly a scientific debate. It was an intensely personal rivalry fueled not by theoretical disagreements but by simple hatred. So the Cope-Marsh conflict is really the odd one out among the debates I have included for study. However, it is so well known that a book dealing with the great dinosaur debates simply could not omit it. I have tried to put that debate in the scientific context of the times.

The other debates I examine certainly all had their share of personal animosity—sometimes to the point of abandoning all dignity and decency. I have tried to depict the scientific aspects of the debates as paramount, as I believe they in fact were. That is, the *science* is the main story that needs to be told. However, I have endeavored to indicate where I think personal feeling and the ambient social and political environment influenced the course of these debates. As I emphasize in Chapter 1, I wholeheartedly repudiate the currently fashionable "social constructivist" views of science. No one can deny, though, that personal feelings and social influences have deeply shaped the course of scientific controversies, so that part of the story has to be told. I go to some lengths, especially in the early chapters, which are set in the nineteenth and early twentieth centuries, to give the historical background to these disputes.

This is a book for students, and for the general educated public. My aim was to write a book that would be appropriate for college freshmen and sophomores. This book is not intended for professional paleontologists or historians, philosophers, or sociologists of science (though, of course, I hope such individuals will find much here to their liking). In writing this book, I have drawn upon my twenty years' experience of teaching to guide my judgment about how deeply to go into issues and how much to tax the reader's attention span. I have not presupposed any previous acquaintance with dinosaurs or the history of these debates. I have attempted to explain all technicalities either in the text or in the two appendixes. Since not every reader was once a six-year-old dinosaur expert, the first appendix is an introduction to the dinosaurs and their world. A glossary is provided for the reader's convenience.

This book is not a comprehensive history of dinosaur paleontology, so many important episodes, debates, discoveries, and persons are not mentioned at all. My intention is to examine some of the most prominent disputes with an eye to showing how debate, controversy, and even personal animosity affect the course of science. The first chapter explains why I consider the study of scientific controversy important.

I have written in a more personal and informal style than is usual for academic books. I have found this style useful in my classroom teaching. The arid "textbook" style of most academic publications is worse than useless. Prose that cannot convey the author's own excitement and enthusiasm is positively harmful—and a frightful bore to read or to write. Therefore I have not tried to write in a detached, neutral manner, or to be careful to hide my own feelings, whether positive or negative.

In particular, when lecturing to my students about any subject dealing with the history of life, I am quickly made aware of the pervasive influence of creationist propaganda in insulating my students' minds against knowledge of evolution. I consider creationism in all of its guises—whether "young earth" fundamentalism or the allegedly more sophisticated "intelligent design" variety—to be pernicious nonsense. I find that before I can instruct my students on evolution, I have to disabuse them of large amounts of misinformation and disinformation spread by creationists. Therefore, I have included a second appendix explaining what evolution is, how it is known to be true, and why the arguments against it are empty.

There are many people to thank for their support while writing this book. Art Stickney and the staff of ABC-CLIO have been very helpful and professional in all of their interactions with me. I particularly appreciate their patience as I struggled to finish this project while shouldering a heavy teaching load. Connie Oehring did an excellent job of copyediting and improved the book's readability in numerous places.

When you write a book, you will inevitably try the patience of people about whom you care a great deal. My wife, Carol, has been most understanding when I had to stay late or go to the office on weekends to get everything finished. My colleagues on the faculty of the University of Houston, Clear Lake, have also had to bear my occasional refusals when asked to attend meetings or partake in committee work. To get anything written, or just to keep my sanity, sometimes I had to just say "no" when asked to take on certain service obligations. I thank my colleagues for their patience and understanding.

In the early and mid-1990s, I was a research associate in the Department of Vertebrate Paleontology at the Carnegie Museum of Natural History in Pittsburgh and a graduate student in the Department of History and Philosophy of Science at the University of Pittsburgh. I would like to thank my friends and colleagues from both institutions, including especially Mary Dawson, Chris Beard, and Jim Lennox.

Their guidance and friendship greatly helped me to get the knowledge and experience I needed to write this book.

Also, I owe many thanks to my mother, Charlotte Blanton Parsons, who always encouraged and supported me in my circuitous and lengthy path to get where I am now. To her this book is lovingly dedicated.

1

Why All the Fuss?

Dinosaurs have always been controversial, and this book is about some of the biggest of the dinosaur debates. Paleontologists have disagreed about the appearance, anatomy, posture, behavior, taxonomy, evolution, physiology, ecology, and reasons for extinction of dinosaurs. These disagreements have often been passionate and sometimes bitter, intensified by personal enmity and resulting in bruised egos and tarnished reputations. If you listen to paleontologists in unguarded moments, you might be surprised at their disdainful comments about some of their colleagues. When I once asked some paleontologist friends about a controversial colleague, the kindest assessment I got was "a total jackass."

So, are paleontologists more cantankerous or peevish than other scientists? Do chemists or physicists have fewer conflicts or resolve their disagreements more quickly and with less acrimony? I have never seen a formal study addressing these questions, but the answer suggested by the history of science is "no." Today's reigning theories had to overcome determined opposition before they were accepted—and sometimes an intransigent rearguard continued the battle even after that. The receptions of the Copernican and Darwinian theories are cases in point. The opposition to these theories was loud and long; with respect to the latter, it has not died out completely even today. Every major figure in the history of science had to face down powerful and outspoken critics. When you look at the writings of those scientists, you find many passages that answer criticisms or attack opponents.

Why do scientists engage in such controversy? Early modern defenders of science such as Francis Bacon thought that scientists, unlike theologians or philosophers, would not engage in doctrinaire squabbles. They viewed controversy as a sign of ignorance. Politicians, lawyers, and theologians debate endlessly because the issues they dispute are truly moot questions—matters of opinion for which there are no hard data and no clearly marked paths to genuine knowledge. Compare this situation to two people who disagree about whether a column of numbers adds up to a certain sum or whether New Guinea lies on the equator. The former opponents pull out a calculator, and the latter get an atlas. In these cases the way to get a reliable answer is clear, and once it is obtained there is no room for further debate—unless somebody just wants to make trouble.

Bacon believed that answering scientific questions should be as straightforward and uncontroversial as calculating or consulting the atlas. Bacon recommended that scientists compile large numbers of facts and then generalize on the basis of observed correlations among those facts. Once we have a set of reliable generalizations at one level, we look for higher-order correlations among lower-level generalizations. If we take care to observe differences also, thereby avoiding accidental correlations, we can ascend to ever more general knowledge. There is no room for subjectivity or bias in making such comparisons and generalizations. The procedure is dispassionate, impersonal, and cooperative, providing no occasion for rivalry or resentment. Once we have Scientific Method and consistently apply it, controversy will wither and rapid consensus will emerge.

But scientists do argue, stridently and at length. We need to address three questions: (1) Why are there scientific controversies, and are they avoidable? (2) What part does controversy play in making science a rational enterprise? (3) How do controversies end?

Why do scientists argue? History has dashed the hopes of scientific methodologists from Francis Bacon in the seventeenth century to Rudolf Carnap in the twentieth. No one has succeeded, or seems the least likely to succeed, in elaborating a single, universal, distinctive Scientific Method applicable to all and only the natural sciences. There is, and apparently can be, no rule book of how to do science. Of course, scientists do frame hypotheses to explain puzzling phenomena and then seek to evaluate those hypotheses as critically and rigorously as possible. But historians, detectives, and even auto mechanics do that. A true Scientific Method would have to give rules specific enough to be of actual use to working scientists in their daily investigations. Those rules would also have to be general enough to be useful in all fields of science from archaeology to zoology. Telling a scientist that his or her method is to frame bold hypotheses and rigorously test them is like advising a stockbroker to buy low and sell high.

Such truisms provide nothing useful to the scientist struggling to comprehend complex and baffling phenomena or a stockbroker grappling with the vagaries of the market.

Now, to say that there is no single, universal, distinctive Scientific Method does not imply, as one philosophical maverick charged, that the lesson of the history of science is that "anything goes" (Feyerabend 1975). There may be no Scientific Method, but there are all sorts of scientific methods employed by real scientists to answer real questions. Nature appears very complex and presents us with deep challenges at many different levels, so the various scientific disciplines must employ methods appropriate for their particular subject matter. For instance, consider the efforts of astronomers to detect planets in other solar systems (Croswell 1997). Such extrasolar planets cannot be seen directly by any telescope because their dim light is swamped by the brilliance of the planet's home star. The only way to detect a planet revolving around a distant star is to watch the star and identify the distinctive effect of the planet's gravitational tug on the star's motion. A planet is so small compared to the mass of its star that the effect of the planet's revolution on the star's motion is minuscule. However, very careful analysis of tiny wobbles in the star's movement can sometimes reveal the presence of a planetary companion.

Again, such measurements are very hard to make; the effects are very subtle, and it is always difficult, when working near the limits of our capacities (which cutting-edge science, almost by definition, always does), to distinguish signal from noise. Pious platitudes about bold hypotheses and rigorous tests are no use at all to an astronomer trying to detect extrasolar planets. A method that such an astronomer could really use would almost certainly be of no use at all to, say, a field biologist trying to detect changes in beak size of successive populations of finches. So the quest for a Scientific Method that would really be useful to working scientists in all fields seems quixotic at best.

The upshot is that if there is no authoritative rule book for doing science, no infallible guide to good scientific practice, it is not surprising that scientists, even within a particular discipline, often disagree on the appropriate way to investigate phenomena. Which methods are better? Which ones give us cleaner, more accurate, or more useful data? How do we weigh the evidence when we apply two different methods and the results disagree? Does a given method really do what it is supposed to do? What are the limitations of a method? These and many other methodological questions crop up constantly in science and are a frequent basis of controversy.

But there are even deeper problems with doing science than merely the lack of a universal Scientific Method. Scientific theories are, of course, based on evidence. A theory for which there is little evidence may be entertained provisionally as a working hypothesis, but it had better get some solid evidence pretty quickly or it is headed for trouble. Yet the relationship between theory and evidence is complicated and subject to numerous limitations and qualifications.

One thing that is clear is that no amount of evidence ever simply entails a theory. Entailment is a logical relation where if certain things are true, then other things *must* be true. For instance, if we know (a) that Senator Claghorn is either a fool or a liar, and (b) that Senator Claghorn is not a fool, then we know (c) that Senator Claghorn is a liar. If (a) and (b) are true, (c) *must* be true; (a) and (b) jointly entail (c). But in science such a relationship never holds between a body of evidence (e) and a theory (T). E may give us excellent reason for believing T (or at least believing T over a rival theory T'), but it is not the case that if e is true, then T *must* be true.

The upshot is that, as philosophers like to put it, a theory is always *underdetermined* by the evidence. When we accept a given theory it is always at least conceivable that a completely different and incompatible theory could explain everything as well as the received theory. This may seem like philosophical nitpicking since the fact that two theories are *compatible* with the evidence does not mean that each is equally well *supported by* the evidence. In fact, to working scientists, the philosopher's worries about underdetermination may seem perverse. Instead of an embarrassment of riches where they have to choose between many theories that explain the data, scientists often have a hard time finding even *one* that will do the job! Still, the evidence for a theory is often ambiguous. As the accompanying illustration shows, what looks like a duck to one researcher may look like a rabbit to another. Even "hard" data often admit various interpretations. Solid, high-quality evidence may strictly constrain our theorizing, but there is almost always some wiggle room for the dissident. So concerns about the relation of theory to data are not merely the philosopher's worry.

In the actual course of science, alternative hypotheses are gradually weeded out until one emerges triumphant. Scientific communities determine that one and only one of the competing hypotheses is the best supported by the evidence. But even if one regards such collective theory-choice decisions by scientific communities as eminently rational and reasonable—as I do—these decisions are still *deci-*

The "duck/rabbit" from Darwin for Beginners by Jonathan Miller and Borin van Loon. New York: Pantheon Books (1982). Illustrations copyright © 1982 by Borin van Loon.

sions; the evidence does not *compel* anyone to accept the theory. In other words, the weight of the evidence for a scientific theory may be more than adequate to *rationally persuade* skeptics, but it seldom if ever can *compel* consensus. When a scientific community moves to consensus, the consensus is seldom 100 percent. There will almost always be holdouts, and a resolute skeptic can almost always finds some rational grounds for doubt. Of course, anything can be taken to absurd extremes. Anyone, for instance, who still denied that the blood circulates would be a crackpot. Still, there is often no clear dividing line between rational dissent and crackpottery.

So controversy in science is inescapable. Once we realize that there is no universally prescribed Scientific Method and that scientific evidence serves to rationally persuade but not to compel, the inevitability of scientific debate becomes apparent. Suppose we have a large batch of data, and two theories, T and T', are offered by rival scientists to explain those data. T and T' both have evidence in their favor, and there is no automatic procedure or set of rules to apply to decide between them. The evidence supporting each theory may be strong, but not strong enough to compel skeptics to give in. Which theory should the relevant scientific community accept? Obviously, there is no alternative except to have the opposing sides argue it out, allowing each side to bring in as many observations, experiments, methods, and techniques as it can adduce in favor of its theory. When the opposing counsels have had their say, the jury—the scientific community—renders a verdict.

Controversy may be unavoidable, but is it an essential ingredient of scientific rationality? Consider the following two sentences: (a) Natural science, and such formal disciplines as logic and mathematics, are the highest expressions of collective human rationality. (b) In no other fields of human endeavor are knowledge claims so well grounded as in the mathematical and natural sciences. A few years ago hardly any educated person would have questioned (a) or (b). Such is no longer the case. In the past two or three decades vocal critics from the "academic left" have subjected science to vigorous and often hostile critique. These critics, including radical feminists, radical environmentalists, "postmodernists," and "social constructivists," are often hard to understand clearly because their prose is frequently nebulous, verbose, or laden with jargon. Yet their aim is clear—to debunk science, to cut it down to size, to expose science as no more "rational" or "objective" than any other field of human inquiry or system of beliefs (see Parsons 2001, 2003).

One of the earliest, and still most influential, of these recent science critiques was *Laboratory Life: The Construction of Scientific Facts* by Bruno Latour and Steve Woolgar (first published in 1979). Bruno Latour, a sociologist, got a menial job at the famous Salk Institute, where he observed scientists in their native habitat—like the anthropologist who lives with a rain-forest tribe to study its odd customs. One of the native practices he frequently observed was that scientists argued a lot. When he and Woolgar wrote their book, they devoted much space to an analysis of the nature and function of scientific controversy.

According to Latour and Woolgar, scientific debate serves a dubious if not downright insidious purpose. When a scientific hypothesis is first formulated, everyone recognizes its provisional status as an unverified and speculative proposal. As debate proceeds, proponents of the hypothesis employ all sorts of tactics to overcome the opposition of skeptics, using many rhetorical devices to persuade, intimidate, or silence opponents. Now, it is important to realize that for Latour and Woolgar scientific methods are not, as usually conceived, reliable means for comparing the predictions of hypotheses to the realities of the natural world. Indeed, for Latour and Woolgar "nature" and "reality" are merely whatever scientists finally agree that they are, not something existing prior to and presupposed by scientific inquiry.[1] Rather, scientific methods are *merely* rhetorical devices useful in the practical business of prevailing over enemies but in no sense providing objective constraint on theories.

---

1. Should anyone think that I am making this up and that no one could possibly seriously affirm this notion, please do read the book.

According to Latour and Woolgar, once supporters of a hypothesis have succeeded in overcoming their opponents through rhetorical devices and other tricks, a curious thing occurs. The hypothesis loses its hypothetical status and is promoted to a "fact." When such an "inversion" occurs, members of a scientific community forget that their agreement on the hypothesis is a social construct, created by rhetorical manipulation and hardball politics, and see the hypothesis as something that has been "out there" waiting to be discovered all along. So scientific facts are not discovered; they are made, and scientific debate is the means of their making. Scientific debate is a smokescreen hiding the constructed nature of "facts" and misleading people into thinking that they stand for something objective.

Latour and Woolgar's view is supported by another influential sociological study of science, *Leviathan and the Air-Pump* by Steven Shapin and Simon Schaffer (1986). Shapin and Schaffer examine the debate in the 1660s between the new breed of experimental scientists such as Robert Boyle and the curmudgeonly philosopher Thomas Hobbes. Boyle and his colleagues devised an air pump that could create a fairly good vacuum. They used the pump to perform a number of striking experiments and concluded that the proper method of science was experimental, that is, that controlled experiment was far more useful for science than uncontrolled and commonsense observation. Hobbes loudly objected, claiming that to constitute genuine and reliable knowledge science must be confirmed by common and public observations, not the word of a scientific elite working with strange equipment only they know how to build or operate.

Boyle won this debate, and experimental methods have been an integral part of science ever since. However, according to Shapin and Schaffer, Boyle won not because his method clearly *was* better but because he played the political game much better than Hobbes. Hobbes had a bad reputation as an alleged atheist and the author of the ferociously authoritarian classic of political philosophy, *Leviathan*. Boyle, on the other hand, was famously pious and the friend of many in high places. Shapin and Schaffer write that it was Boyle's social connections and political allies that led to his victory over Hobbes. The upshot, as they see it, is that the very methods of science, the very means whereby scientists confront hypotheses with data from the real world, are social constructs. So the very basis of scientific rationality—science's methods, techniques, and standards of objectivity and rigor—is a product of the rough-and-tumble of politics, and hence is socially constructed through and through. Scientific methods are historically contingent rules of the game, adopted in response to local social and political conditions. Had the politics gone differently in the 1660s, science might still be following the Hobbesian rather than the experimental model.

We have already seen that science does not proceed by the application of a universal Scientific Method and that no theory is ever entailed by the evidence. Does this mean that we should simply follow Latour, Woolgar, Shapin, and Schaffer in seeing scientific pretensions to rationality as a smokescreen and scientific controversy as the machine blowing the smoke? Not at all. Just because there is no universal Scientific Method does not mean that all scientific methods are merely arbitrary rules of the game. The social constructivist view rests on a *causal* claim. Such theorists claim that scientists adopt a particular method because of ambient social and political factors. Scientists say that they adopt a new method because it is better than the old ones.

Of course, scientists' self-reports do not have to be accepted at face value. Maybe, as Latour and Woolgar think, scientists just fool themselves into thinking that their methods and standards are more than tools of propaganda. Still, science apparently has been quite successful in providing rigorously tested theories that economically explain otherwise incomprehensible phenomena. Even more obviously, science has succeeded in producing some spectacular (if occasionally horrifying) results—from wonder drugs to nuclear weapons. So, at first view science has a pretty impressive track record—or so it seems to many of us who are not social constructivists. It therefore seems that scientists must be doing *something* right and that their methods cannot be just politically expedient hocus-pocus or rhetorical shuck-and-jive. Therefore, social constructivists must bear a heavy burden of proof if they want to convince the rest of us. In particular, to argue that scientific methods are social constructs, that is, that scientists adopt methods that serve social or political agendas—not just because they are better methods—one must show at least one of the following: (a) No method is better than any other. (b) Scientists cannot recognize a better method even if one is at hand.[2]

_____

2. Someone could charge that even if there are better methods and scientists recognize them as better, scientists instead adopt the methods that are more socially or politically expedient. However, even the social constructivists hesitate to accuse scientists of outright dishonesty.

How could anyone argue in favor of (a)? When a scientific method is deemed better than an alternative, it is because, for instance, it appears to provide more or better data, allows for more rigorous or more practical tests of theoretical claims, permits a larger number of such claims to be tested . . . and so forth. It is self-defeating to claim that no methods for gaining knowledge are better than any others. Such a claim is itself a claim to knowledge; namely, it is a claim to know that no method is better than any other. But to make any knowledge claim at all—or at least to claim to know anything beyond the patently obvious—requires that we regard some methods of gaining knowledge as better than others. As the noted philosopher Nicholas Rescher observes, the only alternative to endorsing some methods and rejecting others is to affect total indifference about all methods, which is tantamount to having no convictions about anything (Rescher 1997, 58). On the one hand, judging by their voluminous and highly opinionated writings, the social constructivists have loads of convictions about all sorts of things. On the other hand, maybe the social constructivists are saying that their methods, the methods of the sociology of knowledge, are good and those of the physical sciences are not. However, no social constructivist has yet explained how the sociology of knowledge has reliable methods but particle physics, for instance, does not.

If anything, it would be even harder to argue (b) that scientists cannot recognize a superior method even if they stumble across one. Taken literally, (b) implies that, for instance, scientists cannot know that double-blind studies are better for testing pharmaceuticals than tossing coins or gazing into crystal balls. It is hard to see what, besides a universal curse laid on all human minds, would prevent us from knowing such things. We *do* know that some methods are better than others. To deny this, constructivists would have to resort to universal skepticism about all methods, or examine each particular method and show that there is no rational basis for preferring one to another. I have just noted the problems with the former option. As for the latter, it would be instructive to see a social constructivist debate a working scientist about the efficacy of an accepted method in his or her specialty. For instance, it would be interesting to see Bruno Latour debate the Nobel Prize–winning particle physicist Steven Weinberg about the reliability of methods of particle physics. I have a pretty good idea who would win that one.

If social constructivists cannot make a convincing case for (a) or (b), then they have to admit that at least in principle (and why not in practice?) scientists can adopt a method just because it is better. If they admit this, then they must concede that science is not, or at least need not be, entirely a social construct. They should also concede that scientific methods may not be merely historically contingent rules of the game.

A much more reasonable view of scientific rationality and the role of scientific controversy is given in the book *The Discourses of Science* (1994) by the philosopher and historian of science Marcello Pera. Pera also sees rhetoric as playing a major role in scientific debate. However, by "rhetoric" Pera means the art of producing the best arguments to persuade a given audience. It was in this sense that rhetoric was included among the seven liberal arts in medieval universities. Rhetoric in this sense does not imply anything devious or underhanded; rather, it is essential for the process of rational debate among the qualified parties, which is scientific controversy. What makes the debate rational? Pera agrees that there is no rule book of automatic procedures or universal Scientific Method to definitively settle each issue. However, unlike Latour and Woolgar, he thinks that some arguments really *are* more cogent than others. Darwin and Galileo did have better arguments—better evidence and better logic—than their critics; they were not just more adept at subterfuge and browbeating. Darwin and Galileo did not have to overawe or fool their opponents; their arguments rationally persuaded many of them. The scientific community acts rationally, which it often does, when it awards victory to the debater who really makes the better case.

Pera's view of scientific rationality and the role played by controversy is similar to the one developed by the philosopher Richard Bernstein in his book *Beyond Objectivism and Relativism* (1983). Like Pera, Bernstein rejects the notion that any automatic procedure, rule book, or one-size-fits-all Scientific Method can settle scientific controversies. He views scientific controversy as a rational debate between qualified parties in which each side appeals to broadly shared and deeply entrenched standards. All scientists agree that good theories must meet certain standards of accuracy, fruitfulness, simplicity, and consistency with background knowledge. The controversy is over which of the rival theories is best when judged by those standards. Again, there is no rule book or automatic procedure for picking the winner. Standards are abstract and general; the methods and techniques used to evaluate theories are concrete and particular. Scientific controversy is over the practical means of meeting those abstract standards in scientists' theory-choice decisions.

Apparently, then, the fact that there is no Scientific Method and that theories are underdetermined by data does not mean that the social constructivists are right. Science remains a rational enterprise, and scientific controversy, far from being merely a means of blowing rhetorical smoke, is essential to the rationality of science.

How, then, do scientific controversies end? After his collaboration with Woolgar, Latour went on to write such works as *Science in Action* (1987) and *The Pasteurization of France* (1988). In these works Latour develops what is called his "actor-network" interpretation of science. According to this view, the success of a particular scientist depends on how good an alliance he or she is able to make with other actors in the scientific drama. A scientist with a strong network will have powerful allies and other advantages, such as the support of influential senior scientists, prestigious institutional appointments, the use of state-of-the-art equipment at leading laboratories, or lavish funding from government or industry. When a scientist enjoys the support of such a network, he or she has a great advantage over a scientist with less support. According to Latour, when a scientist with a weaker network of support challenges one in a stronger position, the latter can simply outgun the former. The more powerful scientist is simply able to bring far more resources to bear than the weaker one, who eventually will collapse under the weight of the onslaught. In other words, scientific controversies are decided by might, not right.

Clever half-truths like Latour's always exert a strong appeal. Scientists do form alliances and networks. For instance, Darwin carefully cultivated his friendships with leading scientists of his day such as Joseph Dalton Hooker, T. H. Huxley, and Asa Gray. Alliances with these noted scientists brought their prestige, intelligence, and eloquence into the battle on Darwin's side. Darwin's rapid victory in persuading his colleagues that evolution had occurred (selling his particular mechanism of evolution, natural selection, took much longer) was no doubt due largely to the effectiveness of his allies. It is also true that a strong enough network can silence all opposition to a theory. Consider the case of Trofim Lysenko, the Soviet pseudoscientist whose crackpot theories of genetics were supported by Joseph Stalin. Opposition to Lysenko's theories disappeared, at least in the Soviet Union, when all of Lysenko's critics were sent to starve in slave labor camps in Siberia.

Fortunately, few scientists today are able to appeal to a ruthless dictator to squelch opposing views. Scientific heretics are often tolerated, if only because not much can be done about them. Scientific heretics cannot be burned at the stake, sent to Siberia, or even removed from tenured positions. Many dissident scientists relish their roles as heretics, often living out happy careers in gleeful defiance of scientific orthodoxy. In Chapter 6 we shall encounter the ebullient dinosaur heretic Robert Bakker, who takes every opportunity to thumb his nose at mainstream paleontologists. Sometimes mainstream scientists just stop listening to heretics, but not always. Sometimes a scientific controversy will continue when a whole scientific community is opposed by a few or even one dissenting voice. What matters is whether that minority opinion, even if voiced by only a single dissident, is scientifically interesting enough to challenge the received view. So it is not always true that scientific controversies end when the weaker side is cowed into silence.

In general, it is not true that scientific debates are closed by might rather than right. Any fair-minded study of past scientific controversies will disclose that evidence and logic, the traditionally *rational* factors in scientific debate, really *did* matter for the outcome of those debates (see Parsons, 2001). Of course, as we shall see in examining the following controversies, the contending sides often deployed all the devices of rhetoric and even occasionally stooped to subterfuge and personal attacks. Scientists are merely human, after all, and subject to all the sins and weaknesses that flesh is heir to. Scientists can be motivated by rancor or vindictiveness like everyone else, and though politicians may set the standard for venality, they are not the only ones who sell out to special interests. As the following chapters will show, when a controversy is truly settled, when the tumult and the shouting die, it is because the evidence has spoken most loudly of all.

Further Reading

Professional philosophers of science have never paid scientific controversy the attention the subject deserves. The best recent study of scientific controversy by professional philosophers of science is *Scientific Controversies,* edited by Peter Machamer, Marcello Pera, and Aristides Baltas (Oxford: Oxford University Press, 2000). This volume treats all of the important aspects of scientific controversy from both a theoretical and a historical perspective. The contributors are top-notch, and quite a bit of the book, especially the historical studies, should be accessible to nonspecialists. One of the editors, Marcello Pera, wrote a very interesting book called *The Discourses of Science* (Chicago: University of Chicago Press, 1994). Pera argues, persuasively in my view, that the rationality of science consists not in some unique method available only to scientists but in the quality of the arguments they employ in their debates with other scientists. Therefore, the study of the rhetoric of science, the art of presenting the most persuasive case to one's peers, is essential for the understanding of how science really works. Pera shows that the success of some of the best scientists, such as Darwin and Galileo, depended largely on their skill as rhetoricians.

As noted in the chapter text, Bruno Latour and Steve Woolgar present a very different view of scientific rhetoric in their book *Laboratory Life: The Construction of Scientific Facts* (Princeton: Princeton University Press, 1986, 2nd ed.). Latour and Woolgar see scientific discourse as the means whereby scientific "facts" are constructed and, further, as the means whereby scientific communities achieve collective amnesia and forget the constructed nature of such "facts." Their later writings take even more extreme stands. In his remarkably truculent little book *Science: The Very Idea* (London: Tavistock Publications, 1988), Woolgar defends an extreme, global antirealism asserting that scientific statements can refer only to other statements and not to the world. In my opinion, Bertrand Russell was correct when he said of a similar claim that it is so silly that "only a very educated person could ever come to believe it." Latour also takes a more extreme line about scientific discourse in his *Science in Action: How to Follow Scientists and Engineers through Society* (Cambridge, MA: Harvard University Press, 1987), in which he presents a view of scientific rhetoric as a kind of weapon to bludgeon opponents into submission.

I vigorously criticize the views of Latour and Woolgar on scientific rhetoric in my book *Drawing Out Leviathan: Dinosaurs and the Science Wars* (Bloomington: Indiana University Press, 2001). I also criticize the view presented by Steven Shapin and Simon Schaffer in their book *Leviathan and the Air-Pump* (Princeton: Princeton University Press, 1986). They argue that the methods and standards deployed in scientific debate are historically contingent products of the ambient social and political environments of those controversies. An excellent exposition and critique of Latour, Woolgar, Shapin, and Schaffer is Robert Klee's *Introduction to the Philosophy of Science: Cutting Nature at Its Seams* (Oxford: Oxford University Press, 1997). The views of these "social constructivists" have become a part of the series of academic disputes called "the science wars." My anthology *The Science Wars: Debating Scientific Knowledge and Technology* (Amherst, NY: Prometheus Books, 2003) gives an overview of these debates and representative selections from various combatants. The classic counterblast to the constructivists, and other science critics from the "academic left," is the sharply polemical *Higher Superstition: The Academic Left and Its Quarrels with Science* by Paul R. Gross and Norman Levitt (Baltimore: The Johns Hopkins University Press, 1994).

2

"Fearfully Great Lizards"

Fossils have probably always evoked wonder. They are beautiful, strange, and rare. Remarkably, though, people have not always identified fossils as the relics of organisms of the deep past. For one thing, the idea that the earth has had a deep past is a relatively new one. Only toward the end of the eighteenth century did the new science of geology reveal that the earth was far older than anybody had—or perhaps could have—conceived. Also, what we today call "fossils" were just one type of interesting thing found in the ground. Originally, the word "fossil" meant anything dug up from the earth, including minerals, crystals, gems, ores, rocks, and fossils in the present sense (Rudwick 1976, 1). Some of these dug-up objects did look remarkably like living organisms; others did not look organic at all, and many could not easily be categorized either way. So the first task was to distinguish the remains of once-living organisms—fossils in our sense of the word—from all of those other interesting objects taken from the ground.

Even when fossils (I'll use the word in our current sense from now on) were correctly identified as organic in origin, their nature and relations to the original organisms were often far from clear. For instance, when a seventeenth-century British naturalist named Robert Plot found the broken-off end of a dinosaur femur (big leg bone), he guessed that it came from an elephant brought to Britain by the Romans. In the eighteenth century, the Oxford researcher Richard Brookes was still puzzled by the fossil, but he sought to give it a name in accordance with the new system of binomial classification ("*Homo sapiens,*"for instance). He observed the two bulbous protuberances at the end of the bone fragment and, by reason of resemblance, classified the object as *Scrotum humanum.* Considering the size of the "*Scrotum humanum,*"Brookes must have envisioned a true Goliath (or perhaps a case of elephantiasis). Actually, he did not know what creature could have left such a remnant, so he just gave it a descriptive name (Flannery 2002). Brookes's puzzlement about the fossil and what to call it may be amusing, but we should not scorn him or other naturalists of the past for not recognizing what seems obvious to us. The past is another country, and condescension is inappropriate, just as it is when we regard the (to us) odd customs of the Inuit or Australian Aborigines. Rather than scorn the "foolish" or "absurd" conclusions of past scientists, we should see them as doing what we all must do—attempting to make sense of a stupendously complex cosmos with the concepts, categories, theories, methods, and techniques that we possess at any given time.[1] The point is to ask what kinds of social and intellectual changes had to take place before there could have been much knowledge about the life of the deep past.

One problem in learning about ancient life is that the evidence is almost always piecemeal—usually just a few fragments of fossilized bone. Richard Brookes had only the broken end of a femur. Even by the mid–nineteenth century, when Richard Owen named the order "Dinosauria," he had only bits and pieces of the fossilized bones of a very few species. But even if a complete, articulated skeleton of an *Iguanodon* had been found in medieval Britain, no one would have come up with the idea of a dinosaur. They might have thought it was the skeleton of a dragon or some other monster, but the concept "dinosaur" could emerge only when much other progress had been made in geology and biology, not to mention philosophy and theology. For people to conceive of dinosaurs as enormous, extinct creatures of the deep past, they had to have the idea of deep time. They even had to acquire the idea of extinction. Not until the early nineteenth century did Georges Cuvier, the preeminent anatomist and paleontologist of the day, realize that some species had

---

1. Of course, I do not deny that foolishness and absurdity were just as rife in the past as they are today. Nor am I advocating some form of trendy relativism holding that no perspective is any more true or more objective than any other. I think science progresses in a straightforwardly realist manner, that is, that over time we acquire a surer and broader grasp of the actual structure and history of the universe. We know vastly more about the deep past than Robert Plot did in 1676. I am arguing only that we must evaluate past scientists fairly, meaning that we must not expect them to know what we know or even to investigate things the way we do. All we can expect of them is to have done the best they could have done with the knowledge and intellectual tools available to them.

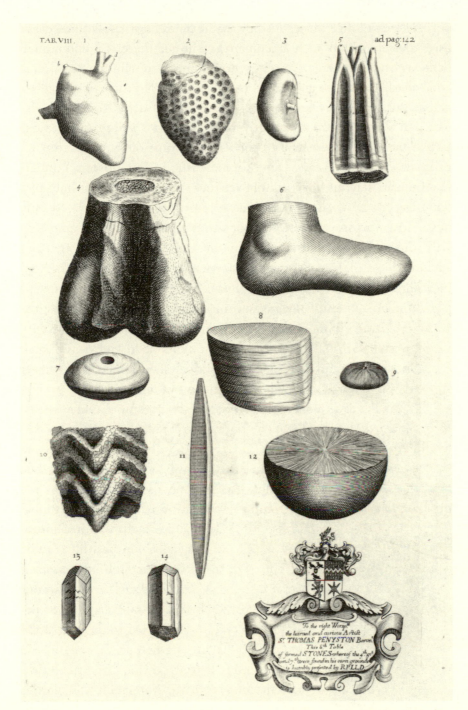

A "human thigh-bone" figured as item 4 on TAB VIII by Dr. Robert Plot in The Natural History of Oxfordshire in 1677 is actually a dinosaur thigh bone, probably Megalosaurus (The Natural History Museum, London).

become extinct. It is therefore not really surprising that we have recognized dinosaurs only for about the last 160 years (one-millionth of the time dinosaurs lived on the earth). Not only did the evidence have to emerge but also humans had to have the background knowledge to interpret the data correctly.

The fossil discoveries leading up to the recognition of dinosaurs as a distinct group were made in Britain in the third and fourth decades of the nineteenth century, a time of enormous intellectual and social change. The industrial revolution had created fabulous wealth for a few magnates, who lived lives of astonishing extravagance. It also created a vast urban underclass that lived in squalor but, unlike the stolid yeomanry of earlier centuries, did not placidly accept subservience. Radical philosophies were in the wind and threatened the ancient authority of church and crown (it did not help that King George III was insane and his successor, George IV, was widely despised for his scandalous behavior). Inevitably, science got mixed up with the social strife of the day. In particular, evolutionary theories such as those of Jean Baptiste Lamarck were identified as atheistic, radical, and (worst of all) French. The establishment therefore supported scientists who opposed evolutionary ideas (the hostility toward evolution partially explains why Darwin waited over twenty years, until 1858, to defend his views publicly). In fact, as we shall see, dinosaurs themselves took on political significance, and strong support was given to antievolutionary views of dinosaurs.

Another consequence of the social and economic change of the early nineteenth century was the rise of the middle class. Prior to that time there was no middle class; there were only the "haves" and the "have-nots." Paleontology would have remained the obscure study of a few specialists had not the public become involved. Members of the middle class had the leisure and the education to pursue avocational interests, and few subjects had more appeal than the new discoveries about past life. Public interest led to the establishment of museums of natural history and displays such as the famous dinosaur exhibit at the Crystal Palace. The burgeoning interest in science led to the rise of the middle-class scientific amateur. Science at the time was much less professionalized than now. Amateurs made many important discoveries and were often welcomed into prestigious scientific bodies.

Gideon Algernon Mantell (1790–1852) was one of those middle-class scientific amateurs. A successful physician by day, he would often work into the wee hours on his fossil collection. A passionate and indefatigable collec-

Gideon Mantell was a British geologist and paleontologist as well as a country doctor. He acquired some large fossil teeth in a quarry and was convinced that he had found fossils of a large, extinct, herbivorous reptile that he called Iguanodon *(The Natural History Museum, London).*

tor, he filled his house with fossilized bones—much to his wife's annoyance. Mantell is an appealing character. His commitment to the healing profession and compassion for the suffering of his patients is evident. His scientific acuity shines through the darkness of a personal life marked by tragedy, heartbreak, and chronic pain. Though historians should be impartial, it is hard not to side with Mantell when Richard Owen criticizes him. Compared to Mantell, Owen, though unquestionably brilliant, seems mercenary, spiteful, and arrogant. Nevertheless, we shall see that Owen's criticisms were necessary for the recognition of dinosaurs as the distinctive and remarkable animals that they were.

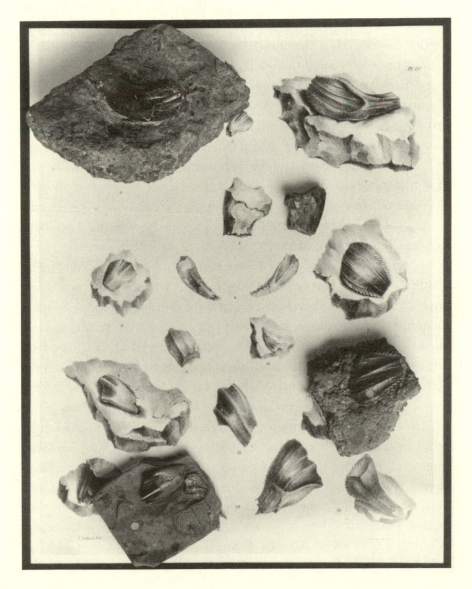

Four Iguanodon *teeth from the Mantell collection alongside their reproduction in* Illustrations of the Geology of Sussex *(1827) by Gideon Mantell (The Natural History Museum, London).*

A pleasant story tells of how Mantell's wife occasionally accompanied him on his rounds. Supposedly, while waiting for him to finish with a patient one day, she walked around the doctor's buggy to scour the ground for fossils. According to the tale, she found a number of large fossilized teeth in some roadside gravel and showed them to her astonished husband. Unfortunately, this is just a story. In 1822 Mantell did acquire some large fossil teeth, which came from a quarry in the Tilgate Forest, near the town of Cuckfield in Sussex (in southern England). In the 1820s, stratigraphy, the study of the deposition and age of stratified rocks, was in its infancy. However, it was known that the quarry that yielded the strange teeth was located in a region of Mesozoic (what was then called "Secondary") aged rock. Mantell knew that the Secondary was the "Age of Reptiles" and was convinced that he had found fossils of a large, extinct, herbivorous reptile. Impressed by the similarity of the fossil teeth to those of the living iguana, he named the fossil creature *Iguanodon*: "iguana-tooth."

Mantell reported his discovery of *Iguanodon* to the Royal Society of London, the most prestigious scientific society in Britain, in a letter that was read to the body in 1825 and published in *Philosophical Transactions of the Royal Society of London* that year. Soon Mantell was invited to join the Royal Society, a singular honor for the son of a small-town shoemaker in those highly class-conscious days.

Mantell was not the first to report on a creature now recognized as a dinosaur. In 1824 William Buckland, the eminent Oxford geologist and one of the oddest in the great tradition of British eccentrics, had described the teeth and jaw of *Megalosaurus,* a big Jurassic carnivore. Mantell compared the bones of *Megalosaurus* with those of *Iguanodon*: "That the latter *[Iguanodon]* equalled, if not exceeded the former *[Megalosuaurus]* in magnitude seems highly probable; for if the recent and fossil animal bore the same relative proportions, the tooth . . . must have belonged [to] an individual upwards of sixty feet long; a conclusion in perfect accordance with that deduced by Professor Buckland from a femur, and other bones in my possession" (Mantell 1825, 185).

In other words, Mantell estimated the size of *Iguanodon* simply by comparing the fossils to the corresponding anatomy of living iguanas and scaling the rest of the creature in the same proportion. Further, Mantell automatically identified *Iguanodon* as a Lacertian (lizard) like living iguanas. His only question was whether they should be classified in the same genera with existing creatures (Mantell 1825, 184). Clearly, Mantell thought of *Iguanodon* as resembling a vastly larger iguana, though he did note some significant

differences in the vertebrac between the living and extinct creatures (Mantell 1825, 185).

Though amateurs like Mantell could be elected to prestigious scientific bodies where they could rub shoulders with the most eminent scientists, changes were under way. Society was just beginning to recognize that there could be such a job as "scientist" (actually, the term "scientist" was not coined until the nineteenth century). By the middle of the century, brilliant and ambitious young men began to plan *careers* in science, and the importance of the gentlemanly amateur waned as professionals took over the most important research. Today, of course, most sciences are thoroughly professionalized, and there is little scope for the contributions of amateurs (astronomy and paleontology remain exceptions to this rule).

Richard Owen (1804–1892) was one of the first of those brilliant, up-and-coming young men who sought a career in science. He

Mantell's Iguanodon. *This image of a restoration of reptiles whose fossil remains were found in Tilgate Forest, Sussex/G Scharf del 1833, served as a sketch for a picture 3 yards long.* Iguanodon *was calculated from the remains to have been 100 feet long (Alexander Turnbull Library, Wellington, New Zealand).*

*Portrait of Sir
Richard Owen,
Professor of
Comparative
Anatomy at the
Royal College of
Surgeons. Oil on
canvas, by Henry
William Pickersgill
(1782–1875),
1844. Original at
The Natural History
Museum, London
(The Natural
History Museum,
London).*

came to London in 1825 to seek his fortune and soon obtained a membership in the Royal College of Surgeons and a position in the Hunterian Museum associated with the college. Hard work, intelligence, and marrying the boss's daughter solidified Owen's position. In 1836 the College of Surgeons appointed him Hunterian Professor, and he eventually succeeded his father-in-law as conservator of the museum. Owen possessed the unwavering self-assurance that fortifies many high achievers. He was unquestionably a brilliant scientist, rightly recognized as the foremost comparative anatomist of his day, but he often engaged in petty controversies and could be ruthless to rivals. Owen's opposition to evolutionary theories endeared him to the conservative establishment and allowed subsequent generations of evolutionists to tar him as an obscurantist.

In the late 1830s Owen began a systematic review of all known British fossil reptiles. In 1839 he reported on the marine reptiles such as ichthyosaurs and plesiosaurs to the British Association for the Advancement of Science (BAAS). In 1841 he spoke again to the BAAS, this time on terrestrial reptiles. It is often said that Owen revealed his term "Dinosauria," meaning "fearfully great lizards," at this meeting as his name for a proposed new order of the class *Reptilia*. However, Professor Hugh Torrens has shown that none of the contemporary accounts of Owen's presentation mention such a momentous occurrence (and science news was much more carefully and fully reported by the press in that day than in ours) (Torrens 1997, 179–180). In that talk Owen did not mention the term "Dinosauria" or suggest that creatures such as *Iguanodon* be placed in a separate order of reptiles. These developments took place in the much-revised printed version of Owen's report, which was published the next year (Torrens 1997, 182). Why then in Owen's published 1842 version, "Report on British Fossil Reptiles, Part II," did he hold that dinosaurs needed to be classed as a distinct order of reptiles?

Owen compared *Iguanodon* and *Megalosaurus* and found that they shared distinctive anatomical features: "This group, which includes at least three well-established genera of Saurians, is characterized by a large sacrum composed of five anchylosed vertebrae of unusual construction, by the height and breadth and outward sculpturing of the neural arch of the dorsal vertebrae, by the twofold articulation of the ribs to the vertebrae" (Owen 1842, 103).

For Owen, the most important feature was the "anchylosed" (fused) vertebrae of the sacrum. The sacrum is the part of the spine that joins the hips. The fusion of those vertebrae greatly strengthens them and permits them to bear the weight of enormous creatures such as dinosaurs. These characteristics were so distinctive that they justified the creation of a new order of reptiles: "The combination of such characters, some, as the sacral ones, altogether peculiar among Reptiles, others borrowed, as it were from groups now distinct from each other, and all manifested by creatures far surpassing in size the largest living reptiles will, it is presumed, be deemed sufficient ground for establishing a distinct tribe or sub-order of Saurian Reptiles, for which I would propose the name *Dinosauria*"(Owen 1842, 103).

Owen did far more than invent a new name for the fossil creatures; he emphasized dinosaurs' uniqueness in a way that was bound to be offensive to Mantell. Owen pointedly referred to the differences between *Iguanodon* and living iguanas. He argued that *Iguanodon* differed from the iguana in its teeth, vertebrae, and ribs: "The important difference which the fossil teeth presented in the form of their grinding surface was afterwards pointed out by Cuvier, and recognised by Dr. Mantell, and the combination of this dental distinction with the vertebral and costal characters, which prove the *Iguanodon* not to have belonged to the same group of Saurians as that which includes the Iguana and other modern lizards, rendered it highly desirable to ascertain, by the improved modes of investigating dental structure, that actual amount of correspondence between the *Iguanodon* and Iguana in this respect" (Owen 1842, 121).

Owen proceeded to make extensive comparisons, showing repeatedly the differences between *Iguanodon* and living reptiles. Mantell was bound to regard these conclusions as a slap in the face, and in his view Owen proceeded to add insult to injury. Mantell thought that Owen had pirated some of his own observations (Cadbury 2000, 243) Owen also displayed his penchant for presenting his views as proven when the scanty evidence admitted of other interpretations.

In a letter to the magazine *Literary Gazette,* which had published a report on Owen's Plymouth talk to the BAAS, Mantell remonstrated that he had proposed the name *Iguanodon* merely because of the general similarity of the teeth to the iguana's and had not intended to imply further affinities. However, it seems clear that Mantell did envision *Iguanodon* as a scaled-up lizard. As noted earlier, in estimating the size of *Iguanodon* Mantell simply compared the sizes of the corresponding bones of the fossil and living creatures (Mantell 1833). Owen sharply criticized Mantell's estimates, even injecting a note of ridicule:

> But it is very obvious that the exaggerated resemblances of the *Iguanodon* to the Iguana have misled the Paleaeontologists who have hitherto published the results of their calculations of the size of the *Iguanodon;* and, hence, the dimensions of 100 feet in length arrived at by a comparison of the teeth and clavicle of the *Iguanodon* with the *Iguana,* of 75 feet from a similar comparison of their femora, and of 80 feet from that of the claw-bone, which if founded upon the largest specimen from Horsham, instead of the one compared by Dr. Mantell, would yield a result of upwards of 200 feet for the total length of the *Iguanodon,* since the Horsham phalanx exceeds the size of the recent Iguana's by 40 times! (Owen 1842, 142–143)

Owen correctly realized that the *Iguanodon* and the iguana were two very different sorts of creatures, and that you cannot estimate the size of the one simply by scaling up the other. Ratios of body-part size to overall length differ considerably among different types of animals. In particular, large animals often have body parts that are disproportionately larger than those of small animals. Owen proposed a much more reasonable length of twenty-eight feet for *Iguanodon* (Owen 1842, 144). An even clearer indication of Mantell's view of *Iguanodon* as a big lizard is seen in the frontispiece to his 1838 book, *The Wonders of Geology.* The engraving by John Martin shows a very lizardlike *Iguanodon* in mortal combat with another great saurian. Mantell approved of the drawing and held that it accurately depicted a long-lost world (Cadbury 2000, 237).[2]

Today we know quite a bit about iguanodontids (see Brett-Surman 1997; Norman 1985). Medium to large (5–10 meters, or 16 to over 30 feet) dinosaurs, they existed from the late Jurassic to the end of the Cretaceous but were most numerous and diverse in the early

---

2. These illustrations remind me of the atrocious "dinosaurs" Irwin Allen created by gluing fake sails and frills on lizards and alligators in his horrid 1960s version of *The Lost World.*

and mid-Cretaceous. In fact, judging by the large number of remains, *Iguanodon* must have been a common denizen of Europe in the early Cretaceous, the age of the Tilgate Forest strata. Fossilized footprint trackways, which show numerous animals walking in the same direction, indicate that it very possibly traveled in large herds, like the bison of the American frontier. Iguanodontids had many interesting features. One genus, *Ouranosaurus,* had a large, sail-like structure on its back, like the better-known creatures *Spinosaurus* and *Dimetrodon.* Iguanodontids were armed with a highly modified first finger that had developed into a stilettolike claw (misidentified as a rhinoceroslike horn by early paleontologists). They must have been capable of inflicting deep puncture wounds on any predator foolish enough to try a frontal attack. Pack-hunting dromaeoraurids, such as *Deinonychus,* may have had more success with slashing attacks to the back and sides, though the bulk and strength of a full-grown iguanodontid would have made it formidable prey for any hunter.

Of course we have the advantage of having far more evidence than Mantell. When anything new is discovered, it seldom is revealed all at once. We begin to make sense of it by comparing it to what we already know. For instance, when Galileo first saw the big moons of Jupiter (at first he saw only three of the four) through his telescope, he initially thought that they might be background stars,

Owen's Iguanodon *is fundamentally different from* Mantell's. Mantell *believed* Iguanodon *resembled living reptile species,* whereas Owen *pointed out numerous differences* (The Natural History Museum, London).

though it was odd that they lined up in one plane with Jupiter. As he watched over successive nights, he noticed that the bodies moved in ways that no background stars could. Drawing on the analogy of the movement of Venus and Mercury around the sun, he correctly concluded that the new "stars" made revolutions about Jupiter (Galileo 1989, 66). Sometimes we are so focused on the familiar that we fail to recognize just how new a new thing is. Christopher Columbus died thinking that he had sailed to the extreme east of Asia, the limits of his known world, and did not realize that he had found a new continent. Similarly, the earliest discoverers of dinosaurs had to interpret scanty fossil evidence in terms of the only model they had—living creatures. Small wonder that Mantell thought of *Iguanodon* as lizardlike. It took a leap of imagination for Owen to see dinosaurs as really different kinds of animals. In fact, we are still learning just how different and extraordinary these amazing creatures were.

Mantell was not Owen's only target, or even his main one. Owen had a much bigger agenda than merely overthrowing the lizardlike *Iguanodon*. Chiefly, he was grinding an ax against evolutionary theories. People often associate the debate over evolution with the publication of Darwin's *Origin of Species* in 1859. In fact, feelings ran high well before then. To see just how high, consider the furor that broke out in 1844 with the publication of *Vestiges of the Natural History of Creation,* an anonymous work that defended a crude but exciting evolutionary view. It quickly became a best-seller and was especially popular with women readers (no doubt prompting many a misogynistic "harrumph!" from scientific gentlemen). The reaction of establishment scientists to *Vestiges* could be described as vitriolic, though even this term would seriously understate the case. Adam Sedgwick, a leading geologist and one of Darwin's teachers at Cambridge, was incensed nearly to apoplexy. He execrated *Vestiges* as a "filthy abortion" and fired off hundreds of pages of bitter polemic (see Ruse 2000, 42).

Owen was oddly sympathetic to the book and refused to join in the general chorus of condemnation. According to Adrian Desmond, Owen agreed with some of the claims made in *Vestiges,* such as the idea that organisms arose through the uniform operation of lawful regularities rather than by particular acts of special creation. However, he was unalterably opposed to the idea that evolution, or "transmutation" as it was then called, of one species into another could be that lawful process (Desmond 1982, 33).

Prior to Darwin, evolutionary ideas were theories of organic progress. For instance, in Lamarck's view, "lower" organisms were always striving to become "higher" ones. Over succeeding generations, organisms within any given lineage achieved ever more "advanced" features, eventually becoming "higher" creatures. By contrast, Darwin held that evolution is a branching process, not a straight-line ascent up a stairway of organic progress. Darwinian evolution does not imply inevitable progress within a lineage; the branching occurs as an adaptive response to local conditions, not in obedience to an innate drive toward perfection (more on this concept in Chapter 4). But in 1842 Darwin's *Origin of Species* was still seventeen years away, and evolution still implied progress. Therefore, Owen was no doubt delighted to find in dinosaurs a splendid counterexample to any such progressionism.

Dinosaurs were in no sense more primitive, or "lower," than current reptiles. On the contrary, they had "advanced" features that were typical of the "higher" mammals. For instance, dinosaurs did not sprawl like lizards, resting on their bellies between periods of activity. Rather, dinosaurs stood upright like elephants, with their great mass supported by legs positioned directly under the body. Indeed, Owen would have had to say that living reptiles had *retrogressed* from the peak represented by dinosaurs! Owen was therefore able to argue triumphantly:

> If the present species of animals had resulted from progressive development and transmutation [i.e., evolution] from former species, each class ought now to present its typical characters under their highest recognized conditions of organization: but the review of the characters of fossil Reptiles, taken in the present Report, proves that this is not the case. No reptile now exists which combines a complicated and thecodont dentition with limbs so proportionally large and strong, having such well-developed marrow bones, and sustaining the weight of the trunk by synchrondrosis or anchylosis to so long and complicated a sacrum, as in the order *Dinosauria*. The *Megalosaurs* and *Iguanodons*, rejoicing in these undeniably most perfect modifications of the Reptilian type, attained the greatest bulk, and must have played the most conspicuous parts, in their respective characters as devourers of animals and feeders upon vegetables, that this earth has ever witnessed in oviparous and cold-blooded creatures (Owen 1842, 200).

In other words, dinosaurs were not crude or primitive reptiles but were anatomically much more sophisticated, with more complex and efficient adaptations, than living crocodiles or lizards. Further, if dinosaurs were on the road to becoming higher organisms, we would expect the even higher mammals to have evolved from them. Owen retorts:

> If, therefore, the extinct species in which the Reptilian organization culminated, were on the march of development to a higher type, the *Megalosaurus* ought to have given origin to the carnivorous mammalia, and the herbivorous should have been derived from the *Iguanodon*. But where is the trace of such mammalia in the strata immediately succeeding those in which we lose sight of the relics of the great Dinosaurian Reptiles? Or where, indeed, can any mammiferous animal be pointed out whose organization can by any ingenuity or license of conjecture, be derived without violation of all known anatomical and physiological principles, from transmutation or progressive development of the highest reptiles? (Owen 1842, 200–201)

The idea that dinosaurs were part of a progressive development leading upward to present creatures was thus thoroughly refuted.

Owen was right that dinosaurs were very different from any living creatures. He recognized, as Mantell did not, that dinosaurs were distinctive creatures, so different that they deserved a whole new, higher-level taxonomic designation. They were not big lizards, nor were they predestined failures, nature's version of planned obsolescence, as they were popularly portrayed not so long ago. They were marvelously diverse and well-adapted creatures, as they had to be to dominate terrestrial faunas for 160 million years. Some may even have been warm-blooded (a possibility examined in Chapter 6). Yet progressive as Owen's views may have been, the modern picture of dinosaurs did not spring into existence with him—far from it. Owen's dinosaurs were unlike anything the world had yet seen, but to us they seem primitive. For instance, we now know that *Megalosaurus* was a lean, mean, bipedal predator like *Allosaurus*. Owen imagined it as an elephantine quadruped. Much of the problem was that Owen still had very scanty evidence, and in the absence of evidence, the scientific imagination runs along tracks of familiarity. Innovative as Owen's ideas were, the idea of a bipedal reptile would have been a wild speculation given what he knew.

But what about Owen's opposition to evolutionary theories? Did this not show that his thought was progressive in some respects and backward in others? Here we have to be very careful. It is bad historical practice, what historians derisively call "Whig history," to judge past scientists by how well they anticipated current views.[3] Clearly, past scientists have to be judged in the context of the assumptions, concepts, methods, and evidence available to *them*. Prior to Darwin, the evidence for evolution was meager and the evolutionary mechanisms proposed were highly speculative. Owen's analyses *were* appropriate rebuttals to the progressionist pre-Darwinian theories of his day.

Still, though evolution may not be progressive, science definitely is, and the story of science is poorly told unless it is shown how such progress has occurred. Historians of science, perhaps due to excessive fear of slipping into the Whig style, have recently been timid about identifying past scientific episodes as progressive or not. Judging by what we currently know (and what other measure of progress could we have?), Owen's dinosaur theories were in some respects progressive and in others not. Things often happen this way in science. New ideas often appear mixed in complicated ways with old ones. For some time it has been fashionable to emphasize the revolutionary aspects of scientific change. This picture is complicated by the fact that, as in Owen's case, surprising new insights often appear in the service of conservative views.

Within a few years, the picture of dinosaurs had changed again. By 1858 undeniable evidence of bipedal dinosaurs had appeared, but by then Owen's reconstructions had literally been set in concrete. The sculptor Benjamin Waterhouse Hawkins had created life-size concrete statues of Owen's dinosaurs. These "antediluvian" creatures were exhibited in the park of the great Crystal Palace exhibition that

---

3. The term "Whig history" was coined by the historian Herbert Butterfield. He noted that members of the British Whig political party wrote history so that all those historical figures who had supported elements of the Whig party line were the good guys and those who had opposed such policies were the villains. It is especially tempting to write Whig history when doing the history of science because science does progress. Those who advocated ideas similar to current ones therefore look amazingly prescient, and those who opposed them come off as obscurantists. However, a historically sensitive interpretation will often show that those who opposed our current views were more rational than those who were not. As a case in point, a Greek astronomer named Aristarchus proposed a heliocentric theory over 1,700 years before Copernicus. But, given the information available in Aristarchus's day, it really would have been unreasonable for astronomers to have abandoned their geocentric model in favor of his theory.

opened at Sydenham, not far from London, in 1854. A cartoon in *Punch,* the *Mad* magazine (or perhaps *National Lampoon*) of the day, depicted a lad being dragged through the exhibit by an elder determined to improve the youngster's mind. The boy's expression shows that he is not amused or edified by Hawkins's glaring monsters. The Crystal Palace burned in 1936, but Owen's dinosaurs remain, recently restored to their former glory, a monument to the remarkable brilliance and inevitable limitations of a great mind.

Further Reading

No one knows more about British paleontology in the nineteenth century than Adrian Desmond—and he's a good writer too. His 1976 book *The Hot-Blooded Dinosaurs* (New York: The Dial Press) contains much excellent information on the beginnings of dinosaur paleontology. On the social and political conditions of the early 1800s in Britain and how they affected the developing life sciences, Desmond's *The Politics of Evolution* (Chicago: University of Chicago Press, 1989) is unsurpassed. As I try to indicate in this and the following chapters, to understand what is going on in science in any given era, you need to understand the political, social, and religious background. All of Desmond's works do an excellent job at this, though in some writings he goes too far in the "social constructivist" direction for my taste.

Martin Rudwick's *The Meaning of Fossils* (2nd ed., Chicago: University of Chicago Press, 1976) is not a comprehensive history of paleontology. By focusing on a number of specific episodes Rudwick is able to show how people interpreted fossils from the middle of the sixteenth century to the middle of the nineteenth century. He wonderfully illuminates the unfolding of scientific understanding of the history of the earth and the life of the deep past. Another of Rudwick's works, *The Great Devonian Controversy* (Chicago: University of Chicago Press, 1985) gives a vivid picture of the community of geologists in Britain in the 1830s and 1840s—precisely the time when the earliest dinosaur discoveries were being made. The cast of characters includes several individuals who were undeniably brilliant but also argumentative to the point of mania or eccentric to the point of lunacy.

Just before this chapter was written, two very interesting books dealing with the earliest dinosaur discoveries came out: Deborah Cadbury's *Terrible Lizard: The First Dinosaur Hunters and the Birth of a New Science* (New York: Henry Holt, 2000) and Christopher McGowan's *The Dragon Seekers* (Cambridge, MA: Perseus Publishing, 2001). Each book tells the story of Mantell and Owen with clarity and style. Both books also tell the story of Mary Anning, certainly one of the most important and appealing of the early paleontologists. She was born into a poor family that lived on the coast of Dorset in the town of Lyme Regis. While still a child of twelve or thirteen, she discovered a nearly complete skeleton of an ichthyosaur, a dolphinlike marine reptile of the Mesozoic. For the rest of her life, before her death of breast cancer at the age of forty-seven, she supported her family with her fossil finds.

The best succinct account of the beginnings of dinosaur science is Hugh Torrens's chapter "Politics and Paleontology: Richard Owen and the Invention of Dinosaurs" in the outstanding anthology *The Complete Dinosaur,* edited by James O. Farlow and M. K. Brett-Surman (Bloomington, IN: Indiana University Press, 1997). Torrens does a fine job of reconstructing those events. The Farlow and Brett-Surman, anthology is the best single work on dinosaurs I have seen. It covers a wide range of topics, including dinosaurs and the media. The contributors are first-rate, and they show that yes, there is real, hard science behind the study of dinosaurs. Also, the book is beautifully illustrated.

3

Huxley Agonistes

By midlife, Richard Owen was sitting on the top of his world. Perhaps the most respected scientist in Britain, he had been offered more honors than he could accept. He even turned down a knighthood, probably so as not to excite the envy of senior colleagues who had not been so honored (McGowan 2001, 187). Then, in November 1859, a book appeared that shook Owen's world, and everyone else's—Charles Darwin's *On the Origin of Species by Means of Natural Selection, or The Preservation of Favoured Races in the Struggle for Life.* Owen, the long-standing critic of evolutionary theories, was confronted with a very serious challenge in Darwin's work. *The Origin* accomplished what no previous work had. It gave a plausible mechanism for evolution—natural selection—and a mountain of solid evidence supporting the theory. It was also written in engaging, at times even inspiring, prose.

In 1860 a review attacking *The Origin* appeared in the periodical *Edinburgh Review.* It was anonymous, but it was clear to everyone that Owen was the author. Darwin regarded the review as spiteful and biased and thought that it even resorted to intentional misrepresentation (Darwin 1860b). "It is painful to be hated in the intense degree that Owen hates me," mourned Darwin (Darwin 1860a). Owen's review was very negative. In his hostility he even went so far as to coach Bishop Samuel Wilberforce,

Title page of Charles Darwin's groundbreaking book on evolutionary theory, On the Origin of Species, *published in 1859.*

ON

THE ORIGIN OF SPECIES

BY MEANS OF NATURAL SELECTION,

OR THE

PRESERVATION OF FAVOURED RACES IN THE STRUGGLE
FOR LIFE.

By CHARLES DARWIN, M.A.,

FELLOW OF THE ROYAL, GEOLOGICAL, LINNÆAN, ETC., SOCIETIES;
AUTHOR OF 'JOURNAL OF RESEARCHES DURING H. M. S. BEAGLE'S VOYAGE
ROUND THE WORLD.'

LONDON:
JOHN MURRAY, ALBEMARLE STREET.
1859.

The right of Translation is reserved.

Portrait of Charles Darwin, the British scientist whose theory of evolution created immense controversy in the mid–nineteenth century (Library of Congress).

the unctuous "Soapy Sam" who attacked Darwin at the famous 1860 Oxford debate (see Ruse 2000, 59–60, for an account of this celebrated confrontation).

Darwin, a gentle soul averse to controversy by nature, refused to confront Owen or his other critics in public debate or published polemics. Not so T. H. Huxley. Thomas Henry Huxley (1825–1895) was an anatomist of incandescent intelligence and fiery temperament. His aggressive defense of evolution earned him the sobriquet "Darwin's Bulldog." His logic, eloquence, astonishing erudition, and mastery of the stinging riposte made him a fearsome opponent in any debate. For forty years he gleefully debunked the claims of scientists, theologians, politicians, and anyone else who had the temerity to cross swords with him.

By 1860 Owen had been the five-hundred-pound gorilla of British science for over two decades. In 1856 he became superintendent of the natural history departments of the British Museum; even his restless ambition could aspire to nothing higher. Huxley, in contrast, was the brash young upstart who differed from Owen in almost every way—except in having the same unbounded confidence in his own opinions. Owen was the very embodiment of establishment science. For instance, he never departed from the tradition of natural theology that saw God's creative hand in the marvelous adaptations of organisms. Huxley represented the new breed of totally secularized scientist; he coined the term "agnostic" to describe his own religious view. Huxley was also a social activist who sought to bring scientific enlightenment to the down-and-out. He worked on the London School Board to bring modern public education to the slums and regularly offered extremely popular evening classes and lectures for the benefit of workingmen. Owen always preferred the company of the up-and-in and took pride

Portrait of Thomas Henry Huxley (1825–1895), English biologist. Undated illustration from a painting (Bettmann / Corbis).

in rubbing shoulders with the rich, powerful, and famous. It was simply inevitable that Huxley and Owen would clash, and the publication of Darwin's *Origin* gave them ample opportunity.

Before *The Origin* Huxley had a low opinion of evolutionary theories, such as the one in the anonymous shocker of 1844, *Vestiges of the Natural History of Creation.* Darwin's book made him a convert, and he acted with a convert's zeal. Although he never completely accepted natural selection to the extent that Darwin did, he believed that *The Origin* had decisively shown that species had evolved, that is, that they had originated by a natural process of descent with modification from earlier species. For Huxley the chief benefit of Darwin's theory was that it at last permitted a completely naturalistic, materialistic approach to nature. Richard Dawkins, a present-day Darwinian, notoriously said that Darwin allowed him to be an intellectually fulfilled atheist (Dawkins 1987, 6). Huxley would have said that Darwin allowed him to be an intellectually fulfilled scientist. By 1859 science had already gone through a long, gradual process of emancipation from oversight by religion. Leading scientists in all fields, even those who were personally pious, had long declared their intention to seek natural, what they called "secondary," causes of phenomena rather than invoking creationist hypotheses (Gillespie 1979). For Huxley, the process of emancipation could now be completed. Supernatural and metaphysical hypotheses could be banished completely from science. Huxley viewed such hypotheses as worse than false. He regarded them as obscurantist, as leading science away from fruitful research programs and into the slough of superstition. His attitude toward the influence of religion on science was summarized in perhaps his most famous remark: "Extinguished theologians lie about the cradle of every new science as the strangled snakes beside that of Hercules" (quoted in Desmond 1997, xv).

Now, Owen was in no sense a biblical literalist, special creationist, or fundamentalist.[1] As noted earlier, he advocated seeking lawful, uniform processes to answer the questions about the origin of

---

1. Actually it would be an anachronism to describe anyone at the time as a "fundamentalist." Fundamentalism was an early-twentieth-century movement that insisted on interpreting scripture as inerrant. Traditionally, both Catholics and Protestants had permitted considerable latitude for the nonliteral interpretation of scripture (see Ruse 2001).

species. Yet he could not abide the complete materialism espoused by Huxley and implied by the theory of natural selection. He was convinced that natural processes had to be grounded in some sense in transcendent principles, that is, that nature must to some extent be understood metaphysically. He expressed this conviction in his theory of "archetypes."

According to Owen, all organisms, however diverse in the details of their anatomy, manifest a small number of basic body plans, or "archetypes." All vertebrates, for instance, embody the primal, generalized vertebrate plan. A newt and a koala bear are very different sorts of animals, but comparison of each with an invertebrate such as a squid shows that they share deep, fundamental similarities in overall structure. A synonym for "archetype" would be "paradigm" (except that this word has been debased by overuse and misuse)—an ideal model or prototype that sets the standard for identifying and evaluating all others of that type. It is important to emphasize that for Owen the archetype was not a flesh-and-blood organism; it was a transcendent or metaphysical entity. The concept of an archetype is an ancient one, grounded in Plato's theory of ideal forms. For Plato (427–347 B.C.E.), all physical things derive their nature and reality from a transcendent world of ideal forms that they "reflect" or in which they "participate." Virtues such as courage, values such as goodness, qualities such as beauty, and physical things such as cats are real for Plato because each is an image of its ideal counterpart—the eternal forms of "courage," "goodness," "beauty," and "cat," respectively.[2] For Christians, naturally, the Platonic forms could be construed as eternal ideas in the mind of God. For Owen, therefore, organisms pointed to archetypes, and archetypes pointed to God.

Evolution eliminates the need for any such concept as the archetype. For the Darwinian, the newt and the koala are more like each other than either is like a squid simply because they share a (much) more recent common ancestor with each other than with squids. There is no need to appeal to transcendent archetypes. The similarity has a much more mundane explanation—the retention of ancestral anatomical features by biological descendants. As far as we can tell today, the first vertebrate arose at some time during the early Ordovi-

2. Read Plato's greatest dialogue, *The Republic,* for his best exposition of his theory of the forms. The concept is illustrated by Plato's famous diagram of the divided line and his haunting myth of the cave.

cian period (about five hundred million years ago). The newt and the koala are descendants of that first vertebrate; the squid is not. Evolution is extremely creative; the fantastic frills, sails, appendages, and appurtenances it has given to so many creatures are evidence of this. Yet it is essentially conservative. If something is a good idea, that is, if it works in a given environment, evolution will keep it and build on it. In other words, evolution does not design; it tinkers. Evolution cannot "go back to the drawing board" and start with a whole new plan. It has to modify what is already there rather than create grand new structures out of nothing. Pigs do not have wings because vertebrate wings are modifications of forelimbs, which pigs and their ancestors put to other use long ago.

The scary thing about evolution is that it proposes a purely mechanical natural process that, operating over geological time, produces effects similar to those that we achieve by intelligent design. The most intuitively appealing argument for God has always been the one that sees the marvelous, intricate adaptations of organisms to their environments as analogous to the work of a human designer in creating complicated mechanisms such as watches. Evolution undercuts this argument by showing how "design" need not imply intelligence, or indeed any cause or principle other than the purely material. For Owen in particular, evolution threatened to explain how organisms could look like embodiments of metaphysical archetypes when, in reality, their similarities are due to a historical process of biological descent with modification. Faced with this threat, Owen refused to back down. In his magisterial *Paleontology*, published a year after *The Origin* in 1860, Owen concludes

> that the phenomena of the world do not succeed each other with the mechanical sameness attributed to them in the cycles of the Epicurean philosophy; for we are able to demonstrate that the different epochs of the earth were attended with corresponding changes of organic structure and that, in all these instances of change, the organs, still illustrating the unchanging fundamental types, were as far as we could comprehend their use, exactly those best suited to the functions of the being. Hence we not only show intelligence evoking means adapted to the end; but, at successive times and periods, producing a change of mechanism adapted to a change in external conditions. Thus the highest generalizations in this science of organic bodies, like the Newtonian laws of universal matter, lead to the unequivocal conviction of a great First Cause, which is certainly not mechanical (Owen 1860, 414).

In other words, Owen equates evolution with the most notoriously atheistic and materialistic doctrine of ancient times, Epicureanism. For the Greek philosopher Epicurus (341?–270 B.C.E.), reality was an infinite number of atoms in motion through an infinite void. Occasional random movements cause these atoms to interact in sheerly mechanical ways. Over infinite time, these accidentally commingling particles produce everything, including us. By the way, Darwinism is *not* Epicurean; Darwin emphasized the selective force of the environment as a decidedly *nonrandom, nonaccidental* element in evolution. At any rate, for Owen the adaptation of organisms to changing environmental conditions, while retaining the fundamental type, was plainly providential and simply could not be explained without ultimately appealing to a transcendent First Cause.

So Huxley supported evolution, and castigated Owen's theory of archetypes, because he supported a materialistic, naturalistic agenda that sought to drive God and metaphysics from science. Owen opposed evolution because he wanted to keep God and metaphysics. Thus, both Owen and Huxley were motivated largely not by purely scientific concerns but by philosophical, or, to put it more bluntly, ideological agendas (Desmond 1982). This conclusion raises the deep question of the relationship of philosophy (or ideology) to science. Ever since the Royal Society adopted the motto *"Nullius in Verba,"* roughly translated as "mere words are empty," science has striven to free itself from the sorts of verbose, stultifying, endless disputes typical of metaphysics and theology. The motto tells us to settle issues by looking and seeing, not by wordy and windy rhetoric, hairsplitting, and logic chopping. In the seventeenth century Robert Boyle pioneered the experimental method because he believed that it would lift science above the morass of politics and ideology and place it on an objective basis. So why does philosophy or ideology enter into science at all? Why not decide between competing hypotheses by performing the relevant empirical tests and relegate philosophy and ideology to the ash heap?

But it is hard to get rid of such "unscientific" considerations, if only because the human imagination is so fecund. The White Queen in *Through the Looking Glass* could think of six impossible things before breakfast. Given any batch of phenomena, clever humans can quickly think of six impossible hypotheses to explain it. The great value of empirical testing is that it provides rigorous, objective constraints on our theorizing. All sorts of theories may sound great, but few, one, or none will survive rigorous empirical testing. However, scientists just do not have the resources to test every theory that could conceivably be tested. Testing is often a long, slow, expensive, and laborious process, so we have to have some criteria to decide which hypotheses are even worthy of testing (Rothbart 1990). Perhaps the best way to put it is that a hypothesis must show considerable *promise* before it can even be a candidate for empirical testing.

How do we rate the promise of hypotheses? How do we tell, of the infinite number of possible hypotheses we could test, which ones are most likely to be worth our effort? A major consideration is track record: Have hypotheses of that sort had a good record when they were tested in the past? But philosophical considerations also inevitably enter into our "pretest" criteria. Consider the case of vitalism, the once very popular biological theory that proposed that life could not be explained in purely chemicophysical terms but only by the postulation of a nonphysical "vital principle" *(vis viva)*. Ernst Mayr describes the fate of such theories:

> This vitalistic school opposed the mechanists, believing that there are processes in living organisms which do not obey the laws of physics and chemistry. Vitalism had representatives well into the twentieth century. . . . However, by the 1920s or 1930s biologists had almost universally rejected vitalism, primarily for two reasons. First, because it virtually leaves the realm of science by falling back on an unknown and presumably unknowable factor, and second, because it became eventually possible to explain in physico-chemical terms all the phenomena which according to the vitalists "demanded" a vitalistic explanation (Mayr 1982, 52).

So vitalism was rejected both because it lost out to physical theories in the process of empirical testing and because it conflicted with *philosophical* convictions about how science should explain things. Philosophical convictions about the nature and goal of science dictate that science should postulate knowable processes or entities and not take refuge in occult principles. For both of the reasons Mayr notes, biologists eventually refused even to consider vitalistic hypotheses as candidates for serious testing. Vitalism became a dead issue in biology (Mayr 1982, 52).

It was therefore quite appropriate for Huxley and Owen to bring philosophical issues into their scientific debates. Huxley explains and defends scientific materialism in his 1868 essay "On the Physical Basis of Life." He carefully explains that the materialism he advocates is methodological, not metaphysical. That is, he argues that science must proceed by using materialist vocabulary and concepts and by accepting only materialist explanations of phenomena. However, he regards this stipulation as a requirement of method and procedure—as governing the way science should be *done*—not as the making of grandiose pronouncements about the nature of ultimate reality. He holds that metaphysical materialism, the claim that all that ultimately exists is matter and physical forces, is just as groundless as the worst theological dogmas (Huxley 1868b, 161–162). Huxley's argument against metaphysical materialism is epistemological; that is, it is based on a view of what humans can know and how they can know it. He colorfully argues that questions about ultimate reality are as unknowable as the politics of extraterrestrials:

> If a man asks me what the politics of the inhabitants of the moon are, and I reply that I do not know; that neither I, nor anyone else, has any means of knowing; and that, under these circumstances, I decline to trouble myself about the question at all, I do not think he has the right to call me a skeptic. On the contrary, in replying thus, I perceive that I am merely honest and truthful and show a proper regard for the economy of time. So Hume's strong and subtle intellect takes up a great many problems about which we are naturally curious, and shows us that they are essentially questions of lunar politics, in their essence incapable of being answered, and therefore not worth the attention of men who have work to do in the world (Huxley 1868b, 162).

The philosopher David Hume (1711–1776), perhaps the greatest ever to write in the English language, argued that giving answers to the deepest questions about ultimate reality simply exceeds human intellectual capacity. Therefore, we should curb the frivolous curiosity that carries us into the cloud-cuckoo-land of metaphysics and focus on the more mundane realities that we can know.

Huxley thinks *all* metaphysical doctrines about the ultimate nature of reality are equally groundless and unknowable. He favors the use of materialist terminology in science for both practical and philosophical reasons:

> With a view to the progress of science, the materialistic terminology is in every way to be preferred. For it connects thought with the other phaenomena of the universe, and suggests inquiry into the nature of those physical conditions, or concomitants of thought, which are more or less accessible to us, and a knowledge

Though a profoundly respected philosopher of the Enlightenment, David Hume is most remembered for his assault on reason as a reliable mode of philosophical inquiry (Library of Congress).

of which may, in future, help us to exercise the same kind of control over the world of thought as we already possess in respect of the material world; whereas, the alternative, or spiritualistic terminology, is utterly barren, and leads to nothing but obscurity and confusion of ideas (Huxley 1868b, 164).

In other words, the goal of science is to *understand,* and materialist hypotheses are the sort that the human mind can grasp. Also, such hypotheses can be empirically evaluated and hence confirmed or disconfirmed with the tools and techniques of science. Further, such hypotheses connect with other materialist hypotheses, and with diverse sorts of observable phenomena, in ways that suggest new and promising lines of inquiry. On the other hand, spiritualistic "explanations"—God, souls, vital forces, etc.—introduce alleged supernatural entities and occult forces that are generally inscrutable and are conjectured to operate in unknowable ways.[3]

Small wonder, then, that Huxley did not think much of Owen's archetypes. For Huxley, such a concept does no work; it is useless metaphysical baggage that explains nothing. It may soothe our spiritual qualms but at the price of intellectual obscurantism. Huxley thought the price too high to pay and urged that we bite the bullet

---

3. An instance illustrating Huxley's point is the special creationist's "explanation" of where birds come from (see Appendix II): God says, "Let there be birds!" and *poof!* The air is full of feathered creatures! There is no way to know how God did it, no way to discover a causal modus operandi, or any way to discover "laws of supernature" that prescribe uniform and predictable rules for divine creation. God, for reasons and in ways unknowable to us, just gives the word, and *poof!* It happens. Such an "explanation" is obviously no explanation at all; it serves merely as a marker for our ignorance. By contrast, the evolutionary explanation in terms of natural selection, whether right or wrong, is at least comprehensible.

A famous cartoon by Sidney Harris also illustrates these points: Two scientists stand in front of a blackboard. The left and right sides of the board are filled with mathematical scribblings. In the middle, which is labeled "step two," the board is blank except for the declaration "and then a miracle occurs." One scientist comments dryly to the other: "I think you need to be more explicit here in step two."

and face the prospect of a universe wholly explicable in godless, materialist terms.

The historian Adrian Desmond thinks that Huxley was far too harsh on Owen's archetypes, and that his opposition to the idea was largely ideological (Desmond 1982, 44, 42). Desmond argues that the concept was scientifically useful and in fact promoted the idea of evolution by providing the notion of a generalized type with more specialized descendants (44). For instance, the modern horse, which runs on one, highly modified, toe, is a specialized development of its five-toed fossil ancestors. This argument raises an important point: All sorts of concepts, even mystical or metaphysical ones, can provide scientifically useful ideas. For instance, despite Huxley's infamous crack about dead theologians lying around the cradle of science, the very idea of a universe governed by unbroken natural law received theological support from the Reverend Baden Powell (1796–1860). Powell argued that a rational God had designed an orderly universe that needed no adjustments or fine-tuning with ad hoc miracles (Desmond 1982, 45).

Still, Huxley had a point. The archetype idea may in fact have been somewhat scientifically useful, and Huxley's opposition may have been ideologically motivated, but this does not mean that his criticisms were unjustified. Criticism of Platonic archetypes goes back to Aristotle, and even to Plato himself (Plato develops the famous "third man" argument against the doctrine of the forms in his late dialogue *Parmenides*). The gist of many of these criticisms is that archetypes constitute an ideal world that merely parallels the physical world without explaining it. The explanatory emptiness of archetypes was one of Huxley's main points too. Many bad ideas have proven useful for a while, but their temporary usefulness should not exempt them from criticism.

Where do dinosaurs fit into all of this? We have seen that dinosaurs can be useful tools in ideological disputes. In particular, Owen thought that dinosaurs were much more advanced and higher animals than any living reptile. Therefore, they serve as counterexamples to theories of progressive evolutionary development. For Huxley, though, dinosaurs were the best evidence for evolution. In fact, he viewed them as evidence of a major evolutionary transition: the transition from reptiles to birds.

As the story is usually told, it was the discovery of *Archaeopteryx lithographica*—the Jurassic bird with claws on its wings, and teeth instead of a beak—that proved Huxley right by providing the missing link between reptiles and birds. As Desmond notes, however, Huxley did not regard *Archaeopteryx* as the missing link because he held that the transition from reptile to bird had occurred long before the Jurassic (Desmond 1982, 128). However, he noted that *Archaeopteryx* had a number of distinctly reptilian features: "The leg and foot, the pelvis, the shoulder-girdle, and the feathers, so far as their structure can be made out, are completely like those of existing ordinary birds. On the other hand, the tail is very long, and more like that of a reptile than that of a bird in this respect. Two digits of the manus have curved claws, much stronger than those of any existing bird; and, to all appearance, the metacarpal bones are quite free and disunited. Thus it is a matter of fact that, in certain particulars, the oldest known bird does exhibit a closer approximation to reptilian structure than any modern bird" (Huxley 1868a, 70).

In fact, with further study and with the advantage of more specimens than Huxley had, we now know that he considerably understated the case. As Christopher McGowan shows, when we consider the anatomical traits that distinguish reptiles from birds, the *Archaeopteryx* had only a few typically avian features, such as feathers and a furcula (wishbone). The remainder of its traits are typically reptilian (McGowan 1983, 117). One could hardly ask for a more perfect candidate for the missing link between reptiles and birds.

Huxley found further evidence in the similarity of dinosaur and bird anatomy. The small dinosaur *Compsognathus,* a bipedal, chicken-sized Jurassic predator, had long legs that looked remarkably birdlike. Huxley mentions a number of such points and concludes: "It is impossible to look at the conformation of this strange reptile and to doubt that it hopped or walked, in a erect or semierect position, after the manner of a bird, to which its long neck, slight head, and small anterior limbs must have given it an extraordinary resemblance" (Huxley 1868a, 74). Huxley did not consider *Compsognathus* the missing link between birds and dinosaurs or hold that birds had descended from dinosaurs. Again, Huxley believed that the transition had occurred far earlier in geological time. Rather, he saw both birds and dinosaurs as retaining similar anatomical features from a common ancestor, so proving that birds had branched off from the reptilian lineage.

Fossil footprints also supported Huxley's hypothesis. In 1835 Edward Hitchcock, president of Amherst College in Massachusetts, found huge, birdlike tracks in the Triassic sandstone near the Connecticut River (Desmond 1976, 43). Hitchcock thought they were the footprints of giant wading birds, but Huxley placed a new interpretation on them: "The important truth which these tracks reveal is, that at the commencement of the Mesozoic epoch bipedal animals existed which had the feet of birds, and walked in the same erect or semierect fashion. These bipeds were either birds or reptiles, or more probably both; and it can hardly be doubted that a lithographic slate of Triassic age would yield birds so much more reptilian than *Archaeopteryx,* and reptiles so much more ornithic than *Compsognathus,* as to obliterate completely the gap which they still leave between reptiles and birds (Huxley, 1868a, 74). Though no uncontroversial instance of a Triassic bird has been found, Huxley's hope for the discovery of more ornithic (birdlike) dinosaurs has been abundantly fulfilled (more on this in Chapter 7).

The evidence that has accumulated since Huxley's day has turned into a truism his once radical hypothesis that birds descended from reptiles (though, as we shall see, whether birds came from *dinosaurs* is still hotly debated). Huxley's own evidence was not overwhelming, but it was just one piece of a growing body of evidence and argument that by 1870 was turning the scientific community toward evolution and increasingly marginalizing and isolating antievolutionists like Owen.

So much then for Owen's ponderous *Iguanodon* that looked like a gigantic scaly rhinoceros! Instead of symbolizing the end point of reptilian design—the ultimate manifestation of the reptilian type—small, bipedal, birdlike dinosaurs illustrate the process of evolutionary diversification. So much, then, for Owen's archetypes. If reptiles can change into birds, then it appears that there are no ultimate divisions in nature. The power of evolution can alter a "type" of organism indefinitely and finally change it into a different "type." Faced with Huxley's onslaught, Owen was still defiant, but, unlike the old soldiers who just fade away, a worse fate often befalls old scientists. A new generation comes along that simply does not listen to them. Not only did the new generation of evolutionists not listen to Owen, they painted him as the archobscurantist who sought to block scientific progress because of ideological prejudice.

Huxley had one more trick up his sleeve. To conservatives like Owen, it was shocking enough for Huxley to suggest that birds and dinosaurs had a common evolutionary ancestor. Even more outrageous was Huxley's suggestion that, like birds, dinosaurs may have been warm-blooded! Reptiles, of course, are cold-blooded. Actually, the terms "warm-blooded" and "cold-blooded" are misleading since they do not have anything to do with the actual temperature of the blood. Rather, they designate how internal temperatures are generated and maintained. Warm-blooded animals (endotherms) generate enough heat through metabolic processes to maintain the body core at a steady temperature. We humans are endotherms, of course; a healthy person has a steady internal temperature of approximately 37 degrees Celsius or 98.6 degrees Fahrenheit. Cold-blooded creatures (ectotherms) do not generate sufficient heat metabolically to maintain high core temperatures. They must bask in the sun or otherwise obtain heat from external sources. Today's reptiles are all ectotherms, so Huxley's suggestion was radical indeed.

The important thing about Huxley's speculation, and at this point it was merely a speculation, is that it marks the beginnings of interest in dinosaur physiology and behavior—topics of much current interest. It may seem a stretch to think that we can know anything about the physiology or behavior of fossil creatures, but bones and trackways can tell us a lot (Chapter 6 will develop these points). Dinosaurs are so interesting that we are not satisfied merely to reconstruct their anatomy. We want to know everything about them— what they ate, how they reproduced, how fast they ran, what happened to them—but we are stymied by the myriad centuries separating them from us. The paucity of fossil materials was a particularly acute problem for Huxley and his contemporaries in the 1860s. The Mesozoic strata of Britain and the eastern United States had provided paleontologists with few and very incomplete specimens. This was about to change. The opening of the American West, symbolized by the completion of the transcontinental railroad in 1869, made available the astonishingly rich bone fields of Wyoming, Utah, and Colorado. The focus of dinosaur paleontology quickly shifted from Europe to America as all the dinosaurs that are most familiar to us today—*Brontosaurus, Tyrannosaurus, Triceratops*—emerged from the great Jurassic and Cretaceous formations. The hunt for dinosaurs soon assumed all of the rip-roaring, shoot-'em-up, claim-jumping rowdiness of the Wild West. In particular, two protagonists emerged to engage in the scientific equivalent of the shootout at the OK Corral. This conflict is the topic of the next chapter.

Further Readings

T. H. Huxley is a fascinating character. Born into a family living in very modest circumstances, he succeeded, by vast labors of pulling himself up by his own bootstraps, in becoming a superstar of Victorian science. It would be hard to find a better exemplar of the typical Victorian virtues of hard work, enterprise, and persistence in the face of adversity. Adrian Desmond's superb biography is titled *Huxley: From Devil's Disciple to Evolution's High Priest* (Reading, MA: Addison-Wesley, 1997). That title pretty much sums it up. Considered purely as a scientist, Huxley was perhaps not quite of the first rank. However, he was the archetype of the scientist as public intellectual, a role more recently played by such scientists as Stephen J. Gould and Carl Sagan. A very good study of Huxley's science is Mario A. Di Gregorio's *T. H. Huxley's Place in Natural Science* (New Haven: Yale University Press, 1984). Huxley's own writings retain their edge in the present day: They are crisp, clear, witty, and incisive. Huxley's learning was simply prodigious. *Man's Place in Nature* was his most important scientific work and is still very much worth reading. His famous essay "On a Piece of Chalk," originally delivered as a talk to working men, remains an engrossing account of the true depths of time discovered by paleontologists.

Richard Owen's complex and disputatious personality makes him a difficult and controversial figure even today. Though he was unquestionably the most brilliant anatomist in Victorian Britain—fully deserving of the accolade "the English Cuvier"—his reputation was irremediably tarnished by his bitter and intransigent opposition to Darwin. Desmond's 1982 book *Ancestors and Archetypes: Paleontology in Victorian London 1850–1875* (Chicago: University of Chicago Press) is a sustained effort to rehabilitate Owen's reputation. Desmond argues persuasively against the traditional stereotype of Huxley as the brilliant young champion of objective science and Owen as the stodgy, ideologically blinded obscurantist. Though Owen's opposition to evolution may have been ideologically based, Desmond notes that Huxley also had his axes to grind, and the evidence did not unambiguously favor his view. However, in my opinion, *Archetypes and Ancestors* goes too far by implying that ideology rules in science. Owen's critique of *The Origin of Species* and expert commentary is provided by David Hull in *Darwin and His Critics: The Reception of Darwin's Theory of Evolution by the Scientific Community* (Chicago: The University of Chicago Press, 1973).

Naturally, it is impossible to understand mid-to-late-Victorian paleontology without putting it in the wider context of the debates over Darwinism. No educated person has any excuse for not reading *The Origin of Species*. The secondary literature on Darwin and Darwinism is astronomically vast. Even the most superficial survey of that literature would be impossible here. My favorite biography of Darwin is *Darwin: The Life of a Tormented Evolutionist* (New York: Warner Books, 1991) by Adrian Desmond and James Moore. The story of the Darwinian revolution is best told by Michael Ruse in *The Darwinian Revolution: Science Red in Tooth and Claw* (Chicago: University of Chicago Press, 1979). Unlike the Copernican or Newtonian revolutions in science, the Darwinian revolution continues to the present day. In *Darwin's Dangerous Idea: Evolution and the Meanings of Life* (New York: Simon and Schuster, 1995), the philosopher Daniel Dennett argues that we are still learning the lessons Darwin taught.

4

Range War

The conflict between Edward Drinker Cope (1840–1897) and Othniel Charles Marsh (1831–1899) has taken on legendary proportions. A recent book lists the Cope-Marsh feud as among the ten "liveliest" in the history of science (Hellman 1998). Actually, "lively" hardly does justice to the sheer rancor mutually expressed by the two protagonists. Though they battled with bones instead of bullets, the mind's eye sees them squaring off like gunslingers against the backdrop of one of John Ford's classic western landscapes. As usual, though, reality was much more complex and interesting than legend. Unlike *High Noon* or *Shane,* it was not simply black hats versus white hats—no simple good guy/bad guy dichotomy does justice to the facts.

The Cope-Marsh conflict did unfold against the dramatic backdrop of the post–Civil War opening of the American West to mining, ranching, agriculture, and commerce. The completion of the transcontinental railroad, with the meeting of the Union Pacific and Central Pacific teams at Promontory Point, Utah, on 10 May 1869, symbolizes the push westward. The period of 1870–1890, commemorated in innumerable pulp novels, movies, and television shows, has entered mythology as the Wild West. The West certainly offered the stuff of myth and legend—vast and

Edward Drinker Cope (1840–1897) (The Natural History Museum, London).

The first vertebrate paleontologist employed by the U.S. Geological Survey, O. C. Marsh is remembered for his four major expeditions into the American West. During these trips he discovered fossil remains of the early horse, primates, winged reptiles, dinosaurs, and birds with teeth. His explanation of these finds helped both the scientific community and the public accept Charles Darwin's then controversial concept of evolution (Library of Congress).

dramatic scenery, desperadoes and frontier justice, and the last valiant but futile resistance of the American Indians. The West is also legendary in the annals of science. It is a geologist's dream and a vertebrate paleontologist's paradise.[1] Uplift and erosion, faulting and folding—all the slow but inexorable forces of geology—have given the West its unique beauty. These forces also exposed vast formations of sedimentary rock that contain some of the world's richest fossil beds. For instance, the Carnegie Quarry near Vernal, Utah, yielded so many fossils that it became pointless to remove any more. Today visitors to Dinosaur National Monument, located at the site of the former Carnegie Quarry, can see dinosaur bones left in the ground where the collectors abandoned them when the museums could hold no more.

The scientific exploration of the West was conducted by appropriately colorful characters, such as John Wesley Powell and Ferdinand Vandiveer Hayden. Powell lost an arm in the Civil War, but the injury did not impair his incredible energy and stamina. A self-taught geologist, ethnologist, and linguist, he led an epic expedition down the Colorado River through the Grand Canyon. In the 1850s Hayden sought fossils in the upper reaches of the Missouri River country as far as the junction of the Yellowstone and Bighorn Rivers in present-day Montana (Lanham 1973, 35). The territory was Sioux country at the time, and the Sioux, incensed by increasing encroachment into their lands, began fighting the U.S. government in 1854. Hayden was safe, though. The Sioux called him "the man who picks up stones while running." They regarded him as an obvious lunatic, and hence a holy person not to be harmed (Lanham 1973, 35).

---

1. Two books by John McPhee beautifully express the wonder and fascination of the geology of the arid west. *Basin and Range* and *Rising from the Plains* are included in McPhee's *Annals of the Former World* (New York: Farrar, Straus, and Giroux, 1998).

Hostilities between the government and Native Americans flared up again in 1870 when Marsh led a group of Yale students on a fossil-hunting expedition. True to form, the meticulous Marsh had gone to the top generals and arranged a military escort (Lanham 1973, 80). "Be prepared" could have been Marsh's motto long before the Boy Scouts adopted it. Like another late bloomer, Charles Darwin, he made up for a slow start with hard work, attention to detail, prudence, and making the right connections. Marsh was twenty-one when he entered prep school at Phillips Academy, but he quickly climbed to the top in all of his classes and swept all of the school's academic honors. With the help of a wealthy uncle, George Peabody, he entered Yale College, where he again distinguished himself, graduating near the top of his class as a member of Phi Beta Kappa. After receiving a master's degree from Yale, he completed his academic training in Germany. Upon returning to Yale, Marsh was appointed professor of paleontology. No doubt he had received excellent academic training in Germany, but he probably helped his case by securing a $150,000 donation from his Uncle George. The money was used to found the Peabody Museum of Natural History at Yale.

The Greek philosopher Heraclitus said, "Character is destiny." Perhaps then the personalities of Cope and Marsh made conflict inevitable. Each had a sizable and rather touchy ego, and neither was modest about his abilities or accomplishments. Unlike the late bloomer Marsh, Cope was a prodigy who in early childhood showed great promise as a naturalist. Compared to Marsh, whose style was slow and steady, Cope could produce a profusion of scientific papers in a short time. His lifetime output was an astonishing total of 1,400 published scientific papers. His contributions to herpetology and ichthyology alone were so great that a leading journal in these fields is named *Copeia* in his honor. His personality was complex. As a young man he suffered from depression severe enough to cause him to doubt his sanity. He was combative and sensitive to slights, yet colleagues and employees frequently commented on his geniality and friendliness. Marsh, by contrast, was recalled as suspicious and crotchety, and his treatment of subordinates was often unconscionable. Privately, Cope was a devoutly religious Quaker whose devotion approached fanaticism (Lanham 1973, 68). Whereas Marsh's fossil expeditions were carefully planned, Cope's were often poorly prepared, but his brilliance, energy, and skill compensated for the haphazardness. Though he was not university educated, he had a sufficiently distinguished reputation to be appointed a professor at Haverford College in 1864. In 1869, at the age of twenty-nine, he sold some property and invested the proceeds, allowing him to resign his professorship and pursue his research independently. His home in Philadelphia became his laboratory and museum.

The conflict between Cope and Marsh has gotten so much publicity that it is easy to overlook their scientific contributions. Each was an outstanding scientist. Between the two of them, they brought to life the world of the dinosaurs that has entered the popular imagination. Consider some of the genera named just by Marsh: *Allosaurus, Apatosaurus, Triceratops,* and *Stegosaurus.* Despite spectacular dinosaur discoveries in Asia and Africa in recent years, the North American dinosaurs of Cope and Marsh are still the most familiar. When most people think of dinosaurs, they think of Cope and Marsh's discoveries. In his lifetime, Marsh described 496 species of fossil vertebrates. Cope more than doubled this number. In 1900 there were 3,200 species of known North American fossil vertebrates. Cope had described 1,115 of these (Lanham 1973, 162).

Yet Cope and Marsh were far more than just bone collectors. The noted physicist Lord Rutherford once snidely quipped that there are two kinds of science—physics and stamp collecting. It is easy to tag paleontology with Rutherford's sneer. Paleontologists spend a great deal of time collecting fossils. Much of their work involves description and classification rather than high-powered theorizing. Yet even cut-and-dried accounts of bones and teeth, presented in austere anatomical jargon, are full of theoretical significance. Small-scale evolution (microevolution) has been observed in laboratories and in the wild hundreds of times, but big-scale evolution (macroevolution) can be seen indirectly only in the fossil record (see Appendix II). Biologists can watch beak size change in successive populations of finches, but they cannot observe, say, reptiles evolve into birds. To get the big picture in evolution, you have to look at the fossil record.[2] When organisms evolve, do they progress in any meaningful sense? Does evolution occur at a steady, gradual pace, or is it long periods of stasis punctuated by rapid bursts of development? Did the dinosaurs flourish until a single cataclysmic blast wiped them out, or did they dwindle slowly until a final disaster delivered the coup de grâce? These and

---

2. Creationists say that the fact that the large-scale events in evolution cannot be directly observed shows that evolution is not real science. But science is full of processes that never have and never will be observed (see Appendix II).

other big questions can be answered only by referring to the fossil record. The exacting, patient, and incredibly detailed field and laboratory work gives substance to the paleontologist's theories. Cope and Marsh both contributed significantly to theoretical discussions. To put their contribution in context, we need to consider the relation of paleontology to the larger issues of evolutionary biology in the late nineteenth century.

As we saw in Chapter 3, the idea of progress was strongly associated with evolutionary thought prior to Darwin. But even for Darwin himself the idea had a strong hold. The notion that the course of evolution must be progressive, beginning with the lowest organic beings and culminating with the highest, has been very persistent. It owes its durability to one of the most powerful ideas in the history of Western thought, the Great Chain of Being (Lovejoy 1936 is still the classic study). The Great Chain of Being postulated that all of reality was arranged hierarchically, with God (of course) at the top and sheer, unformed matter at the bottom. Further, reality was a plenum: Between the extremes of highest and lowest, every possible niche was occupied by something. Just below God were the heavenly beings such as the archangels, cherubim, and seraphim. Just slightly above inanimate matter were the lowest organic beings—pond scum, mold, leeches, politicians. Naturally, humans were the highest organic beings, positioned on the scale slightly lower than the angels.

Believers in biblical creation thought that the Great Chain came into existence all at once, and, like medieval serfs, the lower organisms and all their offspring were at the bottom forever and could not aspire to anything higher. With the stirring of evolutionary thought at the end of the eighteenth century, however, thinkers such as Lamarck and Erasmus Darwin (Charles's grandfather) began to think that organisms could in time be transmuted into something higher. Lamarck believed that inanimate matter was constantly being transformed, by a process of spontaneous generation, into the lowest of living things. These lowest organisms in time are transmuted into higher beings, and this process continues up through apes that are struggling to become human. In fact, for Lamarck, the struggles of individual organisms do promote evolution. By their own efforts organisms can improve the functioning of their organs—like weight lifters working out at the gym—and these self-improvements can be passed on to offspring. So the hierarchy of the Great Chain still existed for the evolutionists, but it became a ladder that creatures could slowly climb. Just as the Enlightenment threw off the last vestiges of feudalism and fostered a philosophy of progress, so evolution made progress possible, indeed inevitable, in the natural realm.

In the nineteenth century, Lamarckism had a double appeal. Social progressives liked the idea that progress was inevitable in nature and were happy to extrapolate the idea and apply it to society. Lamarckism implied that by exerting effort, creatures could improve themselves and their offspring. In turn, this notion implied that since those with humble backgrounds could improve themselves with hard work and initiative, society should permit social mobility and eliminate the rigid class system. Social conservatives found in the idea of organic progress a silver lining to the dark cloud of evolution. Lamarckism kept humans at the top of the chain of organisms, where tradition and religion said they should be, and allowed organic progress to be seen as the unfolding plan of a benevolent deity.

Darwinism threatened the whole idea of organic progress. According to Darwin, natural selection operates on chance variations. The needs of the organism have no influence on which variations arise. For instance, a plant species growing in a dry environment might need deeper roots to tap into the meager groundwater. According to Darwin, the plant's need for water does not in any way make it more likely to develop deep roots. Some seedlings, by chance, might be endowed with genes for deeper roots; others will just happen to get genes for roots that are standard length or shorter. The environment, and competition with other plants of that species, will then cull out the shorter-root ones, leaving only the longer-root specimens to pass on their genes to the next generation. Thus, over time populations of that plant will become better adapted to a dry environment. But this process is not progress in any absolute sense; it is just adaptation to a particular environment. Should the environment become wet, the longer-root plants will be at a disadvantage because they will waste large amounts of energy growing needlessly luxuriant roots. So long-roots are not in any sense intrinsically superior to or higher than shallow-roots; they just happen to survive better in given environments.

For Darwin, evolution was just the gradual accumulation of these adaptations to local conditions, meaning that an organism is not intrinsically higher than whatever it evolved from. Darwin had a long and sometimes painful correspondence with the American botanist Asa Gray. Gray, an admirer of Darwin but also a conservative Calvinist Christian, believed that divine providence arranged for certain favorable variations to arise in populations at the appropriate times. Darwin made a deep and very detailed study of variation and could discern no overall pattern whatsoever. As far as he could tell, organic features varied in all sorts of ways, some few of which just might happen to be advantageous to the organism in a given environment. Hence, there was no evidence that a plan of inevitable progress or improvement was unfolding in the evolutionary process. The evolutionary journey has no planned itinerary. Some might insist that there is a plan nonetheless, but Darwin rightly observes that it is meaningless to say that there is a plan that looks exactly like no plan at all.

Still, the idea of progress had strong appeal even to staunch Darwinians such as Marsh. Although as a Darwinian, Marsh did not believe in automatic or designed progress, he did think that there was a sense in which net progress does occur. For instance, he claimed that the fossil record showed an overall increase in brain size among mammals during the Tertiary Period (Marsh 1876; Bowler 1976, 137). Further, it was the cerebral hemispheres, the part of the brain associated with intelligence, that got larger and more convoluted (Marsh 1876). For Marsh it was clear that bigger brains were more useful and so tended to be preserved by natural selection. Further, it seemed to Marsh that bigger-brained, and hence more intelligent, animals are certainly more highly organized and more advanced than less intelligent ones. This association of primitiveness and stupidity still exists in our culture.

The evidence for progress in evolution seemed much weaker to T. H. Huxley (Bowler 1976, 138). For Huxley, a salient aspect of the fossil record was not progress but "persistence of type." In his view, evolution is essentially a conservative process in which the details may change but basic forms are preserved for vast periods of geological history. In fact, some of Marsh's own discoveries showed how vexed is the question of evolutionary progress. Marsh discovered two toothed birds from the Cretaceous, *Hesperornis* and *Ichthyornis*. Of course, teeth would seem to be a very primitive feature for birds since no modern bird is toothed (though experiments have shown that birds still have the genes for growing teeth). Yet the teeth of *Hesperornis* were more generalized than the specialized teeth of *Ichthyornis*. Since evolution often works to develop specialized characteristics from more generalized ones, generalized characters seem less evolved, and therefore more primitive, than specialized ones. So *Ichthyornis* seemed to have had more advanced teeth than *Hesperornis*. Yet, by applying equally plausible criteria, the vertebrae of *Ichthyornis* seemed more primitive than those of *Hesperornis* (Bowler 1976, 136). One organism could seemingly be both more advanced *and* more primitive than another—so which was more progressive? The fossil record in some ways seems to support our notions of organic progress but in other ways defies those concepts. In fact, clear-cut criteria of progress are hard to formulate. (See Dawkins 1992 for an excellent discussion of the problems of identifying evolutionary progress.) Some evolutionists, notably Ernst Mayr (2002), still endorse ideas of evolutionary progress, but most repudiate the concept.

Cope, the devout Quaker, simply could not accept the Darwinian view that natural selection acting upon chance variation was the driving force of evolution. For Cope, as for many scientists, especially those with strong religious convictions, this process was simply too bleak and uncertain. It implied that untold numbers of organisms were just wasted and that survival was a cruel lottery imposed by implacable nature. In the final sonorous paragraph of *The Origin,* Darwin embraces this terrifying claim: "Thus, from the war of nature, from famine and death, the most exalted object we are capable of conceiving, namely, the production of the higher animals, directly follows" (Darwin 1979, 459). In a number of publications, such as "On the Origin of Genera" (1868), Cope argues that natural selection acting upon chance variation could not be the whole story. He simply could not accept that the variations that prove so useful to organisms could have arisen by chance. Like Asa Gray, he believed that there had to be a divine plan that caused organic characters to vary in beneficial ways. He went considerably beyond Gray in advocating a frankly Lamarckian view whereby advantageous features acquired by the efforts of organisms could be passed on to their offspring (Jaffe 2000, 152). As Mark Jaffe notes, this was a kinder, gentler evolution that religious scientists could more easily accept as God's plan of creation (152). In fact, Cope became a leader of the "American School" of neo-Lamarckian paleontology, along with such noted scientists as A. S. Packer Jr. and Alpheus Hyatt (Roberts 1988, 86).[3]

---

3. Again let me caution against making Whiggish judgments. Because Cope opposed Darwinism, which ultimately triumphed, we might be tempted to see his view as retrograde compared to the more conventionally Darwinian Marsh. But in the late nineteenth century, Darwin's victory was only partial. He convinced most scientists that evolution had occurred but did not convince the majority, not even among his closest allies, that natural selection was the cause. In fact, Darwinism as

How did the famous feud begin? It is always hard for historians to answer such questions. Naturally, the parties involved blamed each other, and since passionate squabbles usually get people near the scene to line up on one side or the other, it is hard to get objective firsthand testimony. One well-known incident seems to have led to a lot of hard feeling. While they were still on speaking terms, Cope invited Marsh to see his reconstruction of *Elasmosaurus,* an extremely long-necked marine reptile of the Cretaceous. When Cope got the skeleton, it was a jumbled mass of bones. He set about reconstructing it, using the anatomical principles that Cuvier had pioneered at the beginning of the century. When finished, he had an imposing thirty-five-foot-long creature, so it was certainly with considerable pride that he invited Marsh to view his monster. It was a monster indeed. Cope had put the skull on the wrong end. Marsh was probably not overly tactful in pointing this out, and Cope was no doubt mortified.

However it began, the feud grew in bitterness year after year until, in the 1890s, it threatened the careers of both men. Much of the conflict grew out of competition for fossils. Vertebrate fossils are rare, and when found they are usually just assorted bits and pieces. Occasionally, though, sites are found where the ground is strewn with fossils, and a little digging will produce many excellent specimens. Como Bluff in Wyoming was such a site. In 1877 Marsh got a mysterious letter from two individuals who offered to disclose the location of a very rich bone field if the price for their information was right. Though the letter writers used assumed names to protect their identity, Marsh soon discovered that the men were employees of the Union Pacific Railroad named W. E. Carlin and W. H. Reed. The site was Como Bluff, where the bones of dinosaurs lay scattered like logs on a forest floor (Lanham 1973, 174). Marsh hired Carlin and Reed and sent his assistant, S. W. Williston, to supervise the digging. Marsh gave the men strict instructions to be on guard to keep Cope and his men away from the fossils, but precisely how they were supposed to do that was unclear since they had no legal claim to the land.

---

an explanatory *theory* of evolution was eclipsed until the emergence of the neo-Darwinian synthesis in the twentieth century (Bowler 1989). Therefore, Cope was not being an oddball or in any sense unscientific when he rejected full-scale Darwinism. In fact, we noted that Marsh, and even Darwin himself, continued to find the notion of progress attractive.

Como Bluff, Wyoming, with Elk Mountain in the background (Jonathan Blair / Corbis).

Inevitably, Cope soon heard of the riches of Como Bluff and sent out his own collectors. Carlin was not the most loyal or reliable of workers, and he soon defected to Cope. Considering Marsh's behavior toward employees, it is remarkable that he did not have more defections. He was legendarily tightfisted. While working in one quarry, Reed was forced to carry heavy bones over a mile of rough country and across a flooded stream because Marsh was too cheap to hire a horse. Carlin and Reed both complained frequently that their paychecks were a month or more in arrears. Marsh was equally stingy with credit for scientific discoveries. Subordinates had to submit their findings to him, and he would then publish them under his own name. Nevertheless, he kept enough loyal workers to defend his claims against the incursions of Cope's men. Such was his rancor that, according to Williston, Marsh actually ordered bones destroyed rather than left where Cope could find them.

For years Cope had fumed because his fortunes had declined while Marsh's rose ever higher. In part this was because Marsh had the right connections to get a lucrative position as paleontologist for the United States Geological Survey (USGS). Marsh received about $150,000 from the USGS, which allowed him to hire a large staff of collectors, preparers, artists, and scientists (Jaffe 2000, 254–255).

Part of Cope's misfortune was his own unwise decision to invest in gold and silver mines, a very risky venture that cost him dearly. Deeply embittered, Cope began a covert campaign to destroy Marsh. He and a friend put together a lengthy attack on Marsh and John Wesley Powell, then head of the USGS, which circulated privately among members of Congress (Lanham 1973, 245). Cope was aware of Marsh's bad treatment of subordinates, so he began to cultivate them to see if they would turn on their employer. A number did; Williston wrote a letter to Cope (an extract of this letter is published in Lanham 1973, 252) that burned with indignation and contained many damning allegations about Marsh's character and science.

In 1889 Marsh struck back. The secretary of the Interior, no doubt with Marsh's encouragement, demanded that Cope turn over his fossil collection to the federal government on the shaky grounds that he had collected the fossils while a member (actually, he was only a volunteer) of the federally funded Hayden survey. Cope escalated the feud by making it public. With the help of a young journalist from the *New York Herald,* his charges against Marsh, Powell, and the USGS became front-page news. The litany of accusations against Marsh was rather long and included claims of plagiarism and professional incompetence. Marsh replied at length and with asperity, even dredging up the story of the front-to-back *Elasmosaurus* to humiliate Cope. In fact, Marsh generally got the better of the exchange. However, though he may have lost the rhetorical exchange, Cope's timing was impeccable. His onslaught against Marsh and Powell got the ear of influential politicians who, with the coming of recession and hard times in 1892, began to question public expenditures on esoteric subjects such as paleontology. One Alabama congressman harrumphed loudly about taxpayers' dollars being spent to find birds with teeth (it is amazing how little some things change in a hundred years). The upshot was that in 1892 Marsh's appropriation was cut off, and he had to resign from the USGS. Cope had his revenge.

John Wesley Powell, explorer, scientist, and head of the U.S. Geological Survey, also became a target of Cope's campaign to destroy Marsh (Library of Congress).

The melodramatic spectacle of two eminent professors engaged in holy war was prime material for the yellow-press journalists of the day. Today it would probably rate an airing on the *Jerry Springer Show*. Such scandalous behavior may titillate, but it cannot really be the focus for those engaged, as we are here, in trying to tell a story about *science*. The important question therefore is how Cope's and Marsh's feud affected the practice of science. As noted in Chapter 1, conflict, even when it is bitter and protracted, is not necessarily bad for science. As Url Lanham notes, "durable, intelligent, well-focused hatred" can be as powerful a creative force as love (Lanham 1973, 271). It is unlikely that either Cope or Marsh would have been nearly as productive if each had not had the strong motivation to outshine the other. Still, when competition is too fierce, the temptation to cheat, or at least to cut corners, can be overwhelming.

Paleontology is not classified as a "hard" science, like, say, particle physics. The hard/soft dichotomy in classifying sciences is based on how rigorously the data constrain hypotheses in that field. To say that the data rigorously constrain our hypotheses is to say that the number of permissible interpretations of those data is strictly limited. Experimental results in any field are prized to the extent that they narrow the range of reasonable interpretations of those results. The ideal, hardly ever actually achieved, is the "crucial experiment" whereby only one of the candidate hypotheses is unambiguously shown to be correct. In paleontology the data almost always leave considerable room for competing interpretations (though much more precise and rigorous methods exist now than in Cope and Marsh's time. See Chapman 1997 and Ward 1998). To minimize this problem, paleontologists are extremely painstaking in their collection of fossil data, excavating with great care and precisely recording all relevant information. To have seen a good paleontologist at work would have given Freud an entirely new definition of "anal retentive." Sloppiness in collecting or classifying fossils is the unforgivable sin in paleontology. Yet the normally meticulous Marsh, no doubt distracted by his conflicts with Cope, was guilty of just such dereliction. To tell this story properly requires getting into the details of another famous incident in the history of American paleontology.

The decade of the 1890s in America has been called "the Gilded Age" because of the massive fortunes and ostentatious lifestyles of tycoons such as J. P. Morgan, John D. Rockefeller, and Andrew Carnegie. As a young boy Carnegie arrived in Pittsburgh as a penniless immigrant, but in 1901, when he sold his steel company for $480 million dollars, he was hailed as the richest man in the world.[4] Yet even the wealthiest often agonize over what they do not have. On a fall morning in 1898, a news item in the morning paper ruined Carnegie's breakfast. "World's Most Colossal Animal Found" shouted the headline, and an illustration showed a gigantic dinosaur rearing up to an eleventh-story window. What galled Carnegie was that the find was made by a collector—Marsh's former employee W. H. Reed—who was working for the American Museum of Natural History (AMNH) in New York. The AMNH was supported by J. P. Morgan, who had soured a railroad deal for Carnegie in 1885 and whom Carnegie had distrusted ever since. It

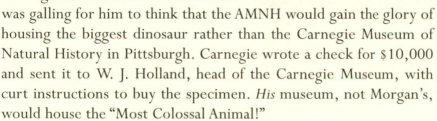

was galling for him to think that the AMNH would gain the glory of housing the biggest dinosaur rather than the Carnegie Museum of Natural History in Pittsburgh. Carnegie wrote a check for $10,000 and sent it to W. J. Holland, head of the Carnegie Museum, with curt instructions to buy the specimen. *His* museum, not Morgan's, would house the "Most Colossal Animal!"

The creature was not available; Reed had found only a single leg bone. Holland hired Reed away from the AMNH and sent him on an expedition to find a big dinosaur for Carnegie. This endeavor was not promising. Dinosaur skeletons do not magically appear in rocks even at the behest of mighty captains of industry. However, on 4 July 1899, Reed succeeded beyond any reasonable expectation. Near Sheep Creek, Wyoming, his party found the nearly intact articulated skeleton of a gigantic sauropod dinosaur. Complete skeletons are extremely rare and hard to find. Soon the Carnegie Museum proudly displayed the *Diplodocus carnegii,* which until recent years reigned as the longest known dinosaur. The local press made much of "Uncle Andy" Carnegie and his fabulous namesake. Messages poured in from emperors, kings, and presidents requesting casts of "Dippy" for their

---

4. Adjusted for inflation, $480 million in 1901 would be approximately $10 billion dollars today.

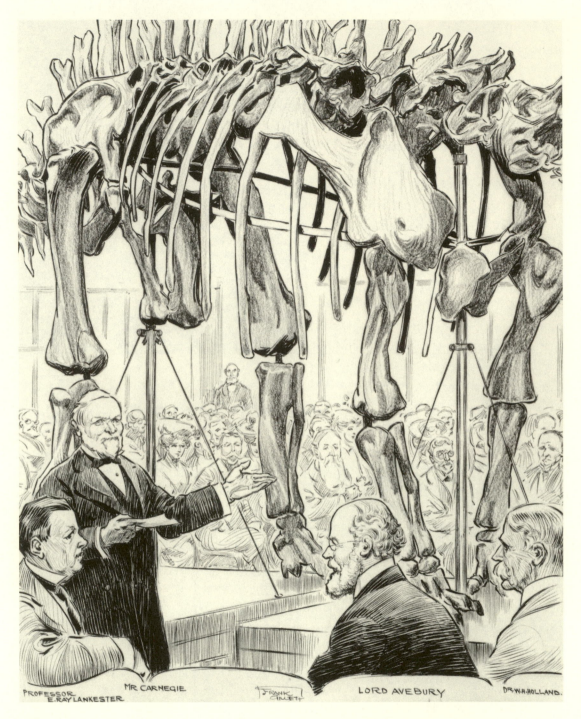

PROFESSOR
E. RAY LANKESTER   MR CARNEGIE   FRANK GILLETT   LORD AVEBURY   DR W.H.HOLLAND.

Andrew Carnegie presenting a plaster cast of Diplodocus carnegii *to The Natural History Museum, London, in May 1905*
(The Natural History Museum, London).

national museums. Basking in the attention, Carnegie liberally funded the museum's paleontology program for years afterward.

Bone hunters for the Carnegie Museum soon found a particularly rich quarry in northeastern Utah. The Carnegie Quarry, later to become the site of Dinosaur National Monument, yielded amazing quantities of fossils. The leader of the excavations at the Carnegie Quarry was Earl Douglass, whose gentle, generous, and introspective personality was a refreshing contrast to the vindictiveness of Cope and Marsh. His letters to Holland at the museum are remarkable documents. They are not the dry, factual reports you might expect but instead vividly describe the dangers and difficulties of fossil hunting, such as the travail of surviving a winter in the wilds of Utah. Douglass also conveys the excitement of discovery and the bitterness of disappointment (paleontologists, like fishermen, have lots of stories about "the one that got away").

In August 1909 Douglass made what promised to be a spectacular find:

> I have found what bids fair now to be an almost complete skeleton of a Brontosaurus (= *Apatosaurus*). I never saw anything that, on the surface, and so far as we have gone, had such promises of a whole thing. . . . We have exposed parts of the bones to the third or fourth vertebra anterior to the sacrum, and the bones are exactly in place, apparently except that the femur appears to have gone down several inches out of its place in the socket. I confidently expect at least to get all of the pelvic girdle, and the two hind limbs. . . . I fully expect to find all the dorsals. Of course, currents may have disturbed the thing before we get to the head but there is no indication as yet that any disturbance has taken place. (Douglas to Holland, 26 August 1909)

A fully articulated skeleton, particularly one with a skull, would have been a very important find. But note the confusion over names. Why did Douglass have to use two different names, *Brontosaurus* and *Apatosaurus,* to identify his specimen? The confusion over these two names, which persisted until quite recently, can be traced back to the Cope-Marsh war (see Gould 1991, 79–93, for an amusing account). In short, the problem was that the unbridled competition between Cope and Marsh made each man eager to name more new species than his rival. This competition inevitably led to sloppiness and misjudgment as descriptions were rushed into print on the basis of inadequate portions of the type skeletons (Berman and McIntosh 1978).

Marsh coined the name *Apatosaurus* and first described its fossils in 1877 (Marsh 1877). Two years later Marsh's collectors discovered two other skeletons of what they thought was a different kind of dinosaur (Norman 1985). Marsh named this supposedly new creature *"Brontosaurus"* and wrote a paper describing it (Marsh 1879). However, *"Brontosaurus"* was soon recognized as the same type of dinosaur as *Apatosaurus* (Riggs 1903; McGinnis 1982, 71). Under the rules of nomenclature the correct name for the genus is therefore *"Apatosaurus."* However, the name *"Brontosaurus"* ("thunder lizard") has always had more appeal than *"Apatosaurus"* ("deceptive lizard"), so *"Brontosaurus"* was the name that stuck. Sometimes with living creatures it is not easy to tell whether two types of creature belong in the same or a different taxonomic classification. With fossil creatures we have to be especially careful. Marsh was not being very careful when he identified *"Brontosaurus"* as a new kind of creature and rushed into print an account of the "discovery." It is hard to believe that Marsh would have been so careless had not scoring points against his rival become more important than scientific accuracy.

The confusion over the names *"Apatosaurus"* and *"Brontosaurus"* was not the most serious consequence of Marsh's slapdash approach. In 1883 Marsh published the first complete reconstruction of *Brontosaurus* (Marsh 1883). However, the *Brontosaurus* skeletons he had found had no skulls. Undeterred, Marsh simply gave *Brontosaurus* a skull from a completely different quarry. No other *Brontosaurus* remains had been found with the skull, and it even came from rock geologically older than the *Brontosaurus* bones (Norman 1985, 81–82; McIntosh, Brett-Surman, and Farlow 1997, 267–268). In fact, paleontologists now know that the skull belonged to *Camarasaurus,* a distinctly different sort of dinosaur—so different that the mistake was like putting a giraffe's head on a horse's body (Norman 1985, 81–82). Marsh may or may not have thought he had found the right head for *Brontosaurus.* Whether he did or not, the wrongheaded *Brontosaurus* was accepted by paleontologists as the real thing, and this mistake was not corrected until the 1970s. In fact, for forty-five years, from 1934 to 1979, the Carnegie Museum displayed its *Apatosaurus (Brontosaurus)* specimen with the wrong head! This story shows just how hard it can be to correct an entrenched mistake.

Holland and Douglass were, of course, aware of Marsh's reconstruction of *Brontosaurus,* and both were skeptical about the skull he had assigned to it. The discovery of the *Brontosaurus* skeleton at the Carnegie Quarry in 1909 gave them high hopes that the issue could be decisively settled by finding a skull in place at the end of the specimen's neck. In September 1909 Douglass reported to Douglass Stewart, the assistant director of the Carnegie Museum, that the skeleton seemed to be oriented with the neck straight down (Douglass to Stewart, 24 September 1909). This position might require digging through thirty feet of solid rock to reach the skull. However, things had taken a turn for the better when he wrote Stewart two weeks later: "We evidently have the most complete of the huge Dinosaurs that ever was found, at least I havent [sic] heard of any other so complete as this appears to be. It is evidently just the thing that Dr. Holland is anxious to get at the present time. I am not sure, but believe now that we will get the head. The cervicals ended abruptly but I find they turn back and the anterior cervicals (with the skull, I hope) lie under the others. Every bone so far was evidently buried entire" (Douglass to Stewart, 12 October 1909).

Ten days later, Douglass's mood remained confident: "So far as I know there is no such specimen of a large Dinosaur in any of the museums. I believe now that we will get the skull. The neck is turned back upon the other part of the neck. I do not dare to go in from this side so will probably have to wait until we get in back of this part before we will know. Some bones are out of place but I believe I have not struck a bone that I am sure belongs to this animal that is not complete from end to end unless it came to the surface" (Douglass to Stewart, 22 October 1909).

However, the *Brontosaurus* excavation ended in crushing disappointment:

Well we went down on the big Brontosaurus until we came to four sternal ribs and true ribs pointing in a direction which indicated that the back bone was bent like a bow, so that the spines of the anterior portion of the back might point nearly vertically downward. We found that we must not go down any farther until we got the upper portions out. We then began to work in back of the left of the pelvis to get below the thing and discovered a series of the vertebrae running to the left (west) and upward. They looked like cervicals and we concluded that the beast had thrown his neck and head back over his back so instead of going down anywhere from ten to thirty feet for the neck and digging for months in a state of suspense for the skull, here it was right before us and if the skull had been buried there and hadn't weathered out we would soon have it. Well we followed it with beating hearts—the neck I mean, and it turned downward a little. I could almost see the skull I was so sure of it for was there not a series of 8 cervicals undisturbed and in natural position, but when I had come to what I thought was the third of fourth cervical I dug ahead and there were no more. Apparently it had gone the way of all Brontosaur skulls. How disappointing and sickening. (Douglass to Holland, 11 November 1909).

Small wonder Douglass concluded his letter: "But I'm tired of Dinosaurs and fain would sleep. So good night."

But Douglass did soon find a skull closely associated with a *Brontosaurus* skeleton. The problem was that the skull he found looked very much like a *Diplodocus* skull, only larger and more robust. Could *Brontosaurus* have actually had a skull so much like that of *Diplodocus*? The problem was that in the twenty-six years since Marsh had published his reconstruction, paleontologists had come to accept a close affinity between *Brontosaurus* and *Camarasaurus,* not *Diplodocus.* Authority matters more in science than most scientists might want to admit, and Marsh had been the dean of American vertebrate paleontologists. So however poorly supported Marsh's claim about the skull of *Brontosaurus* may originally have been, by 1909 the weight of his authority made the claim difficult to challenge. Holland, however, was bold and pugnacious, and an estimable paleontologist in his own right. He presented a paper before the Paleontological Society of America in 1914 claiming that the *Diplodocus*-like skull was the correct head of *Brontosaurus.* He was scathing about Marsh's procedure, noting that if the skull had belonged to *Brontosaurus* it would have had to wash upstream and burrow through eight feet of solid rock to wind up where Marsh found it.

Curiously, when Holland had the Carnegie Museum's *Brontosaurus* specimen mounted, he decided to leave the skeleton headless. He probably hoped that the ongoing excavations at the Carnegie Quarry would eventually turn up a *Brontosaurus* specimen with the head in place, thus removing all doubt. By 1915, though, Carnegie was distracted by world events and ceased funding the museum's fossil digs. A specimen with the head in place was never found. The Carnegie Museum's *Brontosaurus* remained headless until 1934, two years after Holland's death. It was then decided, by whom is not clear, to mount a *Camarasaurus*-like head of the sort that Marsh had given the creature. This head remained on the Carnegie's *Brontosaurus* for the next forty-five years. In other words, for forty-five years one of the world's leading museums of natural history displayed a chimera, a creature that had never existed, as one of its prize specimens. Think of the embarrassment if the Louvre had displayed a fake Picasso for so long. Worse, the whole community of paleontologists accepted this amalgamation; even A. S. Romer's authoritative *Osteology of the Reptiles* (1956) showed *Brontosaurus* with the *Camarasaurus* head.

Getting to the bottom of this mess took a lot of skillful detective work by John S. McIntosh, perhaps the world's leading expert on the sauropod dinosaurs. By reviewing the records of the original *Brontosaurus* discoveries, McIntosh made a remarkable find. He found that at the time of his earliest *Brontosaurus* discoveries, Marsh had good evidence of a close association of *Diplodocus*-like skulls with *Brontosaurus* remains (Berman and McIntosh 1978, 6–7). Williston, Marsh's assistant, had found parts of a diplodocid cranium he associated with a dinosaur that Marsh named *Atlantosaurus*. Williston told Marsh that he believed *Atlantosaurus* actually to be the same species as *Apatosaurus* (= *Brontosaurus*. So at this point there were *three* names for the same creature!). Marsh ignored Williston's suggestion and decided that the cranium had come from a different species found in a different quarry. Sloppy cataloging procedures confused the situation considerably. If Marsh had recognized the close association of a diplodocid cranium and *Apatosaurus* remains, he might have avoided some errors about dinosaur heads.

The upshot was that a series of mistakes and misjudgments on Marsh's part led to long-standing scientific errors that were hard to detect and correct. Why was Marsh, by nature a stickler for detail (as all good paleontologists have to be), so sloppy and prone to poor judgment? The simplest explanation is that in his eagerness to outshine Cope, he rushed to judgment and into print, often naming new species on the basis of inadequate evidence. The resulting errors long outlived Marsh himself, and some were not corrected until quite recently. Cope and Marsh were the two greatest American paleontologists of the nineteenth century, but their legacies are tarnished. When we hear their names, we think more of their feud than their discoveries.

What is the moral? Scientific creativity is like creativity in every field. Great creative acts require the release of primal emotional energies. The philosopher Friedrich Nietzsche realized this when he wrote about the "dionysian" (after the god Dionysus) elements of creativity. These are the untamed, chaotic, even violent or irrational forces that are the wellsprings of imagination and intuition. In other words, great creativity requires deep passion. Yet these primal energies are too strong; by themselves they are as destructive as they are creative. This is why Nietzsche also spoke of the "apollonian" (after the god Apollo) restraints on the raw creative energies. These are the powers of logic, critical thinking, and rationality that channel the primal energies and make them forces for creation rather than destruction. The creative interplay of apollonian and dionysian elements may be obvious in great works of art or literature.[5]

Creativity in science and mathematics may seem entirely apollonian, involving only the dispassionate employment of logic and abstract reasoning. This is a serious misperception. An emotionless Mr. Spock would be a third-rate scientist at best. As should be obvious by now, scientists, especially the good ones, are deeply passionate. Inevitably, such people will occasionally form intense rivalries with one another. In itself, such conflict need not be bad—competition can be the emotional fuel that feeds high achievement. But Cope and Marsh became worse than rivals, and their enmity hurt more people than themselves. At the cost of shamelessly sermonizing, I draw the following moral: The scientist, as a scientist, has one sacred duty—the duty to truth. To the extent that *anything* becomes more important than truth to a researcher in any field of science, to that same extent he or she ceases to be a scientist. Of course, there are times when ethical considerations might rightly interfere with our search for truth. But when personal rancor prevents us from doing all that we can to know the truth, it is hard to see how such a situation can be excused.

---

5. Some years ago I was watching a version of *Hamlet* with my niece, who was then thirteen. About halfway through she looked up and said, "If people knew what this was about, they wouldn't let their kids watch it." She was right. Murder, lust, suicide, madness, treachery, and vengeance are what it is about. Similarly, Aeschylus's *Agamemnon* has the great scene in which Clytaemnestra, reeking with her husband's gore, triumphantly proclaims how she reveled in his murder. The *Iliad,* the founding document of Western literature, tells in excruciating detail just where the spear entered the chest and just how far its brazen head jutted out the back. Much of great literature deals graphically with some of the darkest elements of life. Shakespeare, Aeschylus, and Homer leave absolutely no doubt that they understood intimately feelings of blood lust, vindictiveness, and hatred. Deep, dark passions seethe in many of the great works of art, literature, and music, revealing the primal emotions that enter into the creative imagination. But what makes these works masterpieces and not merely soap opera or kitsch? It is the *intelligence* that shapes raw emotions into timeless words, notes, or images. Anyone can feel emotion, but to make someone reading a book, strolling through a museum, or sitting at a concert feel such emotion—and feel it with much deeper understanding or empathy—requires genius.

Further Reading

Everyone loves a good dustup, and the Cope-Marsh dispute was a bar-room brawl with the lights shot out. Perhaps their feud is a little less famous than the (contemporaneous) clash between the Hatfields and the McCoys, but it has been recounted many times. I have drawn a good deal on Url Lanham's *The Bone Hunters* (New York: Dover, 1973). Lanham sets the dispute in the context of what he calls the "heroic age" of American paleontology. Despite Cope and Marsh's squalid and dishonorable vendetta, it truly was an age of heroes. Many intrepid figures who braved much hardship and danger have been overshadowed and forgotten because of the attention paid to the battling professors. One of these remarkable figures was Arthur Lakes, who worked for Marsh and made many important discoveries (for which, of course, Marsh got the credit). His story is told in *Discovering Dinosaurs in the Old West* (Washington, DC: Smithsonian Institution Press, 1997), edited by Michael F. Kohl and John S. McIntosh. This book includes the field journals of Lakes, which create a vivid picture of bone hunting in the West when it truly was wild. McIntosh, by the way, is a physicist-turned-paleontologist and perhaps the world's leading expert on sauropods.

The journals and letters of Earl Douglass also portray the joys, disappointments, and hardships of fossil digging in the blazing summers and bitter winters of northeastern Utah. Unfortunately, this material has never been published and is available only in the archives of the Carnegie Museum of Natural History in Pittsburgh (see Douglass, E. Correspondence, in the References section). The story of Douglass and his discoveries is told in Helen McGinnis's *Carnegie's Dinosaurs* (Pittsburgh: Board of Trustees, Carnegie Institute, 1982). The story of the headless *Apatosaurus* is told by McGinnis, and in much greater detail in chapter 1 of my book *Drawing Out Leviathan* (Bloomington: Indiana University Press, 2001). The article by David Berman and John McIntosh, "Skull and Relationships of the Upper Jurassic Sauropod Apatosaurus (Reptilia, Saurishia)" in *Bulletin of the Carnegie Museum of Natural History* 8, 1–35 (1978), sets the record straight on how *Apatosaurus* got the wrong head. This essay is mostly detailed anatomical discussion that is incomprehensible to nonspecialists, but it contains an invaluable history of the discoveries and mistakes that led to the mix-up over the skull of *Apatosaurus*.

A recent and comprehensive account of the Cope-Marsh war is Mark Jaffe's *The Gilded Dinosaur* (New York: Three Rivers Press, 2000). Jaffe goes into considerable detail and has done his homework with the original sources. However, the scholarship does not make the work pedantic. It is a great story, and Jaffe tells it well.

The question of whether and to what extent Darwinism will countenance a view of evolution as progressive is still a matter of dispute. Robert

J. Richards in *The Meaning of Evolution* (Chicago: University of Chicago Press, 1992) argues that Darwin himself did see the net tendency of evolution as progressive. Still, Darwin denied the Lamarckian view that there is an inherent and automatic drive to greater perfection in the evolutionary process.

5

Whose Dinosaur? Kaiser Bill's or Uncle Sam's?

In 1914 European civilization self-destructed. For forty-three years, ever since the end of the Franco-Prussian War, the great powers of Europe had been at peace with one another. Of course, there were many colonial wars, as when Britain in 1879 invaded Zululand and defeated Cetshwayo, the Zulu king, in a brief but bloody campaign. But the indigenous peoples of Africa and Asia, lacking European arms or training in lethal Western ways of warfare (Hanson 2001), could be subdued with minor military effort. During those forty-three years of peace, European civilization had achieved unprecedented scientific, medical, and technological progress. Physicists discovered X-rays and radioactivity, new vaccines and antisepsis extended the lives of millions, giant liners sped across the oceans, and the first canvas-and-wire flying machines tested the air. Rapidly expanding economies raised the standard of living so that ordinary people ate better, were housed better, and lived longer than they ever had. The arts flourished. The composers Brahms, Bruckner, Mahler, and Tchaikovsky and such painters as Van Gogh, Monet, Renoir, and Degas did much of their most familiar work in this period. The future was bright. Yet from 1914 to 1918, this civilization that had accomplished so much and had so much promise devoted its enormous intellectual and economic energies to frenzied self-annihilation.

In *All Quiet on the Western Front,* Erich Maria Remarque speaks of the generation of those who had physically survived the Great War but who were nevertheless destroyed by it. The same could be said of European civilization as a whole. Of course, Europe was rebuilt after 1918, went on to endure an even greater war from 1939 to 1945, was rebuilt again, and now (mostly) enjoys peace and prosperity. But the lights that went out in 1914 can never be relit. A civilization died, and the one built on its ashes is very different. The reader of Barbara

Tuchman's classic history of the pre-1914 period, *The Proud Tower*, encounters a society near to us in time but distant in ideas and ideals.

One of the changes evident since 1918 is a greater hesitancy, at least among educated people, to endorse or express nationalistic sentiments. No doubt the sight of millions following streaming banners into the abyss of world war made flag waving passé for many thinking people. By contrast, much of the music and art of the nineteenth century was frankly and unabashedly nationalistic (e.g., though they are enjoyed worldwide, Wagner's operas are emphatically *German*). When Germany went to war in 1914, the intellectual, academic, and scientific communities rushed to pledge support—a phenomenon not always observed in the United States during its recent wars. Dissent from a government's war policies does not necessarily indicate a lack of patriotism, but I think it is safe to say that intellectuals and academics of a century ago freely held and expressed nationalistic sentiments that would be considered maudlin or jingoistic today.

What does all this have to do with dinosaurs? Before the recent work that has brought to light the dinosaurs of Asia, Africa, and South America, most of the great dinosaur discoveries were in North America. The most famous, such as those of Cope and Marsh, were made in the United States. The big discoveries were made by American scientists, funded by American tycoons, and displayed in American museums. American paleontologists led the way in the study and reconstruction of dinosaurs. Under these circumstances, it was hardly surprising that Americans had a proprietary feeling about "their" dinosaurs.

William Jacob Holland was the director of the Carnegie Museum in Pittsburgh during the time of its giant dinosaur discoveries. Holland was a remarkable man, one of those individuals whose accomplishments were so numerous and diverse that it is hard to see how they all could have fitted into a single lifetime. Born in Jamaica in 1848 to parents who were missionaries of the Moravian Church, he was at various times a clergyman (a graduate of Princeton Seminary), chancellor of a university, head of the Carnegie Museum, and the holder of other important offices and positions far too numerous to mention. He was an expert philologist and linguist who taught Greek and Latin; won a prize as a seminary student for his knowledge of Hebrew; and knew many modern languages, including Arabic and Japanese. He was an expert entomologist who authored over a hundred papers and several books on the subjects of moths and butterflies. He gained fame as a paleontologist and was the author of numerous papers on extinct vertebrates.

Diplodocus carnegii and *Apatosaurus louisae* were the pride of the Carnegie Museum during Holland's time. They still are. As a research associate of the museum, I had many opportunities to marvel at these creatures and to think that beasts as long as two city buses or as heavy as a herd of elephants actually walked the earth. Surely Holland must have felt a justifiable pride in himself, his museum, and his nation when he regarded these magnificent specimens.

But what if these specimens were not really so magnificent? Following the practice of every American vertebrate paleontologist up to that time, Holland had reconstructed his sauropods as creatures that stood upright, with their massive, columnar legs positioned directly under the body. We imagine them striding through Jurassic forests with heads held high and powerful footsteps that made the ground tremble. After all, was not *Apatosaurus* also called *Brontosaurus* (thunder lizard) because the earth supposedly thundered with each stride? But what if *Diplodocus* did not stride but crawled on its belly like the lowest lizard? A creeping, crawling *Diplodocus* would not be nearly so magnificent. Yet two paleontologists, an American named Oliver P. Hay and a German named Gustav Tornier, writing in 1908 and 1909, respectively, suggested such heresy. They argued that *Diplodocus,* being a reptile, had a reptile's typically splayed-out posture, resting on its belly most of the time and with limbs projected out horizontally from the side of the body (Hay 1908; Tornier 1909). Their creation looked truly grotesque and made none of the magnificent impression of an upright sauropod.

Holland was furious. The German scholar's arguments particularly provoked him. In an article published in *The American Naturalist* in 1910 he excoriated Tornier and belabored his arguments with an equal mixture of logic and vitriol. Scientific articles are almost always dry in tone and judicially sober. This one seethed. Why did this hypothesis draw such rancor from Holland, especially toward the German scientist? A rationale is not hard to imagine. Sauropods were the pride of American paleontology. Would an *American* dinosaur crawl at the command of a *German* paleontologist? No red-blooded American would tolerate the insult! Worse, *Diplodocus* was *carnegii*. Could a dinosaur named after a captain of American industry, and the benefactor of Holland's museum, be permitted to creep on its belly?

How much of Holland's ire was driven by a desire to defend his nation's pride, or resentment at the perceived insult to his benefactor's namesake? It is hard to sort out all of these elements, but no one can deny that science has often been influenced by nationalism. This seems odd because science is surely the preeminent instance of an international enterprise. Different nations still have many of their own literary and artistic traditions that are little known beyond their national borders, but scientific communities are worldwide. In any given year, an American physicist might consult colleagues in Moscow or New Delhi, or attend professional conferences in Tokyo or Saudi Arabia. But nationalism is still to be found in science, and it was certainly present in the pre–World War I years. In an amusing and horrifying passage from *The Mismeasure of Man,* Stephen Jay Gould tells of how European scientists of the nineteenth century concluded that members of their own nationality had bigger brains than other Europeans (Gould 1981)! Sadly, in the 1930s, when the Nazis extolled "Aryan science" and condemned "Jewish science," this idea was not so aberrant as we would like to think.

In the case of Holland and *Diplodocus,* it is important to recall the context of the times. In the years leading up to 1914, German militarism and expansionism had become a major international worry. The ruler of Germany, Kaiser Wilhelm II, was a swashbuckling figure with a fierce mustache and a steely Prussian gaze. He had overseen an unprecedented buildup of the German military. The vast Krupp arms works supplied the German army with gargantuan artillery pieces (The "Big Bertha" cannon of World War I fame was named after Bertha Krupp, wife of the munitions manufacturer). Naval yards turned out warships of excellent quality and in quantity enough to challenge the British Royal Navy. Society as a whole was militarized: Conscription was imposed, and military parades and musters were a common sight. The young Albert Einstein fled to Switzerland, a fugitive from German militarism.

So it is entirely reasonable to attribute much of Holland's pique to a German's attack on a source of American pride. But when we are studying the history of science, it is not enough to understand the social conditions and personal motivations involved in a controversy. It is also essential to under-

Kaiser Wilhelm II, 1905 (The Illustrated London News Picture Library).

stand the evidence and rational argument involved. As I argued in Chapter 1, however nasty a scientific dispute might become, *reasons* really do matter for the settlement of scientific controversies. Let us take a look at the reasons Hay and Tornier gave for their heretical hypotheses about dinosaur posture.

First, we should recall that when we study extinct creatures, we usually have nothing to work with but fragmentary and incomplete skeletons. Soft tissues almost never fossilize. Even when the skeletons are complete, as with *Diplodocus,* it is hard to answer all questions about the living creature's posture. Bones can be fitted together in various ways, and their precise alignment is not always obvious. For instance, recall that Owen, working with fragmentary evidence, reconstructed *Iguanodon* as an elephantlike quadruped. Later in the nineteenth century, the Belgian paleontologist Louis Dollo, working with complete skeletons, construed *Iguanodon* as a biped, and this was the accepted view for many years. Then, in the late twentieth century, the *Iguanodon* expert David Norman concluded that *Iguanodon* typically walked on all fours (Norman 1985)! Likewise, *Tyrannosaurus rex* was initially viewed as standing upright in a Godzillalike stance. Now *T. rex* is depicted as leaning forward, with its massive head and body balanced over the legs by the long tail. The upshot is that dinosaur stance and posture are not always easy to determine, and heretical suggestions might turn out to be right. So Hay and Tornier were not being patently absurd or irrational.

Hay begins his first essay on *Diplodocus* by noting that many questions about the likely habits or lifestyle of extinct creatures are difficult to answer and are open to diverse interpretations (Hay 1908, 672). One helpful clue about lifestyles comes from the creature's environment. According to Hay, *Diplodocus* lived on a low-lying, swampy plain, drained by sluggish streams, with areas of luxuriant forest and other regions of open savanna—a climate and environment much like the bayou country of southeast Texas. Further, the weak teeth of *Diplodocus* would have prevented it from eating much tough or woody matter. Floating algae and the succulent, loosely rooted plants found in stagnant ponds or bayous would have been its most likely food (675).

Hay writes that this evidence suggests that *Diplodocus* had to be adapted to a semiaquatic lifestyle. One result of the "dinosaur revolution" that has occurred over the last few decades (see Chapter 6) has been to take sauropods out of the swamp and put them onto dry land. As for the sauropod diet, it need not have been restricted to floating algae and succulent water plants. The teeth of sauropods served only to rake in the food, which was swallowed without chewing. The "chewing" took place in a gizzard where the food was ground by swallowed stones. When Hay was writing, though, the consensus was that the sauropods spent most of their time wading in swamps and sloughs. Hay notes that a creature weighing multiple tons and standing fully erect would exert considerable pressure on the ground with each foot and would probably become hopelessly mired in swampy terrain (680). A creature that rested on its belly with limbs projected horizontally from the side of the body would have a body structure like a crocodile's and be similarly well adapted to swampy environments. Hay contends that nothing in the anatomy of sauropods precludes a splayed-out posture (677–679). In fact, he argues that the structure of the sauropod foot, with strongly developed inner digits and reduced outer ones, would have been more efficiently employed in a sprawling posture (679). Finally, the evolutionary ancestors of sauropods must have had a typically reptilian, lizardlike posture, and there would have been no reason for their descendants to have evolved away from that stance (680).

Hay's arguments were hardly conclusive, but they did serve to challenge the accepted restorations of sauropods. Tornier took the argument further. He begins by noting the many anatomical details that have led paleontologists to classify *Diplodocus* as a reptile (my translations): "The *Diplodocus* is without a doubt a reptile with the structure of a typical four-footed lizard on the following grounds: According to the view of all previous authors, it belongs to the reptilian subclass of the dinosaurs. Further, its head has essential reptilian characters; its neck has more than seven vertebrae. Its pectoral girdle shows a typically lizardlike structure and very closely resembles, for instance, that of the chameleon" (Tornier 1909, 195).

Tornier makes numerous other comparisons and concludes that the overall anatomy of *Diplodocus* was much like that of typical reptiles such as lizards and crocodiles (196). Hence, it would not be too surprising if its stance also were typically reptilian.

Tornier's strongest argument was that placing *Diplodocus* upright required it to be posed in a most unnatural and improbable tiptoed stance: "The real reason why the *Diplodocus* forefoot was set up in this tiptoed manner, despite the contrary opinions in the literature, is: The hind limbs of these animals were arranged [by American paleontologists] in such a steep and contorted manner at that time that to bring the animal into correct posture, and not to allow the forepart of its body to appear unnaturally low, and also to erect its forelimbs correspondingly oversteeply, the forelimbs had to be put in tiptoed stance, despite the contrary opinions in the literature" (Tornier 1909, 201–202).

The accepted American reconstruction of *Diplodocus* must be wrong if it requires a multiton animal to go about on tiptoe. Instead, Tornier recommended that *Diplodocus* be given a lizard- or crocodilelike stance of the sort Hay had recommended (see accompanying illustrations).

Holland's reply erupted onto the pages of *The American Naturalist* in May 1910. He begins with a sardonic characterization of Tornier and his claims:

> In the manner of a man who has made a wonderful discovery, Tornier announces at the outset of his paper that *Diplodocus* is a genuine reptile—"Ein echtes Reptil." No student of the sauropoda has ever doubted this. But having predicated the genuinely reptilian character of the animal, Tornier proceeds thereafter to speak of the *Diplodocus* as a lacertilian [lizard]— "ein[e] Eidechse." There are reptiles and reptiles. Having assured himself of the truly reptilian character of the animal, it

Hay's splayed-out crocodilian Diplodocus (Oliver P. Hay, "On the Manner of Locomotion of the Dinosaurs, Especially the Diplodocus, with Remarks on the Origin of the Birds." Proceedings of the Washington Academy of Sciences, vol. 12 [January–December 1910]: 1–28).

THE FORM AND ATTITUDES OF DIPLODOCUS

was a bold step for him immediately to transfer the creature from the order Dinosauria, and evidently with the skeleton of a *Varanus* or *Chameleon* [i.e., typical lizards] before him, to proceed with the help of a pencil, the powerful tool of the closet-naturalist, to reconstruct the skeleton upon the study of which two generations of American paleontologists have expended considerable time and labor, and squeeze the animal into the form which his brilliantly illuminated imagination suggested. The fact that the dinosauria differ radically from existing reptiles in a multitude of important structural points seems not to have greatly impressed itself upon the mind of this astute critic. (Holland 1910, 201–202)

Most notable about this passage are the various instances of personal abuse directed at Tornier—he is a "closet naturalist" who is guided by his "brilliantly illuminated imagination." Perhaps some of Holland's ire was aroused by the differences between German and American styles of academic expression. German academic traditions are deservedly famous for demanding rigorous, painstaking, and thorough scholarship. The downside is that German academics are often perceived as condescending and pedantic. I recall my own experience as a first-year graduate student when a famous German scholar addressed our seminar. With the attitude of a Teutonic Prometheus come to bring us cowboys to culture, Herr Professor opined that all American high school students should be required to recite Part One of *Faust*. I itched to suggest that the history of the twentieth century might have been different had German students been required to pay closer attention to the U.S. Constitution and Bill of Rights. Still, the note of injured national and professional pride is evident in Holland's retort. It sounds as if the most offensive aspect of Tornier's hypothesis for Holland was that it threatened to undo the work of two generations of American paleontologists.

Holland added some logical argument to the brew. He pointed out that experiments with the skeleton of *Diplodocus* showed that the femur (big leg bone) could be positioned according to Tornier's recommendation only by breaking the bones and, further, that had the femur achieved such a position, it would have been locked into place and could not have moved at all (Holland 1910, 263). Holland details other dislocations and contortions that he says Tornier's hypothesis would require (262–264) and proceeds to mock Tornier's "skeletal monstrosity": "As a contribution to the literature of caricature the success achieved is remarkable. It reminds us somewhat of

Tornier's "skeletal monstrosity" (top) and Holland's mounting of Diplodocus carnegii *(bottom) (G. Tornier, "Wie war der* Diplodocus carnegii *Wirklich Gebaut?"* Sitzungsbericht der Gesellschaft Naturforschender Freunde zu Berlin, *April 20, 1909, pp. 193–209).*

those creations carved in wood emanating from Nuremberg, which were the delight of our childhood, and which came to us stuffed in boxes labeled 'Noah's Ark,' and stamped 'Made in Germany'" (264). Holland next argues that a dinosaur's pelvis is not at all like that of other reptiles, and in fact is much more ornithic (birdlike) in structure (265–267).

Holland's most effective move was to show that the rib cage of *Diplodocus* was so deep that had its legs been splayed out to the side lizard-style, its abdomen would have had to project below the ground (see illustration on p. 80)! He concludes with scalding sarcasm: "Of what earthly use the hind limb of the *Diplodocus* could have been to him in such a position, I leave you to determine for yourselves. It has been suggested that kindly nature, to meet the requirements of the case, must have channeled the surface of the earth and provided the *Diplodocus* and its allies with troughs in which they kept their bodies while the feet were employed for purposes of locomotion along the banks. The *Diplodocus* must have moved in a grove or a rut. This might perhaps account for his early extinction. It is physically and mentally bad to 'get into a rut'" (267–268).

As for the "tiptoed" stance that Tornier claimed was so improbable, Holland reproduced an image of a sauropod footprint, the only one known at the time, which showed that sauropods had large, fleshy footpads (280). Thus, they were not walking on tiptoe but stood on large, round pads like those of elephants.

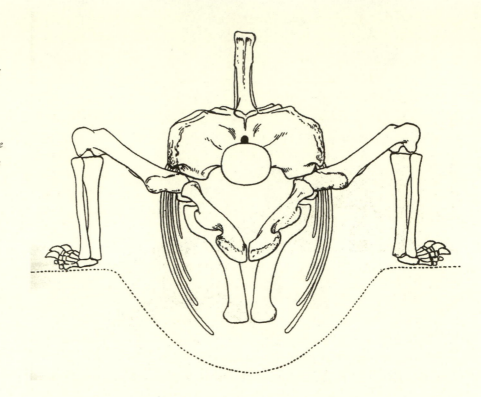

The sprawling stance of Diplodocus *according to Oliver P. Hay and Gustav Tornier (O. P. Hay, 1908. "On the Habits and the Pose of the Sauropodous Dinosaurs, Especially of Diplodocus." The* American Naturalist *43: 672–681).*

What are we to make of Holland's stinging—actually more like a stoning—rebuke? We have already seen that the temperature of scientific debate sometimes gets torrid; later chapters will provide even more instances. A certain sense of propriety or decorum is supposed to govern academic debates. Scholarly disagreement can be firm, but it is supposed to be restrained and respectful and ideally is offered only after conceding the strong points made by the other side. Academic discourse truly does suffer when debate degenerates into mudslinging and scholars act like politicians in a close runoff. But bareknuckled polemic is sometimes appropriate, even in academe (I've penned a few myself). Sometimes the opinions of a purported expert are not only wrong, but perversely so. When an opinion offered by an alleged authority is so inept that it has to be the result of laziness, arrogance, or, worse, ideological bias, then a public chastisement is in order.

It appears that Tornier deserved at least some of Holland's censure. In particular, Tornier seems to have been a "closet-naturalist," as Holland charged. That is, his hands-on experience in reconstructing sauropod skeletons seems to have been slight. In science, and in paleontology in particular, it is not enough to amass vast knowledge of theory and fact. In addition to "book learnin'" there has to be practical knowledge that can only be gained by actual experience such as trying to fit together fossil remains. Those who have actually gotten their hands dirty trying out proposals often realize that many hypotheses that look good on paper or sound plausible to the theorist are in fact completely unworkable. Holland was therefore right to reply: "The critics [i.e., Hay and Tornier] possibly do not realize that weeks and months and years of study have been spent by those who have been charged with the task of assembling these remains, and that the prescriptions, which they now furnish, have been already tried without their suggestion, and have for good reasons been found wanting. It is easy for a knight of the quill, who has never practically attended to the matter, to find fault" (281).

Complete and explicit justifications cannot be given for every sound scientific judgment. Sometimes the practical wisdom of those who have been most involved in the research is our best guide.

Holland's blast did not quite end the debate. In February 1910, just a few months before Holland's article appeared, Hay published a second article in *Proceedings of the Washington Academy of Sciences.* Hay concedes that the sauropod skeleton may have had a digitigrade—i.e., "tiptoed"—stance, with the bones of the foot resting on large pads of flesh, as does the elephant's foot. But he points out that large land tortoises have feet very similar to the elephant's, yet they have the typically reptilian splayed-out posture as well (Hay 1910, 2). Hay notes that even very large crocodiles can move quickly on land when the need arises, despite their reptilian posture (5). Also, some have argued that the sauropods must have stood upright since their femora (big leg bones) were straight. Hay counters with numerous examples of large upright animals with bent femora and sprawlers with straight legs (4–5).

Hay's arguments raise an interesting point. To understand fossil animals, we have to compare them to living ones, and such comparisons are usually risky. To argue that because fossil creatures were similar to living ones in certain ways they must have been similar in others is to base an argument on an analogy. Arguments based on analogy can range from very strong to very weak, and the criteria for distinguishing good analogies from bad are not clear-cut. In general, the more relevant similarities there are between the items being compared, and the more alike those similarities are, the stronger the analogy will be (the real problem is deciding what kinds of similarities count as "relevant" and how "alike" they really are). On the other hand, if the similarities are few and weak, as Hay claimed they were between *Diplodocus* and big mammals, the analogy will be weak and no conclusion can be drawn about sauropod posture.

Later in 1910 W. D. Matthew, an outstanding paleontologist with the American Museum of Natural History in New York, published an article summarizing and commenting on the debate over sauropod posture (Matthew 1910). He contends that Hay misrepresented the argument that cited the straight *Diplodocus* femur as evidence for an upright stance. In fact, claims Matthew, the analogy between sauropod anatomy and that of elephants and other massive upright creatures is much closer than Hay admits:

> No one, so far as the reviewer, knows, has asserted that the straightness of the shaft of the femur of *Diplodocus,* considered alone, proved that the animal walked like a mammal. . . . The argument that Dr. Hay presumably has in mind is this: That in the elephants and several other types of gigantic mammal the femur is relatively long, straight-shafted, with its articulations terminal rather than lateral, the feet short, rounded, heavily padded, and capable of but limited motion, the whole limb being pillar-like and normally held straight under the body. All gigantic mammals show some degree of approach towards this type of limb; and in the Sauropoda the resemblance in form and proportions is very marked. (Matthew 1910, 549)

Matthew concludes: "The straight-shafted femur does not *per se* prove that *Diplodocus* walked in any of the various ways that mammals walk. But taken in connection with the numerous other adaptive resemblances in form and proportions of the bones of the hind limb, feet and pelvis, to the elephants and other gigantic mammals and reptiles cited, it does afford a very strong argument for asserting that *Diplodocus* walked like an elephant as to its hind limbs" (550).

Matthew notes that Hay had a stronger argument in claiming that the sauropods had very different anatomy from dinosaurs that unquestionably did walk upright (Hay 1910, 5–6). Hay points out that the upright posture of the bipedal dinosaurs such as *Allosaurus* must have evolved through all the intermediate stages from the sprawling posture of the reptilian ancestors of dinosaurs (9). In his opinion *Diplodocus* represents a primitive stage in this evolutionary process (10). As Matthew points out, however, Hay backs this opinion with very little argument or evidence (Matthew 1910, 552).

One of the interesting things about studying the history of science is to see how the losing side in a past debate—the defenders of a view now defunct—often supported their losing causes with claims and inferences now regarded as right. Concomitantly, the winning side triumphed in part by accepting what is now universally rejected! For instance, Matthew agrees with Hay that the sauropod limbs did not seem well adapted for the habitual support of the entire weight of the animal (Matthew 1910, 551). He concludes that sauropods generally were waders and that the buoyancy of the water helped hold up their bulk. Thus, an upright posture could be defended even if the animal seemed barely capable of supporting its own weight! Hay, on the other hand, notes that the bones of sauropods seem designed to minimize the weight of the skeleton and argues that such a creature would have been too buoyant to have waded in water (Hay 1910, 25). Paleontologists now agree with Hay that sauropods were not habitual waders, though they did enter wet or muddy areas occasionally, and their lightly built skeletons are seen as adaptations for dwelling on land.

Matthew's careful critique largely ended the debate over sauropod posture. The discovery of a nearly complete juvenile *Camarasaurus* skeleton provided evidence strongly supporting the case for upright posture (Farlow and Chapman 1997, 544). But the debate was not truly settled until 1938, when Roland T. Bird discovered the fossilized trail of a large sauropod in Glen Rose, Texas. These footprints showed unequivocally that sauropods did stand and stride and did not creep or crawl.

So Holland was right and Tornier was wrong. The American *Diplodocus,* the namesake of Mr. Carnegie, could stand proud and tall. Unfortunately, Holland had died six years prior to Bird's discovery and so did not witness the final vindication of American paleontology.

The sauropod posture debate teaches us several things about the nature of scientific controversy. For all intents and purposes, the debate was over nearly thirty years before Bird discovered the fossil footprints that gave the definitive evidence. Such is typically the case; scientific debates tend to dry up long before convincing proof can be offered. Consider the controversy over the head of *Apatosaurus* mentioned in Chapter 4. A specimen of *Apatosaurus* with the skull still in place at the end of the neck has never been found. Only such a find would conclusively show, once and for all, the sort of head *Apatosaurus* had. Yet no paleontologist now doubts that the creature had the diplodocid skull that Berman and McIntosh gave it in 1978.

The problem is that absolutely definitive, knock-down evidence is hard to come by in paleontology, or in any branch of science. Seldom does any single datum, piece of evidence, or particular experiment settle the issue. Generally, the data in a scientific debate will admit of more than one reasonable interpretation, and neither side is simply being obtuse or pigheaded (though they may certainly seem so to those embroiled in the heat of controversy). For a hypothesis to be rejected, the judicial standard of guilt "beyond a reasonable doubt" seldom applies in science. This does not mean that evidence and logic do not matter in scientific debate, or that neither side is really right, or that we can turn the study of scientific controversy over to the sociologists (ugh!). The lesson is simply that in order for science to progress, our reach must always somewhat exceed our grasp. If we always waited until absolutely unquestionable evidence emerged, we would wait a long time, perhaps forever.

Science is like what they used to say about sharks: It can never rest and must swim or die. As I noted in Chapter 1, scientific controversies are closed by the collective decisions of scientific communities. Though these decisions are based on evidence, the evidence is seldom rationally compelling, and such decisions are motivated by the need to get on with the job.

In a sense, the debate over sauropod posture was a continuation of the debates between Mantell and Owen and Owen and Huxley. Those debates were settled in favor of views that saw dinosaurs as different—very unlike the reptiles of our present world. Mantell had seen *Iguanodon* as an overgrown lizard; Owen recognized that it was not a big lizard and gave it an upright, elephantlike posture. Huxley departed even further from the lizard model and noted the affinities between dinosaurs and birds. In general, the more our knowledge of dinosaurs has increased, the more we recognize how unique these creatures were. Hay and Tornier, in interpreting sauropods as lizardlike sprawlers, took a step back from this process. In the next two chapters we shall see how radical thinking about dinosaurs has become, with talk about warm-blooded dinosaurs and even dinosaurs with feathers. In fact, the question might now be whether the emphasis on dinosaur uniqueness has gone too far.

Further Reading

The decades immediately preceding World War I are a particularly fascinating period in the history of Europe, a period of unprecedented achievement that teetered on the brink of cataclysm. The classic account of this era is Barbara Tuchman's *The Proud Tower: A Portrait of the World before the War, 1890–1914* (New York: Ballantine, 1996). This age is fascinating for its contradictions. During an extended period of peace, European nations feverishly prepared for war. International travel and communication were far easier and faster than in any previous age, yet European societies fostered a fervent and even rabid nationalism. In *The Guns of August,* reprint ed. (New York: Ballantine, 1994), Tuchman relates the concatenation of events that led to war in August 1914.

The 1890–1914 period is often called "the Gilded Age" because of the spectacular wealth and ostentatious lifestyles of the captains of industry such as J. P. Morgan and Andrew Carnegie. Like the titled aristocrats of earlier times, these grandees felt a certain noblesse oblige that led them to fund libraries and museums. J. F. Wall's biography *Andrew Carnegie* (New York: Oxford University Press, 1970) tells the story of Carnegie's involvement with the Carnegie Museum of Natural History in Pittsburgh.

W. J. Holland is a particularly interesting individual, but there are few easily available published sources on him or his work. The dinosaurs of the Carnegie Museum are described in detail in *Carnegie's Dinosaurs* (Pittsburgh: The Board of Trustees, The Carnegie Institute, 1982) by Helen McGinnis. The sauropod posture controversy is told in considerable detail in A. J. Desmond's *The Hot-Blooded Dinosaurs: A Revolution in Palaeontology* (New York: The Dial Press, 1976). Otherwise, not a great deal has been written on this debate. I have reconstructed it from the original sources, but these are moldering in library basements and hard to get.

Determining the posture of extinct animals when all you have are the bones is a daunting task. The difficulties are detailed by David Norman in his *The Illustrated Encyclopedia of Dinosaurs* (New York: Crescent Books, 1985). This is one of the best general books on dinosaurs, though it is now slightly dated.

6

Cold-Blooded, Warm-Blooded, or Neither?

During the 1980s anyone visiting a laboratory or a science department at a university was sure to see yellowing copies of Gary Larson's *The Far Side* cartoons festooning walls and office doors. Scientists do have a sense of humor, and a pretty bizarre one at that. *The Far Side* often squarely hit the scientific funny bone. One of my favorites shows a *Stegosaurus* lecturer addressing a dinosaur audience. He says, "The picture is pretty bleak, gentlemen. The world's climates are changing, the mammals are taking over, and we all have a brain about the size of a walnut." This summarizes one very influential image of the dinosaurs—obsolescent, dim-witted hulks unable to adapt to change and doomed to defeat by the smarter, more adaptable mammals. A person is a "dinosaur" if he or she is a holdover from an earlier and more primitive era, with a closed mind and reactionary views.

Stereotypes, of course, usually reveal more about the biases of those who make them than about the nature of the stereotyped subjects. We are mammals, so we regard mammals as obviously "higher" than reptiles. Since dinosaurs were traditionally conceived as big, dumb reptiles, they obviously were inferior to the smarter mammals who eventually beat them in the Darwinian race for survival.

Funny thing, though. Mammals evolved in the Triassic, about the same time as the dinosaurs. Yet for 160 million years, while the spectacularly diverse dinosaur clan covered and dominated the earth, mammals remained small, nondescript bit-players in the drama of Mesozoic life. If they were so much smarter and better, why did it take so long for them to beat the dinosaurs? Why, in fact, did they begin to diversify and fill all sorts of different ecological niches only after the dinosaurs had gone extinct and abandoned those niches? Maybe the dinosaurs were not hapless hulks after all.

Maybe, as dinosaur paleontologist Robert Bakker has insisted in his numerous TV appearances, dinosaurs were not slow, stupid, and in the swamp but were fast, smart, and on the land. Maybe even the great sauropods were active and aggressive foragers of forest and plain—not torpid skimmers of pond scum, up to their necks in stagnant water.

Fighting the stereotype of dinosaurs as evolutionary cul-de-sacs, nature's version of planned obsolescence, has been the life work of Robert Bakker, perhaps the most visible, articulate, and combative of dinosaur specialists. Starting in the early 1970s, while he was still a very young man, Bakker took on the dinosaur establishment and challenged long-accepted views with bold, innovative hypotheses. His most striking claim was that dinosaurs were warm-blooded. Lizards are cold-blooded. Their metabolism does not generate and maintain a steady body core temperature. Instead they are dependent on external sources of heat to maintain core temperatures. Now, there are many advantages to being cold-blooded. You require a lot less food, for one thing. But lizards become sluggish when the weather is too cool and, though capable of short periods of frenetic activity, are generally less active than warm-blooded organisms. Mammals and birds are warm-blooded: They generate their own heat metabolically and do not have to bask in the sun in order to get enough warmth for activity. They are capable of prolonged periods of activity even when the weather is cloudy and chilly.

As we have seen in previous chapters, dinosaurs were all along regarded as big reptiles. Typical reptiles are cold-blooded, so it was simply assumed by everyone (except T. H. Huxley) that dinosaurs had to be cold-blooded. But if dinosaurs were warm-blooded, their behavior must have been closer to that of typical mammals or birds, that is, more continuously active and not restricted to short bursts of energy. Maybe dinosaurs should be completely reconceived—more like super birds than giant lizards. Bakker showed not the slightest hesitation in embracing these and even more radical conclusions. When older or more conservative paleontologists shrank back, he defiantly assumed the role of dinosaur heretic and castigated *them* as the stodgy, out-of-date "dinosaurs." His sharply worded replies to their criticisms soon generated a controversy that quickly boiled over and spilled into the popular media.

In 1978 the controversy led to an American Association for the Advancement of Science colloquium, where Bakker and his colleagues met to debate the warm-blooded dinosaur issue. The papers given at this conference, published under the title *A Cold Look at Warm-Blooded Dinosaurs* (1980), were almost unanimous in holding that Bakker had not proven his case. Bakker, unrepentant and apparently relishing his role as gadfly, has not backed down one iota from his views and instead has publicized them widely in his writings and media appearances.

Bakker has been a big hit with the media . His appearance—hippie-style long hair, full beard, wide-brimmed hat, and jeans—and flamboyant style give him great presence even on the small screen. Movies such as *Jurassic Park* hastened to adopt his view of dinosaurs as warm-blooded, fast, and mean. Recall the scene from that movie in which the *Velociraptor* is looking through a window and its hot breath

Robert Bakker popularized the view of dinosaurs as warm-blooded, fast, and mean. Velociraptor, along with the other Dromaeosaurids, was a very intelligent dinosaur, as calculated from the ratio of its brain to its body weight. Hence it was a very deadly predator and may have hunted in packs (The Natural History Museum, London).

steams the glass. For movie viewers of the '90s, dinosaurs that ran and leaped were expected. Even the *T. rex* in *Jurassic Park* could chase down a jeep. The fat, rubbery, lovable Godzilla of the '50s was out, and the lean, lethal *Velociraptor* was in. Even Godzilla himself suffered outrage when Hollywood made a film called *Godzilla* that replaced the venerable sumolike Japanese icon with a hideous, scrawny monstrosity. By the '90s even dinosaurs, like other Hollywood stars, were supposed to have that svelte, buff, pumped-up, just-left-the-health-club look.

What made Bakker such a zealous advocate of dinosaur revisionism? Why was the popular culture, if not the majority of Bakker's fellow paleontologists, so quick to adopt his views? Here, as in previous chapters, we find two kinds of factors at work, which we may call the social and the scientific. Science is done by communities, and scientific communities are embedded in the larger intellectual community, which in turn is embedded in the society as a whole. Science studies an objective world that exists independently of human cognition. Dinosaurs were real, and would have been just as real had humans never evolved to study them. But science itself is a human creation, and scientists—though they have the incredible chutzpah to think that they can learn something about the hidden nature of this amazing cosmos—are merely human after all. So, even though the self-aggrandizing ambitions of recent sociologists of science are grossly overblown (see Chapter 1), science provides much grist for the sociologist's mill. So understanding the social context of a scientific controversy is essential.

Like me, Robert Bakker is a member of the baby-boom generation. We boomers grew up with dinosaurs. More accurately, we grew up with vivid media images of dinosaurs fleeing, fighting, and running amok. I recall my excitement at age seven when the movie *Journey to the Center of the Earth* featured James Mason and Pat Boone fighting the *Dimetrodon* (not a dinosaur, but close enough). Comic books pitted soldiers and even American Indians against dinosaurs. Best of all were the stop-motion animation dinosaurs (or dinosaurlike beasts) created by Willis O'Brien, Ray Harryhausen, and others for such films as *King Kong, Beast from 20,000 Fathoms,* and *The Giant Behemoth.* These movies, though wildly inaccurate, as every dinosaur fan knew, were great stimuli for the juvenile imagination.

Bakker has mentioned how such films affected him as a child. In particular, he recounts how the scene with the rampaging *Brontosaurus* from *King Kong* shaped his views about sauropods. In this scene, a group of sailors is rafting across a river on Kong's mysterious lost island. They take potshots at a swimming *Brontosaurus*. The infuriated dinosaur overturns their rafts, and the sailors have to swim for it. But dry land affords no safety as the still enraged beast chases them across the terrain. One nitwit tries to escape by climbing a tree but is grabbed by the brontosaur, which avenges itself on the shrieking victim. This is hardly the traditional image of the sauropod as docile and barely capable of moving on land. Bakker says such images convinced him even as a teenager that sauropods could not be swamp-bound sluggards but had to be active and capable of mounting an aggressive self-defense (Bakker 1994, 27).

Bakker also mentions how the 7 September 1953 issue of *Life* magazine affected him. Recently I was lucky enough to find a copy of this issue in an antique shop, and it is easy to see how impressive it must have been to the young Robert Bakker. The featured article by science writer Lincoln Barnett is titled "The Pageant of Life" and gives a fine account of the evolution of life through the Mesozoic. The most impressive thing in the issue is a full-color, six-page foldout of Rudolf F. Zallinger's great mural *The Age of Reptiles* from Yale's Peabody Museum. It begins with the Carboniferous Era on the far left and moves through the three periods of the Mesozoic. Dynamic figures of dinosaurs and other creatures are depicted in a seemingly very realistic landscape. All of the main characters are here—*Dimetrodon, Allosaurus, Brontosaurus, Stegosaurus, Ankylosaurus, Triceratops,* and, of course, *Tyrannosaurus.* Bakker recorded his reaction to these images: "I had discovered an entire world, far, far away in time, that I could visit, whenever I wanted via the creative labors of the paleontologists. And I made my mind up then and there that I would devote my life to the dinosaurs" (Bakker 1986, 9).

Clearly Bakker was deeply influenced by the visual depictions of dinosaurs he encountered as a child, and he has remained loyal to that vision. He is still the wide-eyed kid fascinated by that lost world of heroic creatures that we can visit in our imaginations whenever we like.

Naturally, Bakker went to Yale, the home of the Peabody Museum. The years Bakker spent at Yale, during the mid- to late 1960s, were among the most tumultuous in U.S. history, especially on university campuses. Protests, strikes, and riots shook campuses across the country. Yale did not suffer a major disturbance until the Mayday strike in 1970, but the intensity of that eruption showed that the rage had been building for a long time. Vicious, abusive antiestablishment rhetoric filled the air, and all authority figures were scorned. It was not a time for compromise; intransigence was the norm. When

Bakker, who has called himself "one of the sixties radicals" (Bakker 1986, 9), published work on dinosaurs, his writing reflected impatience and even disdain for the authorities of paleontology—a tone that paralleled the contempt for the political establishment displayed by student activists.

Bakker does not suffer fools gladly. He makes his points with great emphasis, his voice ringing with conviction. He has characterized more conservative colleagues as stodgy defenders of dinosaur "orthodoxy." Here is what he says about Charles W. Gilmore, one of the leading dinosaur paleontologists of an earlier generation (see, e.g., Gilmore 1936): "Between 1909 and 1945 Charles W. Gilmore published extensive descriptive memoirs on dinosaurs, and these studies shaped the new, narrow interpretations of dinosaurean [sic] biology. Gilmore's work had a plodding adequacy devoid of the depth of biological insight shown by Mantell, von Meyer, Huxley, and Riggs. Gilmore described bones as nearly totally inanimate creations and never displayed any firsthand experience with the muscular anatomy or joint structure of extant species. Gilmore's lead-footed reconstructions became the universal standards for textbooks and museums" (Bakker 1987, 44). Against the "plodding adequacy" of Gilmore, whose "lead-footed" dinosaurs trudged through textbooks and museums, Bakker proposed his view of the Mesozoic as a vibrant Age of Dinosaurs, pulsing with life and activity.

Some of Bakker's colleagues have taken issue with the rhetorical tone of his assaults on "orthodoxy." James O. Farlow comments on the tone of the debates over warm-blooded dinosaurs:

> Unfortunately, the strongest impression gained from reading the literature of the dinosaur physiology controversy is that some of the participants have behaved more like politicians or attorneys than scientists, passionately coming to dogmatic conclusions via arguments based on questionable assumptions and/or data subject to other interpretations. Many of the arguments have been published only in popular or at best semi-technical works, accompanied by rather disdainful comments about the stodgy "orthodoxy" of those holding contrary views; what began as a fresh way of considering paleontological problems has degenerated into an exercise in name-calling. (Farlow 1990, 44)

Others have accused Bakker of attacking a straw man in depicting his critics as defenders of a narrow "orthodoxy" (Crompton and Gatesy

1989, 111). Though middle-aged now, Bakker has neither mellowed nor been reconciled to his critics.

It is easy to see the apparent influences of the cultural background of the boomer generation on Bakker's work. The dinosaur images we absorbed as children were depictions not of drowsy browsers but of lively creatures with bad tempers and big appetites. Note that in the earlier quotation in which Bakker recounts his discovery of the *Life* magazine article with the print of the Zallinger mural, Bakker says he decided to devote his life "to the dinosaurs." That is, he dedicated himself not just to the study of dinosaurs but to being their defender and advocate. Small wonder he reacts to any negative characterization of dinosaurs as though it were a personal affront. The intransigent and frequently strident tone of the antiwar and antiestablishment rhetoric of the late '60s is also often apparent in Bakker's speech and writing.

It is even more obvious why the popular culture and entertainment media found Bakker's revisionist views so appealing. Fast, mean dinosaurs are obviously far more exciting than slow, sleepy ones. Also, the jogging and fitness movement that has emerged since the '70s has given us an ultrathin, ultrasleek ideal of physical beauty. Like the modern-day TV heroine who can out-think, outrun, and outfight the bad guys—while still looking as if she just stepped off the cover of *Vogue*—*Velociraptor* is beautiful, smart, fast, and ferocious.

Though it is fun to play the amateur sociologist and trace the possible cultural roots of some of Bakker's ideas, it is, of course, the scientific credentials of those ideas that really matter. Bakker's flamboyance may create the impression that he is more style than substance, but this assumption is far from the case. Bakker has offered a number of ingenious arguments supporting his views. We shall look at three of these and the controversies they generated. First, though, we need to be more precise about the terms of the debate.

Bakker argues that dinosaurs were warm-blooded. "Warm-blooded" and "cold-blooded" are somewhat misleading terms since they seem to imply something about the actual temperature of the blood. But many "cold-blooded" animals employ various devices to keep their internal temperatures quite high, whereas many "warm-blooded" animals sometimes have quite low body temperatures—during winter hibernation, for instance. The real issue is whether body temperature is regulated metabolically or in some other way. "Metabolism" refers to those internal chemical processes, such as digestion, necessary for maintaining life. A high rate of basal metabolism, which we typically find in birds and mammals, produces considerable internal heat. "Endothermy" is the pattern of thermoregulation whereby core body temperature depends on the high and internally controlled generation of heat by metabolic processes. "Ectothermy" is the pattern of thermoregulation whereby body temperature depends upon the uptake of heat from the environment. Reptiles are typical ectotherms, which is why those who keep reptiles must often use heating pads to keep their pets warm.

For over thirty years Bakker has argued that dinosaurs were endotherms, like mammals or birds and unlike lizards or alligators. Why is this categorization important? Because endotherms are typically capable of much more prolonged periods of high activity. Lizards and alligators can move very fast when necessary, but they are active in spurts. A snake's strike or a crocodile's lunge can be faster than the eye can see, but such ectothermic hunters cannot pursue prey for mile after mile, as do endothermic wolves or cape dogs. TV's "Crocodile Hunter" likes to provoke crocodiles and run away before they can snag him; I doubt he would try the same trick with lions. Clearly, then, if you envision dinosaurs as active and lively, showing that they were probably endotherms strengthens your case.

In the early 1970s Bakker developed an argument for endothermy as the best explanation of the fact that dinosaurs were the culmination of an evolutionary trend toward ever more upright postures. Like present-day lizards, the original reptiles were sprawlers. In the Triassic, dinosaurs evolved from a kind of reptile called a "thecodont." Thecodonts were not sprawlers; they had achieved a semi-erect stance. The earliest dinosaurs had made the transition to a fully erect stance with the legs positioned directly under the body and not splayed out to the side. An erect stance is much more conducive to an active lifestyle than a sprawling one. A belly-resting sprawler has to lift itself from the ground each time it moves, which requires considerable effort from the shoulder and hip muscles (as anyone who has ever done a pushup knows). Also, holding the body up places considerable stress on the muscles of an animal with a sprawling posture. With an erect stance, much less muscular effort is required to keep the body up because force is transmitted directly to the ground through bone and joint.

In an important article in the journal *Evolution* Bakker argues that posture and physiology are closely correlated (Bakker 1971). He notes that short-legged sprawlers among contemporary lizards and amphibians have a low core temperature while they pursue their ordinary activities. For such creatures, which are often active only at night or dusk, low activity temperature is a useful adaptation. Since the earliest reptiles of the Carboniferous were short-legged sprawlers, Bakker infers that they also had low activity levels and were ectotherms. However, the thecodonts, the immediate ancestors of the dinosaurs, had a stance and body proportions very much like those of large predatory lizards such as the Komodo dragon. Komodo dragons are ectotherms, but they are active hunters that maintain a high internal temperature. The earliest dinosaurs already had the fully erect posture that Bakker associates with high activity levels. Bakker's argument therefore makes a double inference: From the evolutionary trend toward ever more upright posture, culminating in the fully erect stance of dinosaurs, he infers a trend toward increasing activity levels. From the inferred high activity level of dinosaurs, he further infers that they must have been endotherms.

Bakker's argument is suggestive but hardly conclusive. It is based upon analogies with living organisms, but as some of Bakker's critics have pointed out, the correlation of erect stance with endothermy among living organisms is imperfect (Feduccia 1973, 167). True, all living ectotherms have a nonerect stance, but the converse does not hold—not all creatures with a nonerect stance are ectotherms. Seals and sea lions, for instance, do not have an erect stance, but they are good endotherms. If the correlation were perfect, the analogy would be stronger, say the critics. Even if among today's fauna all and only endotherms had an erect stance, the inference would still be shaky. It could simply be an accident of evolution that no erect-standing ectotherms have survived to the present. Among living primates, only *Homo sapiens* has an erect stance, so there is presently a perfect correlation among primates between having an erect stance and having a large brain. However, this does not mean that all primates with an erect stance have had large brains. *Australopithecus* had evolved an erect stance over three million years ago, but its brain was the size of a chimpanzee's.

Bakker's argument could be taken as the best explanation of the trend toward uprightness in the line leading to dinosaurs: that these developments were adaptations to an increasingly active lifestyle powered by an increasingly endothermic physiology. But this argument is also risky. Some critics proposed that the trend toward more erect posture could instead have been an adaptation for the large size of dinosaurs (Feduccia 1973, 167; Bennett and Dalzell 1973, 171–172). Bakker could reply that the earliest dinosaurs of the Triassic were not yet gigantic but had already developed the erect stance. Still, adaptationist explanations are fraught with danger. It is often hard to distinguish a legitimate explanation from what Stephen Jay Gould called a "just so" story (after Rudyard Kipling's *Just So Stories*). An evolutionary "just so" story proposes a plausible scenario suggesting the adaptations that an evolutionary development might have served. Painting an attractive and even plausible picture might entertain the armchair biologist, but it is not good enough for science. Plausible hypotheses crash and burn every day in science, and it often turns out that the real explanation is something nobody would have thought of. So there needs to be some way to acquire independent confirmation of even the most plausible-sounding hypothesis, and such evidence is precisely what is usually missing when adaptationist stories are told.

Realizing the need for additional evidence, Bakker soon proposed two additional arguments, one based on the microscopic structure of dinosaur bone and the other based on the ratios of predators to prey in fossil ecosystems. Each of these ingenious arguments deserved the close and careful attention that it got from other paleontologists.

Bakker notes that there are distinct differences in the fine structure of the bones of endotherms and ectotherms (Bakker 1972). The bones of endotherms such as birds and mammals are highly vascularized, with many densely packed structures called haversian canals. Bone stores various minerals important for an organism's physiological functions. Endotherms continuously adjust their metabolic rates as they go from rest to high levels of activity. These changes require rapid and easy access to the minerals stored in the bone, and Bakker claims that the numerous haversian canals facilitate the transfer of needed minerals from bone to blood plasma. Some mammals also feature a bone architecture called laminar bone that permits an even closer association of capillaries and bone cells, allowing for an even more rapid absorption of minerals.

The bone of living and fossil reptiles, with very few exceptions, is poorly vascularized. Apparently, the physiological needs of ectotherms do not require rapid access to bone mineral. Bakker reports that all of the dinosaurs he has examined have haversian or laminar bone, or both (Bakker 1972). With a bone architecture much more like that of mammals than reptiles, and with an apparently clear connection of that structure to endothermic physiology, Bakker infers that dinosaur thermoregulation was much closer to that of mammals than reptiles.

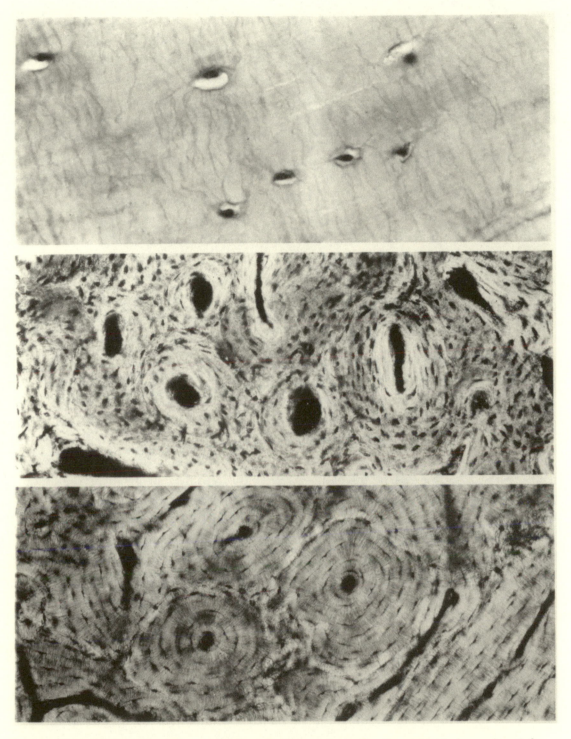

Bone tissue from a living ectotherm, a living endotherm, and a dinosaur, showing that dinosaur bone had many channels for blood vessels, like the bones of endotherms (The Natural History Museum, London).

Perhaps Bakker's most original and interesting argument is based on the ecology of predator-prey relations. Endotherms have a number of advantages over ectotherms, as we have seen, but endothermy has one very big drawback. Endotherms have to eat a prodigious amount of food to maintain such high levels of basal metabolism (which is why the phrase "eating like a bird" is so inapt). Small animals lose body heat faster than large ones, so small endotherms have very high energy budgets, often having to consume two or three times their body weight in a day (which is why birds' excretory functions are so copiously productive). But in the natural world, food is often scarce, and often hard to get even when abundant. This is especially true for predators. Prey animals are wary, fast, and often capable of violent resistance. They have to be located, stalked, caught, and subdued, and these are very energy-expensive activities.

The upshot is that endothermic predators require a lot of prey, and the "standing crop" of predators—the total biomass of predators in a given region—will be strictly limited by the availability of prey animals. For instance, 30 tons of prey carcasses a year will sustain a standing crop of only 2.5 tons of lions. That's a lot of wildebeest and zebras to sustain a very modest pride of lions. Compare this to the situation in the Komodo Islands of Indonesia, where the famous Komodo dragon, the world's largest lizard, is the top predator. Thirty tons of prey carcasses a year will sustain a standing crop of 40 tons of dragons. So the biomass ratio of lions to prey in any given region cannot be higher than 1:12, whereas the ratio of the standing crop of Komodo dragons to the biomass of their prey could theoretically be as high as 4:3.

Given this information, a very interesting pattern emerges when we look at fossil ecosystems. Bakker notes that in the Permian Era, prior to the age of the dinosaurs, the sprawling, presumably ectothermic *Dimetrodon* was the top predator (Bakker 1975, 64). The ratio of the standing crop of *Dimetrodon* to its prey in Permian ecosystems was quite high, from 35 to 60 percent, says Bakker (1975, 68). However, when we look at the biomass ratios of predators to prey in dinosaur ecosystems, we find a ratio of only 1 to 3 percent, a ratio as low as that found in the mammalian communities of the Cenozoic Era (Bakker 1975, 71). In other words, *T. rex* was quite rare compared to the numbers of late Cretaceous herbivores such as *Edmontosaurus* or *Triceratops*. Apparently it took a lot of carcasses of these latter creatures to maintain a quite small standing crop of tyrannosaurs. Standing crop ratios of predator to prey in dinosaur ecosystems therefore were much closer to ratios such as those found in the Serengeti rather than on Komodo Island. It seems then that *T. rex* must have been a warm-blooded predator like a lion instead of a cold-blooded hunter like the Komodo dragon. Maybe *Velociraptor's* hot breath could have steamed glass.

Ingenious as they are, Bakker's arguments have not convinced the majority of his colleagues. The 1980 volume *A Cold Look at the Warm-Blooded Dinosaurs* contains a number of responses to these arguments. The French scientist Armand J. de Ricqlès, perhaps the world's leading authority on the microstructure of fossil bone tissue, wrote a response to Bakker's argument that dinosaur bone indicates mammalian-style endothermy. De Ricqlès agrees with Bakker that bone microstructure correlates closely with physiology, and he notes that dinosaurs possessed highly vascularized "fibro-lamellar" bone, which is always associated with rapid bone deposition and high growth rates (de Ricqlès 1980, 117). Endotherms grow rapidly, reaching their adult size soon after sexual maturity and then ceasing to grow. Ectotherms grow slowly throughout their lives, and their bones show lines of arrested growth (LAGs)—lines indicating that growth speeds up when it is warm and slows when it is cool. De Ricqlès remarks that the function, or functions, of haversian canals is not yet fully clarified, but he does note that among living animals only endotherms have dense masses of haversian bone like the dinosaurs (de Ricqlès 1980, 124). He could find no LAGs in dinosaur bone. However, he notes that dinosaur bone does show that dinosaurs grew more slowly after sexual maturity but did not stop growing like today's endotherms.

De Ricqlès concludes cautiously that though bone microstructure does provide good evidence that dinosaurs had high metabolic levels, they probably had a thermoregulatory pattern intermediate between the reptile and mammalian models (de Ricqlès 1980, 132–133). More recent examinations of dinosaur bone microstructure have supported de Ricqlès's hesitancy. For instance, researchers have shown that dinosaur bone does have LAGs, like that of ectotherms and unlike endotherms (Chinsamy 1993). On balance, therefore, the bone evidence indicates that dinosaurs did not have a typically reptilian, ectothermic metabolic rate, but neither did they seem to match mammalian physiology.

The leading dinosaur paleontologist James O. Farlow has cogently criticized Bakker's argument from predator-prey biomass ratios. Farlow notes that there are many uncertainties when we attempt to compare such ratios in fossil communities with those in present ones. He agrees that when we look at smaller predators of approximately the same size, the endotherms do seem to require much more food than the ectotherms (Farlow 1980, 57). But it just is not clear that such an extreme disproportion would hold for very large predators. In fact, the differences in these ratios appear to decrease with larger animals. In other words, the food needed to sustain an extremely large (e.g., *T. rex*–sized) ectothermic predator would not be that much less than the amount needed to sustain an endothermic predator of the same size (Farlow 1980, 73–75). The problem is that there is little basis for comparison of fossil with living faunas since there are so few terrestrial communities with really large ectothermic predators. Bakker therefore has to extrapolate from small- to medium-sized ectothermic predators to make predictions about the expected standing crop of gigantic Mesozoic predators.

The largest Komodo dragon is about one-tenth the size of the biggest tyrannosaurs. Population turnover occurs much more rapidly in populations of small animals than large. Mice live only a couple of years, and they reproduce prodigiously; elephants live a long time and reproduce slowly. Populations of prey animals with high turnover rates will support a larger standing crop of predators than a population of slowly reproducing, long-lived prey. Since the huge Mesozoic predators preyed on animals as big or bigger than themselves, the turnover rates in those prey populations must have been slow. Therefore, basing estimated standing crop sizes for gigantic ectothermic predators by extrapolating from the data for small- to medium-sized ectothermic predators might well result in a considerable overestimate.

Farlow also notes that Bakker assumes that the standing crop of predators would be limited only by the availability of prey (Farlow 1980, 67). However, *T. rex* may have faced other challenges that kept its population small compared to its prey species. For instance, *T. rex* was probably not a very picky eater, and probably not very friendly to members of its own species. It may well have practiced cannibalism, which would have kept its population well below the maximum that the availability of prey animals could sustain. Injuries inflicted on *T. rex* skeletons show that they viciously attacked each other, so intraspecific conflicts might well have kept the population down. Even with living creatures, we often cannot sort out the factors that control their population size. Clearly, we should be even less confident when speculating about such factors in fossil populations. An even more basic problem is that the biomass of a population obviously depends on how big the individual animals were. But nobody knows just how big an adult *T. rex* was. Soft tissues do not fossilize, and a wide range of body weights is compatible with the skeletal evidence.

A fundamental problem with making the kind of estimates Bakker needs is that the proportions of organisms found in a fossil assemblage might poorly represent their actual ratios in the living community. Like telescopes, fossil assemblages have a certain resolving power. One thing that makes a particular telescope better than another is that it provides images with better resolution. A telescope with better resolving capability can, for instance, split a pair of stars in a close binary system, whereas a weaker telescope will see them as a single star. So a higher-resolution telescope will show the universe in much greater and truer detail. Since fossilization is a *very* chancy process and is dependent on many imponderables, fossil assemblages have a rather low resolving power. That is, from the raw numbers of remains of animals in a fossil assemblage, we cannot tell very well how many of those animals were in the living community. Our estimates might be off by a large margin. If, as Bakker thinks, the predicted difference in ratios of endothermic to ectothermic biomass is as large as a factor of ten, we probably could reliably detect that difference in the fossil record. However, if we predict that the standing crop of ectothermic dinosaur predators would only be three to six times as large as that expected for endotherms—as Farlow estimates—the fossil record will not have sufficient resolving power to verify this theory (Farlow 1980, 75).

Bakker has offered vigorous rebuttals to all of the criticisms leveled at his arguments (Bakker 1980), but the bulk of his colleagues remain unconvinced. In science, the burden of proof is always on the maverick, the one making the new and surprising claim. The consensus has been that Bakker's hypothesis has not borne the burden of proof.

After the AAAS symposium, the dinosaur physiology controversy quieted down but did not go away. Since 1980 various intriguing lines of evidence have been offered both for and against the dinosaur endothermy hypothesis. Endotherms have a high rate of respiration, and since some moisture is lost with every exhaled breath, there is a danger of excessive evaporation of water from the lungs. Endotherms compensate by developing complex nasal turbinates, folded bones within a nasal cavity that provide a surface for mucous membranes. These mucous membranes function to reclaim moisture from the breath. Ectotherms, because of their much lower respiratory rates, need no such turbinates and therefore do not have them. Researchers have pointed out that dinosaurs had no nasal turbinates and so must have been ectothermic, or nearly so, while pursuing their ordinary activities (Ruben et al. 1997).

Other researchers have countered by examining the ratios of oxygen isotopes in dinosaur bone as a measure of how much the body temperatures of dinosaurs varied with the seasons (Barrick, Stoskopf, and Showers 1997). Bone contains phosphate, and phosphate contains oxygen. The oxygen in bone phosphate comes in two different isotopes, and the ratio of these isotopes in an animal's bone phosphate is influenced by the body temperature of the animal when the phosphate was formed. These studies show that the body temperature of dinosaurs, unlike lizards, did not vary greatly from season to season. So these data indicate that dinosaurs maintained a consistent body temperature, like endotherms and unlike most ectotherms. However, the picture is complicated by the fact that extremely large animals, by virtue of their sheer bulk, gain or lose heat slowly and so tend to maintain more stable core temperatures than a small animal.

So were dinosaurs warm-blooded? The answer indicated by the mass of evidence is a definite "maybe." The great dinosaur endothermy debate remains unresolved. It is frustrating not to know, but we may have to resign ourselves to ignorance. I began this chapter with a description of one of Gary Larson's *Far Side* cartoons, and I'll finish with another. In this cartoon, a scientist has stepped out of his time machine and is approaching the backside of a very large sauropod with an enormous thermometer. The caption reads, "An instant later both Professor Waxman and his time machine were destroyed, leaving the coldblooded/warmblooded dinosaur debate still unresolved." Alas, we have no time machines, and even if we did, any researcher attempting to use a rectal thermometer on *Brachiosaurus* would likely meet Professor Waxman's fate. Hence, the dinosaur endothermy controversy may well remain unsettled. Those who love dinosaurs would like to know everything about them, but the myriad centuries separating us from them make that impossible. Sometimes in a scientific controversy we have no choice but to agree to disagree.

Further Reading

Anyone who has ever watched a dinosaur program on the Discovery Channel has probably seen Robert Bakker. Scientists are usually terrible on TV, coming across as the quintessential nerds. Bakker is so animated and enthusiastic that the small screen can barely contain him (small wonder a character in the second *Jurassic Park* movie was a Robert Bakker look-alike). He has always been eager—too eager, other paleontologists think—to take his case to the public. His article "Dinosaur Renaissance" written for *Scientific American* in 1975 (232: 58–78) presented his case for warm-blooded dinosaurs to a wide audience. His 1986 book *The Dinosaur Heresies* (New York: William Morrow and Company) is very well written and lively. Bakker is also an excellent illustrator, and his works are enlivened by his own drawings. Bakker's writings are opinionated, and even polemical, but they are never dull.

As I note in this chapter, Bakker admits that, like many of us Baby Boomers, his view of dinosaurs was strongly stimulated by the entertainment media. A terrific chapter on "Dinosaurs and the Media" by Donald F. Glut and M. K. Brett-Surman closes *The Complete Dinosaur* volume edited by Brett-Surman and Farlow. The chapter is wonderfully illustrated with many of my favorite images, including stills from *King Kong*, *The Lost World* (the good 1925 version), and *Beast from 20,000 Fathoms*.

The papers presented at the 1978 AAAS conference on dinosaur endothermy were published in the volume *A Cold Look at the Warm-Blooded Dinosaurs*, edited by R. D. K. Thomas and E. C. Olsen (Boulder, CO: Westview Press, 1980). This is *the* source for anyone who wants to find out about the dinosaur endothermy debate of the '70s. Unfortunately, it is long out of print and hard to find. Interlibrary loan would be your best bet. This volume contains excellent articles by Bakker's critics and his lengthy and vigorous reply. Popular accounts of the warm-blooded dinosaur controversy are given in John Noble Wilford's *The Riddle of the Dinosaur* (New York: Vintage Books, 1985) and Don Lessem's *Dinosaurs Rediscovered* (New York: Simon and Schuster, 1992). I give a more detailed and critical view of the controversy in Chapter 2 of my book *Drawing Out Leviathan* (Bloomington: Indiana University Press, 2001).

Reconstructing the biology of extinct creatures is a complex and fascinating study. Farlow and Brett-Surman's anthology *The Complete Dinosaur* (Bloomington: Indiana University Press, 1997) has thirteen articles by leading experts on various topics of dinosaur biology. One of the most interesting issues is how such stupendously huge creatures actually managed to get around. When you consider that even a moderate-sized dinosaur could be as big as the largest African elephant, and that the biggest were as massive as a herd of elephants, it is a wonder that they could move at all, much less fight, mate, etc. R. McNeill Alexander's chapter "Engineering a Dinosaur" is particularly interesting. Alexander's book *Dynamics of Dinosaurs and Other Extinct Giants* (New York: Columbia University Press, 1989) applies the insights of an engineer to questions about how dinosaurs lived. Of particular interest is his formula for determining how fast dinosaurs ran from a study of their fossilized footprints.

Other chapters in the Farlow and Brett-Surman anthology provide a variety of perspectives on dinosaur physiology. I found R. E. H. Reid's "Dinosaur Physiology: The Case for 'Intermediate' Dinosaurs" to be particularly persuasive. Reid argues that dinosaurs were probably not like classical warm-bloods or cold-bloods but had a unique physiology intermediate between these. Naturally, everybody wants to know how dinosaurs acted, and Scott Sampson's chapter "Dinosaur Combat and Courtship" is particularly interesting. Maybe even more interesting is the evidence of dinosaur disease and injury reported in Bruce M. Rothschild's chapter "Dinosaur Paleopathology." I cannot think of anything crankier than a *T. rex* with gout.

7

The Raptor and the Hummingbird

People love to honor the superlatives in almost any category, whether it is the NCAA Division 1-A football national champion, Miss America, the best actor in a leading role, or the Nobel Prize in Physics. Vicious arguments break out over who really is number one. Among vertebrate paleontologists, it is a safe bet that one particular specimen would get the most votes for the most important fossil find of all time—the Berlin specimen of *Archaeopteryx lithographica*. Pat Shipman expresses the awe paleontologists feel toward this specimen: "The importance of the Berlin specimen cannot be overstated. It is more than a stony record of an extinct species. It is an icon—a holy relic of the past that has become a powerful symbol of the evolutionary process itself. It is the First Bird" (Shipman 1998, 14).

Paleontologists have dropped to their knees in wonder upon first seeing this specimen. It *is* awesome and inspiring. In a sheet of the beautiful, fine-grained Solnhofen limestone from Bavaria lie the fragile, fossilized bones of a small creature. Its neck is arched backward by rigor mortis. It seems to be the skeleton of a small reptile, perhaps a small dinosaur. But wait—surrounding the bone in the rock is the unmistakable impression of a thick coat of feathers. The creature is a bird. It is 150 million years old.

Seemingly misplaced in a world of giant sauropods and terrifying allosaurs is this delicate creature. When we look at it more closely we see that its skeleton really is more reptile than bird. *Archaeopteryx* had a few distinctly avian features, such as wings, feathers, larger eyes and brain, and a furcula (wishbone). Otherwise, its anatomy is almost totally reptilian. In fact, the collectors of some specimens of *Archaeopteryx* initially identified their finds as *Compsognathus,* a small dinosaur. It is so perfect a missing link between reptiles and birds that

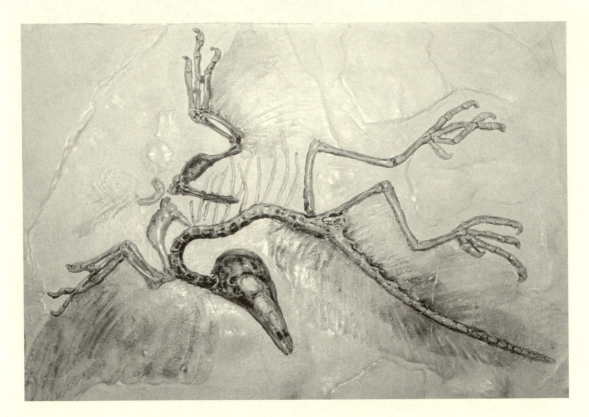

antievolutionists, desperate to deny its transitional status, have been driven to declare it a hoax.

No legitimate scientists now deny that birds descended from reptiles. However, it is one thing to say that birds descended from some nondescript reptile; it is quite another to say that they descended from *dinosaurs*. It is a startling suggestion; if it is true, then the dinosaurs are not really gone. It is amazing to think that the red robin bobbing along might be an evolutionarily modified dinosaur or that last night you may have dined on Kentucky Fried Dinosaur. Lowell Dingus and Timothy Rowe are defenders of the theory that dinosaurs are the evolutionary progenitors of birds; they titled their 1997 book *The Mistaken Extinction*. It should come as no surprise at all that this theory has had its detractors as well. The debate has been spirited, and, though it may be moving toward closure, it is not yet over.

Why think that birds descended from dinosaurs? As we have seen, T. H. Huxley attacked Richard Owen's archetypes and antievolutionary stance by adducing similarities between birds and dinosaurs. However, Huxley stopped short of asserting that birds had descended from dinosaurs. The majority of paleontologists came to believe that birds had evolved from reptiles called "thecodonts." Paleontologists no longer use the term "thecodont"; it was really a sort of "wastebasket" term to group various creatures. The theory that birds descended from thecodonts was widely accepted after 1927, when the leading ornithologist Gerhard Heilmann defended this claim in his highly influential book *The Origin of Birds* (Heilmann 1927). Thecodonts were also thought to be the ancestor of dinosaurs, so birds and dinosaurs were thought to be cousins that had descended along parallel tracks from the same type of ancestor.

It was a dinosaur discovery that got people thinking about birds and dinosaurs once again. In the early and mid-1960s, John H. Ostrom of Yale University, along with Peabody Museum staff and students, conducted a series of explorations of Early Cretaceous sediments in Montana and northern Wyoming. There are many rich sites of Late Cretaceous fossils but very few from the Early Cretaceous. Therefore, the fauna of that period is not nearly as well known as that of earlier and later times. Ostrom and his team were gratified to find the remains of eight new dinosaurs. One of these creatures was very remarkable. Ostrom called this dinosaur *Deinonychus*, which means "terrible claw." The second, or inside, toe of each foot had been modified into a long, sharp, sicklelike claw. This claw could be raised from the ground as the creature ran on its other toes and then snapped out like a switchblade when its leg raked down in a powerful stroke. Clearly, this was a weapon that could inflict terrible injuries.

Deinonychus was small for a dinosaur, only about six feet long and weighing probably between 150 and 175 pounds—about the size of a leopard. It was built for speed, with powerful legs and a tail stiffened by ossified tendons. This rigid tail could be used as a counterbalance to stabilize the creature as it ran and maneuvered rapidly. In addition to its terrible rear claws, *Deinonychus* had strong, mobile forearms with clawed hands to grip prey. It had big jaws with sharp teeth serrated like steak knives. *Deinonychus* was a killing machine. It was built to run down its prey, seize it with hands and jaws, and slice it to bits with its clawed feet. Worst of all, it may have hunted in packs. *Velociraptor*, of *Jurassic Park* fame, was a smaller cousin of *Deinonychus* (actually, Steven Spielberg made his velociraptors the size of *Deinonychus*). Perhaps the most frightening land predator of all time was *Utahraptor*, a twenty-foot version of the same sort of beast. This whole group of nightmarish predators is called the "dromaeosaurs."

Ostrom noted that *Deinonychus* could not have had an upright, kangaroolike posture, as other theropods such as *Allosaurus* and *Tyrannosaurus* were traditionally depicted. Its stance had to be much more birdlike: "Several important features preserved in the neck and back vertebrae of *Deinonychus* indicate that the posture of this little biped was very much like that of an ostrich or a cassowary, with the trunk held in a near horizontal attitude and the neck curving upward . . . in my opinion this is a much more natural-looking posture than the 'Kangaroo' pose that is commonly illustrated for other carnivorous dinosaurs such as *Allosaurus* or *Tyrannosaurus*" (Ostrom 1969, 8). In fact, no one had previously noticed, but the well-known theropods also had these back and neck features, indicating that they too had a birdlike stance: "What is particularly significant, though, is that these same anatomical features are present in the neck and back vertebrae of many of the other well-known theropods. Because none of the specimens of *Allosaurus*, *Tyrannosaurus* or *Gorgosaurus* were as perfectly preserved as our *Deinonychus* material, these features have not been recognized before. The similarity to *Deinonychus* is very close, and I am now convinced that even the large theropods like *Tyrannosaurus* had an ostrich-like rather than kangaroo-like stance" (Ostrom 1969, 9).

Paleontologists now concur that the theropods did have an ostrichlike stance, with the body tilted forward and balanced over the legs. Over the next several years Ostrom authored a number of studies making detailed comparisons between theropods and *Archaeopteryx*. He concluded that the similarities were very close, especially between *Archaeopteryx* and dromaeosaurs such as *Deinonychus* (e.g., Ostrom 1976). Over the past quarter century, a considerable number of paleontologists have come to believe that Ostrom's and other evidence does indeed show that birds evolved from small theropod dinosaurs. What is the evidence for this claim?

First of all, just what is a bird? What are the anatomical features that distinguish birds and set them apart from other creatures? Most obviously, birds have feathers, which are among the most remarkable structures in the natural world. Like fingernails, horns, or hoofs, feathers are made of material containing a tough protein called keratin. Each feather consists of a central hollow shaft that tapers toward the end. From the shaft radiate a number of barbs that are linked together with small hooks called barbules. The linked barbs form the sheet of feather material called the vane. The reason feathers are so important is that it is most unlikely that such a complex and unique structure would have evolved more than once. So anything with feathers must either be a bird or closely related to birds.

Birds are also distinctive in what they lack. Modern birds, of course, have no teeth. They have only four toes, three in front and one in the rear. This rear toe, which is homologous to our big toe, gives perching birds an opposable digit to grip branches. Birds also lack a long tail; the tailbones are fused into a small pointed structure called a "pygostyle." Birds do not have separate hand and wrist bones; these are fused into a unique structure called a "carpometacarpus."

Birds also have remarkable skeletons. Flying birds must have bones that are strong yet light. This is achieved by bone that is thin walled and hollow but is braced inside with buttresses. Such bone is called "pneumatic" because most of its volume is taken up by air spaces. Birds have a number of other unique skeletal features, most notably the furcula, or wishbone, mentioned earlier.

Paleontologists have conducted careful studies showing many similarities between the birds and the theropod dinosaurs: "A great many characters classically considered 'avian' apply to more general levels within Theropoda. Basal theropods have lightly built bones and a foot reduced to three main toes, with the first usually held off the ground and the fifth lost. Closer to birds, the fourth and the fifth digits of the hand are progressively reduced and lost, the skeleton (especially the vertebrae) becomes lighter. . . . In coelurosaurs . . . the forelimbs become progressively longer until they are nearly as long as the hindlimbs in some dromaeosaurs . . . the first toe (hallux) begins to rotate behind the metatarsus" (Padian and Chiappe 1997, 73).

In other words, like birds, theropods developed more lightly built bones, feet with three main toes in the front and the big toe starting to rotate to the rear, and the fifth toe lost. The third and fourth digits of theropod hands were gradually reduced, then lost, coming closer to the complete fusion of the hand and wrist bones found in birds.

These similarities are striking, but how important are they? Merely comparing skeletal similarities and dissimilarities seems to be a largely subjective procedure. Which similarities are important for determining evolutionary relationships? How similar or dissimilar do they have to be? Too often in the past, paleontologists trying to answer these questions were thrown back on their personal judgment and could hardly be surprised when many colleagues disagreed. In recent decades biologists have developed a method called "cladistics" that makes anatomical comparisons a much more objective and useful procedure for determining evolutionary relationships.

First some definitions: A category used in the classification of organisms—species, genus, or family, for instance—is called a "taxon" (the plural is "taxa," and the adjective is "taxonomic"). Taxa are not all of the same level. For instance, in the traditional Linnaean (after Carolus Linnaeus, Swedish scientist of the eighteenth century) scheme of classification, genus is a higher (more inclusive) taxon than species. A "phylogeny" is the evolutionary history of a species or other taxon. In other words, phylogeny is "the evolutionary relationships within and between taxonomic levels, particularly the patterns of lines of descent, often branching, from one organism to another, i.e., the relationships of groups of organisms as reflected by their evolutionary history" (Allaby and Allaby 1991, 280).

Now, imagine that you are at a family reunion and that you find these affairs as confusing as I do. You are introduced to a gentleman you have never met, and you try to find out how you are related. To find out how closely you are related to each other, you try to identify your closest common ancestors. If you discover that you both had the same parents, then he is your long-lost brother. If you share a pair of grandparents, then you are first cousins; if you share a pair of great-grandparents, then you are second cousins; and so on. In other words, how closely related you are depends on how recently you shared a common pair of ancestors. The same sort of thing applies in evolutionary relationships. You are more closely related to your pet cat than your pet boa constrictor because you and your cat share a much closer common ancestor than either of you does with the boa constrictor.

Given any three groups of organisms, cladistics will allow us to determine which two had a more recent common ancestor than the third, and therefore which two groups are more closely related. This is done by distinguishing traits of the organisms that are "primitive" from those that are "derived." When we compare humans, cats, and snakes, we see that members of all three groups have lungs, so lungs must come from a distant ancestor that lived before the line leading to snakes split from the line leading to the cats and us. However we share with felines but not with snakes the trait of having hair or fur, so that trait must have first evolved in a more recent ancestor that we share with cats but not with snakes. So in comparing humans, felines, and snakes, having hair or fur is a derived trait because only two of the three groups have that novel characteristic. Also, human and feline females have mammary glands, but snakes do not, so having mammary glands is another derived trait shared by cats and humans. If we included sharks in our comparisons, having lungs would be a derived characteristic shared by humans, cats, and snakes but not sharks. Having a backbone would be a primitive trait shared by humans, cats, snakes, and sharks.

The many derived traits such as hair and mammary glands that we share with cats and not with snakes must have first evolved in more recent ancestors after their line of descent branched off from the line leading to snakes. The guiding principle is that evolutionary novelties (such as feathers) probably did not evolve two or more times independently. If two kinds of organisms have feathers, they probably got them from a common ancestor. This conclusion is not always true. There is a phenomenon called convergent evolution that occurs when two different lineages sometimes do independently evolve very similar traits. For instance, humans and cephalopods (such as squid and octopuses) have advanced and similar eyes. However, if there are enough novelties shared between two groups, the chances that so many instances of convergent evolution would have occurred are slim. It is a much simpler hypothesis to attribute the similarities to common ancestry. Therefore, if two groups of organisms share a number of derived characteristics compared to a third group, those first two groups very likely have a most recent common ancestor that they do not share with the third group. Therefore, members of those first two groups are more closely related to each other than to members of the third group.

By repeated application of such cladistic analyses, extensive phylogenies can be reconstructed. In particular, cladistic analysis attempts to identify "monophyletic" groups. A monophyletic group consists of an ancestral group and *all* and *only* those groups that share that ancestral group as their most recent common ancestor. Birds are a monophyletic group: They all share a most recent common ancestor that they share with no other group of creatures. Reptiles are not a monophyletic group unless mammals and birds are added to the group, since mammals and birds also descended from the most recent common ancestor of reptiles.

Extensive cladistic analyses have allowed paleontologists to reconstruct detailed phylogenies of birds, indicating that they descended from a group of theropods called "coelurosaurs." These phylogenies indicate that the dromaeosaurs also split from the coelurosaurian line, thus accounting for the many similarities between birds and dromaeosaurs. In fact, paleontologists supporting the dinosaur ancestry theory have identified dozens of shared derived characteristics in the skeletons of birds and theropods (Padian and Chiappe 1997, 75). These comparisons, made less subjective by the application of cladistic principles, have convinced many paleontologists that birds did in fact evolve from theropods sometime in the Jurassic.

As noted earlier, it should come as no surprise that some paleontologists sharply disagree. For over twenty years Larry D. Martin, a paleo-ornithologist and curator for vertebrate paleontology at the University of Kansas Natural History Museum, has been a vocal critic of the dinosaur ancestry theory. He admits that the list of anatomical features supporting the connection between birds and coelurosaurs looks quite impressive at first glance (Martin 1983, 116). However, he thinks a close examination of the proposed anatomical links shows that practically none of them hold up. Some of these discussions of anatomy get quite technical and detailed, so let's take a close look at just one disputed point. First, though, let's remind ourselves of some basic anatomical terminology. Your finger bones are called "phalanges." At your knuckles they join the bones of the hand called "metacarpals." In your wrist, the metacarpals join the bones called "carpals."

In his studies of *Deinonychus* and other dromaeosaurs, John Ostrom noted that these dinosaurs have an unusual half-moon-shaped bone in their wrists called the semilunate carpal. This bone may have helped the ferociously predacious dromaeosaurs snap their hands forward to grab rapidly fleeing prey. The developing bird also has a semilunate carpal that fuses to the adjacent bones of the hand in the adult. For advocates of the dinosaur ancestry theory, the fact that both birds and dromaeosaurs possess this distinctive bone is one of many pieces of anatomical evidence linking the two groups.

Martin disagrees. He notes that advocates of the dinosaur ancestry theory place great trust in cladistic analysis, but he argues that these comparisons are high in quantity but low in quality. That is, these analyses compare many features, but little effort is made to ensure that the compared traits are really homologous. The semilunate bone is a case in point:

> Ostrom says *Deinonychus* has a wrist "almost identical" to *Archaeopteryx*. But that statement cannot be true. *Deinonychus* has only two bones in the wrist, though its half-moon bone may be the result of a fusion within another bone. *Archaeopteryx* resembles all other known birds in having a total of four bones arranged in two rows. One row of two bones is connected to the arm, the other row to the hand. . . . Ostrom saw only half as many bones in the wrist of *Deinonychus*. He reported that both of them, including the half-moon bone, lay in the row connected to the arm. (The row connected to the hand, he believed, did not exist.) Hence by Ostrom's own logic the half-moon bone in

Deinonychus occurs in a different part of the wrist from where it occurs in birds. Therefore, though the half-moon bones in the wrists of birds and dinosaurs look alike, they develop from completely different wrist bones during the animals' growth. Birds and dinosaurs must have evolved the trait separately (Martin 1998, 43).

If the half-moon bones of dinosaurs and birds develop from different bones, they cannot really be the same trait derived from a recent common ancestor and instead must be a product of convergent evolution.

Kevin Padian and Luis M. Chiappe are defenders of the view that birds descended from theropods. They think that *Deinonychus* did have two rows of carpals like the birds, one row (the "proximal" carpals) contacting the radius and ulna of the forearm and the "distal" carpals connecting to the metacarpals of the hand (Padian and Chiappe 1997, 76). Why don't we find both rows of carpals in *Deinonychus* and its kin? Padian and Chiappe think that the proximal row simply was not preserved in the extant fossils. This theory sounds like special pleading, but they have a point. They note that carpal elements are often missing or misidentified and may not even have been composed of ossified (bony) material. They think Ostrom incorrectly identified the carpal bones of *Deinonychus* as the proximal ones that connect to the arm, when instead these bones were the distal carpals, the ones adjacent to the metacarpals of the hand.

Padian and Chiappe point out that early theropods such as *Coelophysis* did have carpals in two rows like birds, one row contacting the radius and ulna of the arm and the other row overlying the metacarpals of the hand (76). In the wrists of those early theropods, two distal carpal bones have fused and overlap the second metacarpal bone. Padian and Chiappe point out that the semilunate bone in later theropods, including the dromaeosaurs, overlaps the second metacarpal in precisely the same way. They conclude that the semilunate bone of later theropods such as *Deinonychus* is an evolutionary modification of the fused distal carpals of the early theropods. The semilunate carpal in birds seems to be a homologous structure, and hence is apparently a derived trait shared with the dromaeosaurs. For Padian and Chiappe, therefore, the semilunate bone shared by birds and some dinosaurs is good evidence of a close evolutionary relationship.

The debate of bird ancestry does not turn merely on these esoteric and (for the nonspecialist, anyway) rather dry anatomical points. Martin notes that another major problem for the dinosaur ancestry theory is that it has the burden of explaining how birds could ever have begun to fly. Theropods were ground dwellers. They made their living running along the ground, often in pursuit of prey. How could ground dwellers ever have taken to the air? A much more plausible scenario for the origin of flight would seem to be that it started among creatures that already lived in the trees and perhaps glided among the branches. After all, we are quite familiar with creatures such as the flying squirrel, which is not a true flyer but can glide quite efficiently from branch to branch. So part of the argument about dinosaur ancestry is over whether a "ground-up" or "trees-down" theory best accounts for the origin of flight.

Alan Feduccia is another well-known opponent of the theory that birds descended from theropods and an eloquent defender of the trees-down account of the origin of flight. His 1996 book *The Origin and Evolution of Birds* gives a detailed and comprehensive statement of his views. He begins by noting that in addition to the three types of vertebrates that developed true powered flight—birds, bats, and pterosaurs—various reptiles and three different orders of mammals evolved gliding ability (95). After surveying the many different kinds of flyers and gliders, both living and extinct, he concludes: "As far as we know, all these living forms, vertebrate or extinct, parachuting, gliding, or truly flying, have one feature in common: an arboreal (or other elevated place) origin. None began to fly from a purely ground dwelling habitat. This fact alone argues strongly against the theory that birds began flight as runners and jumpers" (97).

The most obvious advantage of the trees-down (arboreal) theory is that gravity provides free energy to power takeoff for both gliders and flyers, whereas running is very energy expensive. Intuitively, it would be easier for evolution to exploit a source of energy available to every creature than to develop the cursorial capabilities of a few. As Pat Shipman points out, the theory has other advantages as well:

> One great strength of the arboreal hypothesis derives from living animals who move through the air. This theory relies on a logical sequence of adaptations involving an increasing mastery of the air: plummeting, parachuting, gliding, and flapping. Each stage is a viable locomotor mode in and of itself, one that is practiced successfully by living species. There are no awkward "almost flying" stages along the hypothesized evolutionary transition to raise the question of how a creature moving in this way could have survived for a week, much less long enough to leave successful progeny. It is a simple mental leap from recognizing

this ordered sequence of living, but unrelated species, to postulating an evolutionary trajectory for a single lineage that moved
through these stages over time, eventually producing a modern
bird (Shipman 1998, 182–183).

In other words, known animals move through the air by simply drop-
ping, parachuting, gliding, or flying. Further, the anatomical and behavioral changes needed to move from one form of aerial locomotion
to the next do not seem great, and it is easy to imagine a lineage of
creatures passing through these stages. Most importantly, at no point
in such an evolving process will organisms be stuck with nonadaptive
anatomy or behavior—or so it seems.

However, things are not as simple as they seem. Gliding and flying are really two very different kinds of activities. Gliding is a pretty
simple process and seems fairly easy for evolution to achieve. It has
evolved independently at least seven times among the mammals
alone. Basically, gliding requires the development of a membrane
called the "patagium" that the animal can stretch out and manipulate
with its fore- and hindlimbs. Flying is a much more complex activity
requiring many specialized anatomical adaptations:

> In contrast [to gliding], flapping flight is an anatomically demand-
> ing adaptation that has nonetheless evolved in animals of a wide
> range of body sizes and ecological habits. Flapping requires many
> distinct skeletal specializations, such as changes in forearm and
> shoulder anatomy, an enlarged and reinforced sternum, the fu-
> sion of the clavicles into a furcula, to anchor flight muscles. . . .
> The marked differences between gliders and flappers mean that
> evolving a flapper from a glider might involve almost as many
> anatomical changes as would evolving a flapper from a terrestrial
> biped or quadruped (Shipman 1998, 187).

But if getting from a glider to a flapper involves as many evolutionary difficulties as getting from a runner to a flapper, a chief apparent advantage of the arboreal hypothesis is negated. Maybe so, defenders of the trees-down theory might reply, but we do know that
flyers have descended from gliders at least once. Bats unquestionably
evolved from gliding animals. However, bat wings are very different
from bird wings, and what worked for the bats very likely would not
have worked at all for bird evolution.

Another problem is that if flight had to originate from the trees
down, there had to be trees in the surrounding landscape. But *Archaeopteryx* lived next to a lagoon. The Solnhofen area where its fossils

were found was a large, wet area with salty soil and vegetation of low brush and scrub. Nothing taller than about ten feet seems to have grown there. Nor were there any cliffs or even a decent hill in this low-lying, flat country. If *Archaeopteryx* was learning how to fly, it was not doing so by swooping from the trees down. Feduccia replies to this objection in the spirit of the aphorism "absence of evidence is not evidence of absence." He notes that there are many fossil sites that presumably had trees, though no large woody remains are found there (Feduccia 1996, 109). True, says Shipman, but it is hard to explain why fossils of plants and shrubs are preserved in the Solnhofen limestone, but not trees (Shipman 1998, 190). After all, large, heavy logs should be more likely to fossilize than delicate plants. So the attractive trees-down theory has more problems than are apparent at first glance. What is the alternative? Supporters of the dinosaur ancestry theory argue that flight could have started from the ground up.

An illustration of a possible "ground-up" evolution of birds from dinosaurs. Left to right: Compsognathus, Avimimus, Archaeopteryx, *pigeon (The Natural History Museum, London).*

The story goes that a philosopher once defined "human" as "a featherless biped." Humans are bipedal: We walk on our two legs. The birds are the only other living true bipeds of the animal kingdom.

Birds' wings are forelimbs greatly modified for flight. If the immediate ancestor of birds had been a terrestrial quadruped, birds would have had to evolve bipedality and also develop their wings for flight. It is much simpler to assume that they evolved from creatures that were already bipedal, thereby saving a very major step in the process.[1] Since the theropods had evolved bipedality long before the first birds, and since there are no other known bipeds that could have been the ancestors of birds, this automatically makes theropods plausible bird ancestors. Since theropods were ground dwellers, if birds evolved from theropods, they had to start flying by going from the ground up. Still, this theory leaves the basic question unanswered: How could a runner become a flyer?

The basic problem for any ground-up theory of the origin of flight is how to achieve takeoff. Defenders of the ground-up theory ask us to imagine a small theropod that runs along the ground chasing insects or other small prey. Suppose further that this creature already had feathers, which is not as far-fetched as it sounds. Feathers need not have originally evolved for flying, but perhaps for thermal insulation or some other purpose totally unrelated to flying. Natural selection can take organs that have one use and adapt them (or "exapt" them, as some biologists put it) for a completely different function. (Perhaps the best-known example are the malleus, incus, and stapes bones of the middle ear that were developed from reptilian jawbones). As the feathered theropod ran and leaped after prey, it could achieve greater balance and maneuverability by extending its arms. Natural selection would favor any changes that enhanced such functions, and eventually these feathered forearms would start to achieve lift. Arms would have become rudimentary wings. The first flights really would have just been exaggerated leaps of short duration and low altitude. Surely it would have taken a very long time for these glorified jumps to have evolved into the eagle's soar . . . but one thing evolution has plenty of is time.

---

1. Besides the birds, the only other vertebrates to have achieved powered flight are the bats and the pterosaurs. The immediate ancestor of bats was a gliding quadruped, and bats have retained the quadrupedal form. Their wings attach to both their front and rear limbs, and when they move on the ground they crawl on all fours. Pterosaurs were traditionally depicted as batlike quadrupeds, but Kevin Padian argues that they were in fact birdlike bipeds (Padian 1987). However, a number of other eminent paleontologists disagree and consider the pterosaurs to have been quadrupeds, so it is still not clear how pterosaurs fit into the picture.

This is a nice story, but how do we know it is not just another evolutionary "just-so" tale? The solid evidence to support it would be the discovery of feathered dinosaurs. Amazingly, these have been found. As we have seen in this book, the fossil finds that drove the first wave of dinosaur discovery, the teeth and fragments known to Mantell and Owen, were found in England. Then the scene shifted to North America and the fantastic western bone fields dug by Cope, Marsh, and Douglass. Lately China has been the site of the most dramatic dinosaur discoveries as its rich and exceptionally well-preserved fossil deposits have been extensively mined just in the past few years.

In 1996 a small dinosaur was found in the early Cretaceous formations of Liaoning Province of China and given the name *Sinosauropteryx*. It was very similar to the well-known small theropod *Compsognathus,* but it seemed to have hairlike fibers running down its neck, spine, and tail. Controversy immediately ensued over whether these fibers were downy protofeathers. Opponents of the dinosaur ancestry theory dismissed the fibers as muscle tissue or a reptilian frill. In 1997 a much better-preserved specimen was found that showed that *Sinosauropteryx* had these hairy fibers over its whole body, so they were not a frill or sail.

In the 25 June 1998 issue of the journal *Nature,* the paleontologists Ji Qiang, Philip Currie, Mark Norell, and Ji Shu-An (Ji et al. 1998) reported an even more dramatic find—an Early Cretaceous dinosaur with unmistakable feathers. This creature, called *Caudipteryx,* had long feathers on its arms and a fan of plumage at the end of its tail, and there was evidence that it had a downlike covering over its whole body. The feathers of *Caudipteryx* were identical to those of birds, with a central shaft and parallel barbs sticking out of opposite sides. Opponents of the theory of dinosaur ancestry could not deny that these were feathers, so they offered a different critique. The magazine *Science News* quoted Feduccia as saying that *Caudipteryx* was not a dinosaur at all, but was really a bird that had descended from a flying ancestor (Monastersky 1998, 404). However, even more recent discoveries have made it hard to defend that line of argument. For instance, in 2003 the paleontologist Xing Xu and his colleagues described *Microraptor gui,* a tiny theropod with flight feathers on its fore- and hindlimbs (Xu et al. 2003).

Recently, though, there was a major embarrassment for defenders of the dinosaur-to-bird theory. The November 1999 issue of *National Geographic* reported that an amazing creature called *Archaeoraptor* had been discovered. *Archaeoraptor* looked like a true missing link between birds and dinosaurs, with an avian body but a long, stiff, dromaeosaurlike tail. It seemed too good to be true. It was—*Archaeoraptor* was a fake. An enterprising Chinese farmer had put together parts of unrelated animals to produce a chimera. Overly eager fossil buyers have been duped in this way more than once. Naturally, when the hoax was revealed, some of the opponents of dinosaur ancestry hooted, "Told you so!" However, the hoax does not really undermine the recent discoveries of feathered dinosaurs. Professional paleontologists were not fooled. The professional journals *Science* and *Nature* turned down the article reporting the "discovery" of *Archaeoraptor.* You should always beware when alleged discoveries are reported in the popular press rather than in professional, peer-reviewed journals.

The genuine feathered-dinosaur discoveries do not force opponents of the dinosaur ancestry theory to give up. As I noted in Chapter 1, there is almost always some wiggle room for dissidents. However, it seems that most paleontologists regard these discoveries as backing the dissidents into a corner, or at least placing a heavy burden of proof on them. The objections to a theropod ancestry for birds sound increasingly strained and ad hoc to the ears of other paleontologists. An ad hoc hypothesis is one that is invented for the special purpose of saving a cherished theory from falsification. When a theory can be preserved in the face of mounting contrary evidence only by piling up ad hoc insulators, that theory loses status in science. For instance, in the eighteenth century the chemist Antoine Lavoisier performed a famous series of experiments that undermined the traditional view of combustion. Defenders of the old view were forced to resort to bizarre ad hoc hypotheses to preserve their theory in the face of Lavoisier's results. Perhaps opponents of the dinosaur ancestry theory are heading in the direction of Lavoisier's critics.

The debate over bird ancestry may be moving toward closure. Those who say that birds descended from theropod dinosaurs seem to be winning. Perhaps, after all, the elegant hummingbird sipping nectar is not too far from the bloodthirsty, ultraviolent *Deinonychus.* Dinosaurs are fascinating, and birds are fascinating, and if they are closely related, then each group is made even more interesting. So the idea that birds are the evolutionary offspring of dinosaurs is an exciting thought. Far more importantly, it may even be true.

Further Reading

The idea that birds descended from dinosaurs sounds too good to be true, but we have seen in this chapter that it might be true after all. One of the best popular books on the subject is Lowell Dingus and Timothy Rowe's *The Mistaken Extinction: Dinosaur Evolution and the Origin of Birds* (New York: W. H. Freeman, 1998). Dingus and Rowe make a strong case for the theropod ancestry thesis. The book is very well written and beautifully illustrated. A parallel volume with the opposite view of bird evolution is Alan Feduccia's *The Origin and Evolution of Birds* (New Haven: Yale University Press, 1996). This book is more technical and detailed, but it is clearly written and accessible to a nonspecialist audience. It is also copiously illustrated and contains much information that anyone with an interest in birds would find fascinating, whatever his or her views on the question of avian ancestry.

Deinonychus and the other fearsome dromaeosaurs are certainly some of the most interesting dinosaurs. They are described in David Norman's *The Illustrated Encyclopedia of Dinosaurs* (New York: Crescent Books, 1985). Unfortunately, Norman's book came out before the discovery of *Utahraptor,* which was the biggest and baddest of the whole clan. Robert Bakker, of course, is very friendly to the idea that birds descended from dinosaurs, since this idea supports his view that dinosaurs had already evolved an avianlike physiology. His account of Ostrom's discovery of *Deinonychus* and how this is related to bird ancestry is given in his *The Dinosaur Heresies* (New York: William Morrow and Company, 1986).

Larry Martin's long-standing opposition to the theropod ancestry theory has gotten him spots on TV specials—most notably on PBS's *Nova* program "The Case of the Feathered Dinosaur." His 1983 book chapter "The Origin of Birds and Avian Flight" in *Current Ornithology,* Vol. I, edited by Richard F. Johnston (New York: Plenum Press, 1983), is a very clear statement of his views on these topics.

Overall, the best book available for a general audience on the evolution of birds is Pat Shipman's *Taking Wing:* Archaeopteryx *and the Evolution of Bird Flight* (New York: Simon and Schuster, 1998). Shipman gives the whole story, with fair discussions of all points of view. Her account of the controversies over the "trees-down" and "ground-up" theories of the origin of flight is particularly noteworthy. Shipman understands the science very well and has a knack for explaining things clearly.

As of this writing, many of the discoveries of feathered dinosaurs are so recent that few accounts exist outside the professional literature. The Internet has a number of very good sites (along with some creationist junk). Just put "feathered dinosaurs" in your search engine and take a look.

"Sometimes You Have a Really Bad Day, and Something Falls out of the Sky"

The quotation that titles this chapter is from geologist Walter Alvarez. In 1980, with his father, the Nobel Prize–winning physicist Luis Alvarez, and two other coauthors, he published one of the most controversial articles in the history of science (Alvarez et al. 1980). By 1994 over 2,500 articles in scientific journals had contributed to the controversy ignited by this one piece (Glen 1994, 2). The article appeared in the 6 June 1980 issue of the journal *Science;* its title was "Extraterrestrial Cause for the Cretaceous-Tertiary Extinction." This essay proposed a startling new hypothesis to explain one of the most intractable mysteries of science: the extinction of the dinosaurs.

Had the dinosaurs been the clumsy, obsolescent hulks of popular caricature, there would be no real mystery about their disappearance. The real mystery would be how such dysfunctional creatures could have lasted so long. But as we saw in Chapter 6, the dinosaurs were not dysfunctional; they were among the most successful types of large animals ever to inhabit the globe. They increased in diversity until late in the Cretaceous, living through many changes in climate and topography as the continents drifted. They may have begun to decline a few million years before the end, but there were still many thriving species in the latest Cretaceous, including such spectacular types as *Triceratops* and *Tyrannosaurus rex*.

Then there were none. They vanished, along with the great marine reptiles and the flying pterosaurs in one of the five great mass extinctions that have afflicted life over the last half billion years. Extinction is usually piecemeal—a species here, a species there. Such "background" extinctions are a normal part of the history of life on Earth. Of all the species that have ever lived, 99 percent are now extinct. Generally, there is nothing terribly mysterious about extinction.

Luis and Walter Alvarez, father-and-son research team, view a sample of an iridium layer deposit, which is found worldwide. Based on this layer, the Alvarezes postulated that a giant asteroid hit the earth at the end of the Cretaceous period, approximately sixty-five million years ago, with cataclysmic results that likely led to the extinction of dinosaurs (Roger Ressmeyer / Corbis).

All organisms live under stress from their physical environment, from predators, from parasites and disease, and from competition with other species or members of their own species. Should these stresses become too great, a species that cannot adapt to the harsher conditions will become rare. Sometimes a species will pass through a "bottleneck" when it reaches the brink of extinction and then recover to become quite numerous again. Other species will not be so lucky and will disappear completely. Occasionally the last member of a species will be found in captivity, like the last passenger pigeon, which died in the Cincinnati Zoo in 1914.

Because background extinctions are caused by many very complex and unpredictable factors, they are scattered randomly through the history of life. Sometimes, by sheer chance, they will bunch up. Everyone occasionally has a bad day in which, purely by happenstance, several unpleasant things happen at once—you forget your wallet, spill lunch over your pants, get stuck in traffic, and arrive home to find that the dog has left a large "accident" on the carpet. Sometimes, though, you can have a *really* bad day, and it is owing to one major thing (such as a heart attack) rather than lots of minor irritations that just happened to pile up. The same holds in the history of life. On five occasions in geological history, truly mass extinctions have taken place, occasions when the mortality was far too great to have been due to the accidental conjunction of independently caused extinctions. These mass extinctions have been so great that they are genuinely frightening. It is as though five times in the last half billion years some extremely potent and malignant force has tried to sterilize our planet.

Because a human life, or even the whole of human history, is such a tiny and unrepresentative sample of the history of the earth, we have been lulled into a false sense of security. It seems to us that we live on a fairly congenial planet (given to occasional outbursts) in a relatively safe and stable corner of the cosmos. Certainly compared to our celestial neighbors Venus and Mars, Earth is a veritable paradise of abundant water, equable temperatures, and breathable air. Also, Earth's protective layer of atmosphere shields us from cosmic rays and brutal radiation. But appearances are misleading. The universe is a dangerous place, and even Mother Earth is not as benign as she appears.

We know from experience that dangerous things lurk in space. On 30 June 1908 something—a piece of comet perhaps—fell with tremendous violence into a remote area of Siberia. This so-called Tunguska Event destroyed thousands of square miles of forest. Had the object followed the same trajectory but arrived a few hours later, the rotation of the earth would have brought the great city of St. Petersburg into the target area. We also know that the earth itself is capable of great violence. The final, paroxysmal explosion of the Indonesian island of Krakatau in August 1883 was so loud that people all the way across the Indian Ocean thought they heard heavy guns firing in the distance. The eruption of the volcano Tambora in 1815 ejected so much ash into the atmosphere that sunlight was dimmed all over the world and 1816 was so cold it was called "the year without a summer." Yet, violent as these events were, they may have been only firecrackers compared to the massive cataclysms that have sometimes wracked the earth. The truly great events—great enough maybe to threaten all life on Earth—might occur on average only every so many hundreds of millions of years. In that case, since human records go back only a few thousand years, the disasters we have experienced might not be representative of the full extent of the violence that can rain down from space or erupt from the earth. We might be like the people who survived the hundred-year flood and rebuilt their home ten feet above the high-water mark. Little did they know that the thousand-year flood would hit the next year.

Surely the disappearance of the mighty dinosaurs invites a catastrophic explanation. Surely only some literally earth-shattering disaster could account for their apparently sudden demise. However, until the Alvarez article in 1980, practically all the proposed hypotheses of dinosaur extinction postulated nondisastrous causes. Some of these hypotheses were whimsical, if not comical, such as the one that blamed the dinosaurs' demise on chronic constipation (no kidding!). (Maybe chronic constipation did not kill them, but it may explain why they were so cranky.) Other hypotheses were much more plausible, such as the ones postulating gradual climate change as the shallow epicontinental seas receded at the end of the Cretaceous. Some proffered explanations were catastrophic, such as the theory that the explosion of a nearby supernova could have killed the dinosaurs. What all these hypotheses had in common was that there was little evidence in their favor.

Why, in general, were there so few attempts to suggest catastrophic causes for the dinosaurs' extinction when by their very nature mass extinctions seem to invite such hypotheses? To see why, we have to consider the history of the earth sciences. What Newton's *Philosophiae Naturalis Principia Mathematica* was to physics and *The Origin of Species* was to biology, Charles Lyell's *Principles of Geology* (1830) was to geology. When his book first came out, Lyell was a young

The British geologist Charles Lyell made geological expeditions to various regions of Europe and opposed catastrophic theories of great geologic changes. He is regarded as the father of modern geology (Bettmann / Corbis).

(thirty-three years old), ambitious scientist who had been trained as a lawyer and had a lawyer's knack for presenting a persuasive case. In the late eighteenth and early nineteenth centuries there had been an explosion of interest in geology, but Lyell was concerned that geology had not yet been put on a truly *scientific* basis. The main problem, as Lyell saw it, was that earlier geologists had been far too quick to invoke catastrophes to explain the major changes in the history of the earth. For instance, Georges Cuvier, the renowned French scientist discussed in Chapter 2, explained the episodes of turnover and extinction in the history of life by invoking a series of catastrophes. The last of these corresponded to the great flood of Noah as recounted in the Book of Genesis in the Bible. After each of these catastrophes had wiped out the old fauna, new types were created, which then spread over the earth. Cuvier thus accounted for the succession of faunas that we observe in the fossil record.

For Lyell, the main problem with such "catastrophism" is that it is just too easy. Any time we are faced with an enigma about the earth's history, say a mass extinction or some great change of topography or climate, it is just too convenient to invent an ad hoc catastrophe to explain things. Science cannot be too easy: You cannot just sweep inconvenient facts under a rug by appealing to an attractive but insubstantial scenario. In the book *Chariots of the Gods?* a best-seller of the 1970s, Erich von Däniken said that ancient astronauts from other worlds had helped our ancestors build the pyramids, the great stone heads of Easter Island, and many other monuments of antiquity (von Däniken 1987). Unfortunately, von Däniken could not tell us who these alleged ancient astronauts were, how they got here, how they helped build the pyramids, and why they did it—or suggest any independent means of testing or checking his story. The ancient-astronauts theory remained an interesting scenario that appealed to the imagination but was based on no solid evidence. Such scenarios may be fun—and they do have all the advantages of theft over honest toil—but in the end they only make things more mysterious rather than less.

This is just how Lyell viewed catastrophist geology—as obscurantist because it introduced new mysteries instead of clearing up old ones. What were these catastrophes? What caused them? Why have we never observed catastrophes even approaching such magnitude? Instead of invoking catastrophes, Lyell argued that geological method must stress uniformity. Lyell's view has therefore been called "uniformitarian." The precise nature of Lyell's methodological prescriptions is still debated by historians, but I think we can identify three theses:

1. Uniformity of law: Geological theories must presuppose the same basic laws of nature recognized in other fields of science.
2. Uniformity of process: Geological theories can postulate only the same kinds of geological processes observed occurring in the world today, for example, erosion, deposition, uplift, subsidence, or volcanic eruption.
3. Uniformity of magnitude: Geological processes of unprecedented magnitude must not be postulated, for example, no floods or volcanic explosions bigger than any ever recorded.

In short, all geological theories should attempt to explain the earth's features in terms of geological processes identical in kind and magnitude to those that we observe in the present day. As a consequence, geological theories explain things in terms of very gradual processes working over immense time. Gradualism must be the norm in geology; sudden catastrophes that achieve big effects in a short time will not be acceptable.

Lyell believed that only by obeying these rules in our geological theorizing could the earth sciences be put on a truly scientific basis. Further, he showed how these methodological prescriptions could be followed in doing good geology. Vast changes in the earth or in the history of life could be explained by theories postulating slow change brought about by observable types of processes operating at observed intensities. Mighty mountains could be eroded down to plains grain by grain and then lifted up again inch by inch. All it took was time—vast, incomprehensible amounts of time.

By 1980, 150 years after the initial publication of the first volume of *Principles of Geology,* earth scientists were thoroughly imbued with Lyell's methodological principles. This meant that most earth scientists were strongly disinclined to take seriously any hypothesis postulating catastrophes, especially catastrophes far bigger than any disaster previously recorded. Worse was a hypothesis postulating an extraterrestrial source for the catastrophe, since the only types of causes Lyell would countenance were earthly processes. The paper by the Alvarezes and their coauthors proposed that the cause of the dinosaurs' extinction was both extraterrestrial and spectacularly catastrophic. Their hypothesis was that sixty-five million years ago, at the end of the Cretaceous, a gigantic object—an asteroid or comet—hurtled from space and crashed into the earth. This object was about ten kilometers (six miles) in diameter, weighed ten billion tons, and was moving many times faster than the muzzle velocity of a rifle bullet. Such an object striking the earth would release the energy of about 100 million megatons of TNT—ten thousand times the combined thermonuclear weapons arsenals of all nuclear powers at the height of the cold war. The atomic bomb that destroyed Hiroshima was about .02 megatons. The largest thermonuclear explosion ever detonated by humans was the Russian "Monster Bomb" of fifty-seven megatons.

Needless to say, the immediate effects of such an impact would be cataclysmic in the extreme. Vast amounts of superheated rock would be blasted from the earth's crust at the point of impact. Launched into ballistic trajectory, these incandescent ejecta would fall

all over the earth, even on the side opposite the blast. Raging fires would start in forests worldwide, and in many places the air itself would be heated to broiling temperatures. Should the object land in the ocean, which is likely on a planet that is two-thirds water, tsunamis much larger than those created by earthquakes would devastate the surrounding land areas. The long-term effects would be even more severe. After the fires had burned out, the smoke and soot would remain in the atmosphere, blocking the sunlight and bringing temperatures below freezing worldwide, even in the tropics. Photosynthesis would stop, and herbivores, starving and freezing in the dark, would quickly die, as would the carnivores that feed on them. Here then we have a mechanism sufficient to cause the demise of even the mighty dinosaurs.

Such a scenario certainly appeals to the imagination. Imagine the last day of the Cretaceous: On this final day of the 160 million–year reign of the dinosaurs, summer warmth creates strong updrafts, which the big pterosaurs negotiate with perfect skill. Under the blue sky, a large herd of *Triceratops* grazes placidly in an open field bordered by underbrush and scrub oaks. The scrub is not very high, but it is high enough to conceal the perfectly camouflaged bulk of the *T. rex*. The tyrannosaur, with infinite patience, waits and watches a young *Triceratops* to see if it foolishly wanders away from the herd and toward her hiding place. Attracted by a patch of lush vegetation, the young dinosaur moves toward the waiting predator. The tyrannosaur's great thigh muscles, so huge and strong that they can accelerate a five-ton body to attack speed in two seconds, tense like a steel spring. Suddenly the sky is ablaze as an object far brighter than the sun streaks toward the horizon. Dazzled by the blinding light, the *Triceratops* herd begins to mill about in confusion, and the confusion turns into a panic as the herd stampedes. Minutes later the tyrannosaur, also dazzled and wandering in the scrub brush, is knocked into the air and slammed to the ground as a boulder of flaming rock smashes into the ground a few yards away. With legs and back broken, she cannot escape the raging brush fire that soon engulfs her.

Clearly the impact theory gives us a vivid and exciting image of the destruction of the lords of the Mesozoic in a Götterdämmerung (tale of catastrophic collapse and destruction) that not even Wagner could have envisioned. Scientists get excited like everyone else, but what excites them is a good theory. Is the impact account a good theory, or is it an exciting piece of fiction like the ancient-astronauts scenario?

Unlike earlier extinction theories, the impact story seemed to make testable and, in fact, verified predictions. Like any good geological theory, the impact hypothesis said we should see what the rocks say. If the extinction was literally sudden (not just geologically "sudden," which could be a million years), the dividing line between the Cretaceous period, the last period of the Mesozoic Era, and the succeeding Tertiary, or Cenozoic, Era should be pencil thin. The ash knocked into the atmosphere by the impact should have settled relatively quickly, leaving a very thin layer in the rock. That rock should also have a very unusual composition. Among the rarest of elements in the earth's crust are the so-called siderophile, or iron-loving, elements such as platinum, iridium, and osmium. These elements bond readily with iron, and since most of the earth's iron is in the planet's core, most of the earth's supply of these iron-loving elements is down there with it, leaving only trace amounts in the crust.

Things are different in space. Asteroids, meteoroids, and comets (which have rocky cores) contain a much greater abundance of these iron-loving elements because they do not have a massive iron core to draw these substances inward. Therefore, if the layer of rock marking the Cretaceous/Tertiary (K/T—the word for "Cretaceous" in German is "Kreide"; hence the "K") boundary, since it is largely composed of pulverized asteroid or comet core, should be much higher in iron-loving-element content than the surrounding rock. This was exactly what Walter Alvarez found. Samples of K/T boundary material were taken from the clay of Gubbio in Italy and carefully analyzed. In the rock from that layer, and not in the rock above or below that layer, was an iridium "spike"—a much higher concentration of the element iridium than is generally found in the earth's crust. A sample of rock from the K/T boundary at another site, Stevns Klint in Denmark, found an even higher concentration of iridium—160 times the concentration found in the surrounding rock.

That was not all. Quartz is one of the most common elements in the earth's crust, and it takes a number of different forms. When quartz crystals are subjected to extreme pressures, they take on a characteristic "shocked" structure. Crosshatched bands called "lamellae" appear in grains of quartz that have been subjected to very high pressures. The pressures required to produce the characteristic "shocked" structure are so high that some forms of shocked quartz had previously been found only at sites of known meteorite impact and in underground caverns where nuclear weapons had been tested. Starting in 1984, a number of geologists found shocked quartz gran-

A plastic coating encases a sample of K/T boundary. Strata, from the top down, are coal, iridium layer with shocked quartz, Tonstein boundary clay, and coal (Jonathan Blair / Corbis).

ules in the K/T boundary rocks at various locations around the world, thus apparently lending strong support to the impact hypothesis (see Frankel 1999, 27–31).

Still, there was no "smoking gun." If such a massive body had hit the earth sixty-five million years ago, it would have blasted a really big hole in the ground, and we should still be able to see the crater. After all, numerous other impact sites had been found that were far older than the end of the Cretaceous sixty-five million years ago. Where was the crater?

By the late 1980s, advocates of impact hypothesis were chagrined not to have found an impact site. Several sites were proposed, but for one reason or another they did not work out (see Frankel 1999). There were some tantalizing clues, though. In southeast Texas the lazy Brazos River winds through swampy lowlands until it enters the Gulf of Mexico. A team of sedimentologists found some strange structures in a layer of sediment in an area of riverbank worn away by the Brazos. At the end of the Cretaceous that particular layer was seafloor one hundred meters (about sixty yards) deep. The group reported that the seafloor at this level had been greatly disturbed by oscillating currents and that a rippled layer of clay had been deposited over the sand (Frankel 1999, 80). They interpreted these intrusions as a tsunami deposit and were able to estimate the size of the wave. From their estimates they were further able to say that if the tsunami had been caused by an impact, that impact could not have been more than five thousand kilometers (about three thousand miles) away. It must have occurred in the Caribbean, the western Atlantic, or the Gulf of Mexico.

Intrigued by this and other evidence, the geologist Alan Hildebrand began the careful examination of geological maps of the Caribbean and Gulf of Mexico region to look for anomalous circular structures. He found a promising candidate, a circular anomaly 180 kilometers (112 miles) in diameter buried under hundreds of meters of sediment and centered on what is now the coast of the Yucatán Peninsula. In the 1950s geologists had discovered this structure while prospecting for oil. They thought it was a caldera, the collapsed remnant of a volcano. They dug a few exploratory wells, took rock samples from various depths, and kept careful logs of what they found (alas, no oil). Hildebrand got access to the samples and the logs and found rocks that contained shocked minerals, indicative of extreme pressures. Also, the logs of the oil geologists described a great sheet, several hundred meters thick, of what had once been molten rock. An impact would create enormous heat, which would melt large amounts of rock. Putting all the evidence together, Hildebrand was convinced that he had found his K/T impact crater. He named it "Chicxulub" after a fishing harbor near what might have been ground zero. Impact theorists believed they had found their smoking gun.

Apparently, then, the impact hypothesis provided a cause sufficient to wipe out the dinosaurs, and unlike the earlier speculative scenarios, this one was backed by solid evidence—even a "smoking gun." However, many scientists were unimpressed, to say the least. In fact, the controversy generated by the impact theory was so widespread and intense that the controversy over warm-blooded dinosaurs paled in comparison. The rhetoric on both sides of the debate descended to depths of character assassination that had hardly been plumbed even in the Cope-Marsh conflict.

Why was there such an intensity of bad feeling? Naturally, personalities figured into the mix. Unlike Charles Darwin, Luis Alvarez was not willing to endure the slings and arrows of hostile criticism without responding in kind, and some of the critics were definitely hostile. Maybe one reason for the mutual disdain was disciplinary chauvinism. Luis Alvarez was a physicist, and physicists often see themselves as the purest of scientists pursuing the hardest of sciences. Such is their, at times, overweening confidence that they sometimes undertake to tell other scientists their own business. In turn, scientists in fields such as geology or paleontology are likely to feel that a physicist who butts into their domain does so out of ignorance and arrogance.

One of the most persistent and articulate critics of the impact hypothesis has been Charles Officer, a research professor of Geology at Dartmouth College. In technical articles in the professional journals and in more popular writings for the educated public, he has tirelessly criticized the hypothesis, offering point-by-point rebuttals of each major and supporting claim. His book *The Great Dinosaur Extinction Controversy*, coauthored with Jake Page in 1996, provides a clear summary of his critique.

First of all, did the last dinosaurs die off all at once in a period of just weeks, months, or a very few years, as the impact hypothesis implies? Officer argues that the evidence shows a gradual decrease in dinosaur diversity and in total numbers during the last several hundred thousand years of the Cretaceous (Officer and Page 1996, 66–67). For instance in the famous Hell Creek formation in Montana, one of the most productive fossil fields of the latest Cretaceous, dinosaur bones start to become scarcer well below the level containing the iridium anomaly cited by the Alvarezes and their coauthors. In fact, one study of the Hell Creek formation showed that dinosaur remains disappeared *completely* from the rock well below the iridium-rich layer (66). J. David Archibald further argues that there simply is no evidence that the dinosaurs went extinct simultaneously all over the earth (Archibald 1996, 13–18). He notes that sites containing dinosaur fossils from the very end of the Cretaceous are rare, and the known ones are concentrated in North America. There is therefore simply not enough information to conclude that a worldwide, virtually instantaneous extinction occurred at the end of the Cretaceous.

Well, what about that iridium-rich layer and the shocked quartz? Officer notes that large volcanic eruptions often eject enhanced-iridium airborne particles into the stratosphere, thereby distributing them globally (Officer and Page 1996, 112). So iridium anomalies may be associated with large eruptions as well as impacts. Further, he cites several studies, including one by Helen Michel, one of the coauthors of the original 1980 *Science* article that proposed the impact hypothesis, showing that the actual distribution of the iridium-enriched material indicates that it collected over a period of 400,000 years or more (117–121). That is, the iridium-enhanced material seems to have been deposited gradually over a long time rather than as the fall-out of a single event. As for shocked quartz, Officer claims that the particular kind of deformed quartz associated with impacts is distributed not in K/T boundary layers worldwide but only in North America (125). He infers that a small impact may have occurred in North America, causing limited and localized extinctions. Elsewhere in the world, the shocked quartz is of a different type that is typically produced by volcanic eruption and is deposited in a thicker layer than in North America, indicating that it was laid down over a longer period. So the shocked-quartz evidence is ambiguous at best, Officer thinks.

What about the "smoking gun": the Chicxulub crater? Not an impact structure at all, says Officer. He writes that the shocked quartz found there is of the type typically produced by volcanic or tectonic forces (155). As for the alleged melt sheet, Officer says such a sheet would be chemically homogeneous: The very high temperatures of an impact would very rapidly melt all of the target rock into a chemically uniform aggregate (154–155). However, the rock from Chicxulub shows a wide range of chemical compositions and is deposited in a series of layers indicating that at least six major volcanic events had produced the various kinds of rock. Also, a massive impact would have eradicated the upper Cretaceous rock, blasting a hole ten kilometers (six miles) in depth. Yet the upper Cretaceous rock is still there, claims Officer, as indicated by the fact that limestone (a sedimentary, fossil-bearing rock) interbedded with the volcanic material contains characteristic Late Cretaceous microfossils (154). Officer concludes that the original petroleum geologists were correct in identifying the Chicxulub structure as volcanic.

If Officer is correct that the evidence for the impact hypothesis is so weak, why has it been so widely accepted, not just by the public but by many reputable scientists? One of the most interesting chapters in the book by Officer and Page is titled "Media Science." The authors note that scientists distance themselves from science as reported in the popular media. The scientific community disdains scientists who circumvent evaluation by their peers and appeal directly to the public. Yet, say Officer and Page, scientists seem easily swayed by the hype in their *own* media. They accuse the editors of *Science,* a highly respected and widely read publication of the American Association for the Advancement of Science, of having a strong bias in favor of the impact hypothesis (96–99). They charge that submissions promoting the impact hypothesis are readily published, but critical ones are snubbed. Further, news reports published in *Science* appear to give the impression that the impact hypothesis is well established. Officer and Page argue that such misleading impressions given by publications such as *Science,* and to a lesser extent the similarly respected British journal *Nature,* have swayed scientists to jump on the impact-hypothesis bandwagon.

What did kill the dinosaurs, in Officer's view? "The earth did it," he says (Officer and Page 1996, 157). The theory of plate tectonics has shown that volcanoes are associated with three different types of geological processes. There are those that occur in places such as the mid-Atlantic ridge, where two plates are spreading apart, driven by magma welling up from below. Iceland sits atop this ridge, which is why it is so volcanically active. Subduction zones, where plates are colliding and one is overriding another, are other areas with many active volcanoes. Indonesia and other locales in the Pacific "ring of fire" are located over subduction zones. Then there are oddball volcanoes that occur in the middle of a crustal plate, such as the Hawaiian volcanoes. These "hot-spot" volcanoes are caused by mantle plumes, vast domes of molten material that rise from deep in the earth and punch right through the overlying crust. Hot spots have produced the most intense and prolonged episodes of volcanism in the earth's history. The Siberian Traps are a vast area of Siberia covered with basalt, a volcanic rock, produced by massive hot-spot eruptions over a 600,000-year period between 244 and 252 million years ago. The amount of basalt that poured from these eruptions is truly staggering—between two and three *million* cubic kilometers (480,000 to 720,000 cubic miles)! These eruptions coincided with the end of the Permian Period, and of the whole Paleozoic Era, the time of the greatest of all mass extinctions, when 70 to 80 percent of all living species went extinct.

Second only to the Siberian eruptions were the volcanic episodes that produced the Deccan Traps, vast, steplike ("Traps" derives from the Dutch word meaning "staircase") basalt structures in what is now India. These events occurred over a 500,000-year period at the end of the Cretaceous and the beginning of the Tertiary. These episodes of intense eruption coincided with two other major changes in the environment: the disappearance of the inland seaway that had covered much of what is now North America during the Cretaceous and a severe regression of sea levels worldwide. These events occurring together put tremendous stress on many types of organisms. The loss of shallow-water habitat would have doomed many creatures. The disappearance of the lush environments around the inland sea would have left a much harsher, more arid landscape for the dinosaurs. The volcanic eruptions would have polluted the air with large volumes of noxious substances that had various deleterious effects. Officer thinks the combination of these factors is the best explanation of the extinctions that occurred at the end of the Cretaceous (Officer and Page 1996, 158–177).

Needless to say, defenders of the impact hypothesis did not take Officer's criticisms lying down, nor have they refrained from criticizing his account of the K/T extinctions. Charles Frankel summarizes their response to Officer's interpretation of the Chicxulub evidence:

> The arguments put forth by the volcanists were not convincing. The so-called interstratification of "volcanic and sedimentary" layers was probably nothing more than the mixing of pods of impact melt with slumped sediment at the edge of the crater. Neither did the chemical variability of the igneous rocks speak against an impact origin. On the contrary, impact specialists had long shown that in large astroblemes [impact remnants], melt lenses were chemically diverse, because of the variety of rocks that are fused and mixed together. As for the shocked minerals, their deformation in criss-crossing planes was symptomatic of impact and not of volcanism, an interpretation

The Deccan Traps in India (The Natural History Museum, London).

which no longer suffered any contest among qualified specialists. (Frankel 1999, 98)

In short, defenders of the impact hypothesis believed that they
had cogent replies to each of Officer's criticisms of their identifica-
tion of Chicxulub as the Cretaceous-ending impact site. They also
believed that other evidence greatly strengthened their case. They
claimed that material from the Chicxulub structure chemically
matched microtektites found in K/T boundary layers around the
Gulf of Mexico. A tektite is a small, distinctly shaped fragment of silicate glass that is formed when an explosive impact liquefies and
ejects material, which then cools and solidifies in flight. They claim
that careful dating of the Chicxulub material and the microtektites
shows that they are the same age and date from the end of the Cretaceous (Frankel 1999, 96–97).

The geologist James Lawrence Powell devotes a whole chapter
to rebutting Officer in his book *Night Comes to the Cretaceous* (1998).
Officer argues that the iridium "spike" at the K/T boundary was actually more of a "hill" indicating that the iridium had been concentrated
over hundreds of thousands of years, contrary to the impact hypothesis. Powell replies that independent measures using blind tests, that is,
where the testing laboratories are not told the origin of the material
they are testing, confirm a single iridium peak with adjacent "molehills" of much lower concentration (Powell 1998, 77). Powell argues
that any number of processes could have caused the dispersion of the
iridium-rich material into surrounding rock, accounting for the
"molehills." The material could have been eroded and redeposited,
chemically dissolved and reprecipitated, or disturbed by burrowing
worms before it solidified. He writes that the "spike" is so pronounced that it appears to be a primary feature of the rock later modified by various secondary processes (Powell 1998, 77). As for
shocked quartz, Powell asserts that quartz crystals deformed by tectonic or volcanic forces do not show the multiple crisscrossing sets of
lamellae that are diagnostic of rock subjected to the extremely high
pressures of impacts (81). Therefore, Officer is wrong to think that
earthly forces could have produced shocked quartz of the type found
in the K/T boundary layers.

As you can see from this very small sample of the many arguments and counterarguments offered by proimpact scientists and
their opponents, the evidence is complex and turns on many small
details. I think it is safe to say that the impactors have not given any

knock-down arguments, and neither have their critics. As I argued in Chapter 1, it should not be surprising that Officer and other skeptics still find grounds for principled disagreement.[1] Some former skeptics have been convinced that a major impact did occur at the end of the Cretaceous. Perhaps one factor that weakened dissent was the occurrence of the spectacular collision of comet Shoemaker-Levy 9 with the planet Jupiter in July 1994. The enormous explosions caused by large comet fragments as they hurtled into the Jovian atmosphere dramatically confirmed that major impacts do occur in the solar system and that the energies they release are sufficient to cause catastrophic effects. Perhaps even Charles Lyell himself, had he witnessed these collisions, would have had to admit that major extraterrestrial impacts do occur and could have greatly influenced the earth's history. What is left of old-fashioned gradualism and anticatastrophism when we can *see* such spectacular events take place?[2]

Still, no consensus has emerged in the dinosaur extinction debate. Neither volcanists nor impactors have carried the day. In fact, *both* hypotheses still face many problems. For instance, neither theory really explains the selectivity of the extinctions. Why did some groups, such as the dinosaurs, go extinct, while others, such as crocodiles, sailed right through and are with us to this day? It would seem that anything powerful enough to eliminate the mighty dinosaurs would also have doomed other types that survived. A number of interesting speculations

---

1. Defenders of the impact hypothesis have expressed doubts about just how principled Officer's dissent is. For instance, James Lawrence Powell asks how far Officer is willing to go to oppose the impact hypothesis (Powell 1998, 216–217). Powell charges that Officer stoops to stratagems and subterfuge more typical of the polemics of creationists than scientific debate (216). He concludes that Officer's animosity is so great that he is willing to abandon scientific principles to attack the impact hypothesis (216). Needless to say, Officer fully reciprocates these sentiments and describes the impact theory as "pathological science" (Officer and Page 1996, 178). Speaking as one who stands outside this debate, it seems to me that both sides have engaged in rhetorical excesses, straw-man arguments, and ad hominem abuse. Further, though it seems to me that the weight of evidence has accumulated in favor of the view that a major impact leading to mass extinctions did occur at the end of the Cretaceous, the evidence does not seem so solid as to preclude rational dissent.

2. Astronomers have provided additional evidence. They have detected over two thousand objects in our solar system one kilometer or larger in diameter that have an orbit that crosses the earth's orbit. The collision of any of these objects with the earth would result in a major catastrophe.

have been offered to explain the selectivity of the extinctions (see Frankel 1999, 136–140), but so far no definite answers.

As I noted earlier, the debate over dinosaur extinction has been especially prolonged and nasty. Are the two sides so polarized that there is no hope for consensus to emerge in the foreseeable future? Well, perhaps consensus is still a long way off, but progress has been made. During the 1990s impactors and volcanists discovered a good deal of common ground. Two debates, one published in 1990 and the other in 1997, show how much agreement was gained in those few years. In that short period impact theorists and their critics found that they had much to agree on and, for some individuals at any rate, that they could respectfully disagree where they still differed.

The first of these debates appeared in *Scientific American* in 1990 under the title "What Caused the Mass Extinction?" Walter Alvarez and Frank Asaro argued for an extraterrestrial impact, and Vincent E. Courtillot defended volcanism. These essays clearly and cogently present the opposing views. Perhaps because the debates of the 1980s were so vicious, both essays express hope that some mutuality will be found. For instance, Alvarez and Asaro think that a synthesis might ultimately emerge from the clash of these diametrically opposed views: "In the past few years the debate between supporters of each scenario has become polarized: impact proponents have tended to ignore the Deccan Traps as irrelevant, while volcano backers have tried to explain away evidence for impact by suggesting that it is also compatible with volcanism. Our sense is that the argument is . . . one . . . with an impact thesis and a volcanic antithesis in search of a synthesis whose outlines are as yet unclear" (Alvarez and Asaro 1990, 84). In other words, perhaps a synthesis will someday combine elements of both theories (see Archibald 1996 for such an attempted synthesis).

Courtillot likewise seeks some ground for agreement between the two sides. He notes that in a sense both the volcano and impact theories are revivals of catastrophism and require that some of the old gradualist and uniformitarian theses be rejected:

> Both the asteroid impact and volcanic hypotheses imply that short-term catastrophes are of great importance in shaping the evolution of life. This view would seem to contradict the concept of uniformitarianism, a guiding principle of geology that holds that the present state of the world can be explained by invoking currently occurring geologic processes over long intervals. On a qualitative level volcanic eruptions and meteorite impacts happen all the time and are not unusual. On a quantitative level,

however, the event witnessed by the dinosaurs is unlike any other of at least the past 250 million years. (Courtillot 1990, 92).

In other words, both theories imply the rejection of the Lyellian prohibition against processes acting with a degree of intensity never previously observed. No human has ever witnessed an impact as big as the one the impactors postulate or episodes of volcanism as intense as some of those that occurred while the Deccan Traps were forming. Perhaps in the long run the similarity between these hypotheses—that each entails the rejection of a time-honored and enshrined methodological precept—will seem more important than the many differences between them.

By the late 1990s impactors and their critics *had* discovered considerable common ground. Even where they continued to disagree, the dispute was not over fundamental matters of principle such as uniformitarianism versus catastrophism but over how best to evaluate the various lines of evidence. The excellent recent compendium *The Complete Dinosaur* (Farlow and Brett-Surman 1997) contains a very interesting exchange between Dale A. Russell, a proponent of catastrophic extinction, and Peter Dodson, who still maintains a more gradualist view (Russell and Dodson 1997, 662–672). Their exchange indicates the broad areas of agreement that have emerged in the extinction debates of the last twenty years. Their continued disagreement over extinction hypotheses despite substantial agreement on the evidence is best explained by the inevitable uncertainties involved in scientific judgments about how to weigh different sorts of data and degrees of evidence.

Russell and Dodson begin by listing points of agreement and the concessions each side has made. For instance, they agree that the dinosaur record presently indicates a peak in global dinosaur diversity early in the Maastrichtian stage (the final stage of the Cretaceous period), several million years before the end of the Cretaceous (Russell and Dodson 1997, 664). This is an important concession since one of the issues dividing the two camps has been whether dinosaurs had begun to decline well before the end of the Cretaceous or whether they had flourished until catastrophically destroyed. Both sides now also agree that it is reasonable to postulate an extraterrestrial impact coinciding with the final disappearance of the dinosaurs (Russell and Dodson 1997, 665). These mutual concessions show that the extinction debate, for all its acrimony, did succeed in bringing two initially polarized factions closer together.

Russell and Dodson continue to disagree about how to weigh the various lines of evidence. Russell argues for abrupt changes in the end-Cretaceous biota: "Apparently in both marine and terrestrial environments, the Cretaceous ended with an abrupt collapse in green plant productivity associated with the bolide [i.e., asteroid or comet] trace-element signature. Marine and terrestrial animals belonging to food chains based on organic detritus tended to dominate the post-extinction assemblages. However, those dependent directly or indirectly on living plant tissues (e.g., dinosaurs on land and planktonic foraminifera and mosasaurs in the sea) are postulated to have died on time scales consistent with starvation" (Russell and Dodson, 1997, 666).

He concludes: "A relatively parsimonious interpretation of the foregoing points is that dinosaur dominated assemblages prospered in the Western Interior of North America until they were altered by regional topographic and climatic changes which began in the middle Maastrichtian time. Several million years after they had achieved a new balance regionally, these assemblages were decimated by a catastrophic environmental deterioration resulting from the impact of a comet" (Russell and Dodson, 1997, 666).

Dodson sees more continuity than disruption across the K/T boundary and draws a different conclusion: "The record of non-dinosaurian terrestrial and freshwater aquatic vertebrates, including fishes, amphibians, turtles, lizards, champosaurs, crocodiles, multituberculates, and placental mammals, shows substantial continuity across the Cretaceous-Tertiary boundary. . . . Plant communities also show continuity, although significant disruptions have been noted. These observations suggest that terrestrial communities did not suffer a devastating catastrophe, but responded to changing environmental conditions" (Russell and Dodson 1997, 666–667).

Together, Russell and Dodson conclude:

> We have been pleasantly surprised to discover a broad area of common agreement. We concur that the latest Cretaceous dinosaurian record is far too incomplete to support either the catastrophic or the gradualistic model. . . . We differ in our assessment of which data are of greater significance. As we have described them, the two extinction models surely exist only in our imaginations. The truth lies in nature, which through the scientific method continually reveals ever-fascinating constellations of data which render the pursuit of scientific knowledge so enjoyable. (Russell and Dodson, 1997, 669)

In short, when the data are complex and conflicting and insufficient for decisive tests, each scientist is thrown back on his or her own judgmental skills. For Russell, the indicators of catastrophe seem paramount; for Dodson, the continuities seem more striking than the disruptions. It is refreshing to see Russell and Dodson express themselves so modestly after the very strong rhetoric that has flowed from both sides in this debate.

It is appropriate that the exchange between Russell and Dodson should end this book, for it shows how science should work—and, in fact, how it often does work. Controversy is inevitable in science, and sometimes it will get ugly. When it does, there is a saving grace that, most of the time, raises scientific controversy above the level of the polemics of creationists or Holocaust deniers. That saving grace is that many of the people doing science are, like Russell and Dodson, people willing to be rationally persuaded. Being willing to be rationally persuaded may not sound like much of a virtue, but it is. Such willingness indicates a degree of humility before the evidence, an attitude that sharply contrasts with the invincible certainty of the ideologue. To be susceptible to rational persuasion one must admit, as a genuine possibility, that one can be wrong, and this is hard for anyone to do, especially in the heat of controversy. The fact that so many scientists are able to make that admission—at least eventually—is part of the reason why science is the glory of the human race.[3]

For controversy to be of value, there has to be a principled way for it to end. If controversies went on forever, or ended only when one side was tricked into acquiescence or cowed into silence, scientific debate would merely be a form of obfuscation (as we saw in Chapter 1 that some think it is). But I think that in following the controversies examined in this book we have seen that even those controversies that emitted the most heat did not snuff out the light. That is, in even the most inflamed controversies we saw that logic and evidence—the traditionally *rational* factors in scientific debate—*really did matter* for the outcome of the debates.

---

3. Of course, there are cynics who say that scientists do not change their minds; rather, the proponents of outmoded theories just die off eventually. Other cynics say that when scientists do change their minds, it is not a matter of rational persuasion but owing to a semireligious "conversion" experience. Both cynical claims are demonstrably false (see Chapter 3 of my 2001 book *Drawing Out Leviathan: Dinosaurs and the Science Wars*).

Even where the questions are still undecided, progress has been made. The kind and amount of progress differed from one debate to the next. The dinosaur physiology debate—whether dinosaurs were endotherms, ectotherms, or something in between—has not and maybe never will be settled to everyone's satisfaction. Still, the debate over these points has taught us a great deal about what we can and cannot know about the biology or behavior of fossil creatures. Also, if any paleontologists really ever did conceive of dinosaurs as obsolescent hulks, it is safe to say that all now recognize them for the marvelously adapted and successful creatures they were. The bird/dinosaur controversy shows how spectacular discoveries can affect a sharply polarized debate. The dramatic new discoveries of feathered dinosaurs might not settle the debate, but they do place a very heavy burden on those who still deny that birds evolved from dinosaurs. We have seen that the most divisive debate of all, the one over dinosaur extinction, has moved to the point where the two sides can now share significant amounts of common ground. The most significant outcome of that debate may well be a fundamental change in how the earth sciences are done; that is, catastrophism has been made respectable again (see Huggett 1997).

Contrary to the fantasies of the Whig historians, scientific progress has not been a triumphal march from past ignorance to present enlightenment. Neither has the history of science been a darkling plain where ignorant armies clashed by night, as some of the recent antiscience zealots would have it. It has been a difficult ascent up a steep and slippery slope, with many obstacles and detours. Sometimes we have lost our grip and slipped back; sometimes we have found our footing and made rapid progress. Scientific progress can also be compared to those tests of visual perception where random bits of an image are gradually revealed to test subjects (Young 1992, 220). At first the visual information is so scanty that nothing can be made of it. Eventually there is enough of a picture to see something, but different people will have different hypotheses about what they are seeing. Finally enough of the picture is revealed that everyone can see what it is.

So it has been with our picture of dinosaurs. During the past two centuries we have seen them first as overgrown lizards; then as erect, elephantine quadrupeds; and now as spectacularly diverse and successful creatures, perhaps warm-blooded, perhaps the ancestors of birds, and perhaps snuffed out in an era-ending cataclysm. Such progress is not due to any one thing or any one kind of scientist. Science needs its bold, heretical innovators such as T. H. Huxley, Robert Bakker, and Luis Alvarez. It also needs its staunch conservatives such as Richard Owen, Larry Martin, and Charles Officer. It needs its fiery polemicists such as W. J. Holland as well as patient diggers such as Gideon Mantell and Earl Douglass. Last, but not least, science needs its cooler heads such as James O. Farlow, Dale A. Russell, and Peter Dodson—those who uphold the virtues of rigorous argument and rational debate even when the winds of controversy blow most fiercely. We are fortunate that there have been all these kinds of scientists. They have enriched our lives and made the history of science the fascinating pageant that it is.

Further Reading

The best account of the debate over the K/T extinctions from 1980 to 1994 is *The Mass Extinction Debates: How Science Works in a Crisis* (Stanford: Stanford University Press, 1994) edited by William Glen. Glen's anthology brings together pieces by some of the leading participants in the debate as well as overviews by historians and sociologists. Glen opens the volume with two essays that survey the controversy and give his account of the history of the debates. To my mind, Glen rather too readily adopts the line associated with the philosopher/historian/sociologist of science Thomas Kuhn: that in a "crisis" scientists on opposite sides engage in "incommensurable" discourse. That is, they do not share enough common ground even to argue rationally with one another; they simply talk past one another. Further, according to this view, scientists who do switch to the other theory undergo a sort of semireligious "conversion." In *Drawing Out Leviathan* I take a close look at David Raup, a leading paleontologist and one of the chief participants in these debates. I show that at no point were his debates with his opponents incommensurable and that when he changed his mind, he apparently did so for traditionally rational reasons. By the way, Raup tells his own version of the story in his entertaining book *The Nemesis Affair: A Story of the Death of Dinosaurs and the Ways of Science* (New York: W. W. Norton, 1986).

One of the unanticipated consequences of the debate over dinosaur extinction has been the surprising reemergence of catastrophism as a respectable principle in the earth sciences. Both volcanists and impactors have had to invoke events of a magnitude never observed by humans. Richard Huggett in *Catastrophism: Asteroids, Comets, and Other Dynamic Events in Earth History* (London: Verso, 1997) gives a good account of the resurgence of catastrophism.

Luis Alvarez told his side of the extinction controversy in his autobiography *Alvarez: Adventures of a Physicist* (New York: Basic Books, 1987).

Undeniably brilliant and completely unhampered by feelings of modesty, he was a force to be reckoned with in science for more than four decades. His son Walter wrote his own lengthier account of the controversies in his book with the exciting title *T. rex and the Crater of Doom* (Princeton, NJ: Princeton University Press, 1997). The best book for a general audience defending the impact theory is James Lawrence Powell's *Night Comes to the Cretaceous: Dinosaur Extinction and the Transformation of Modern Geology* (New York: W. H. Freeman and Company, 1998). Powell goes into the science in much greater detail than either of the Alvarezes, but he is careful to explain things in a clear and insightful way. He makes a very good case for the impact theory.

Charles Officer's book, cowritten with Jake Page, *The Great Dinosaur Extinction Controversy* (Reading, MA: Addison-Wesley, 1996) gives a very clear account of his grounds for dissent from the impact theory. Officer is pugnacious and unyielding. He regards the impact theory as so unreasonable that it really is no longer science. In his book, Powell returns the compliment and says that Officer goes so far in his intransigence that he abandons science. For those tired of the mud-slinging, J. David Archibald's *Dinosaur Extinction and the End of an Era* (New York: Columbia University Press, 1996) is a good antidote. Archibald thinks that impacts did occur but that the story of dinosaur extinction is more complex, multiply caused, and gradualistic. He gives considerable fossil evidence for his claim that the dinosaurs were declining well before the terminal K/T events and that the extinctions did not take place at the same time worldwide.

Appendix I
Dinosaurs and Their World

The first thing to grasp about dinosaurs is just how long ago they lived. One of the greatest discoveries of science has been that the earth is very old. Just a few centuries ago, people were quite happy to think of the world as only a few thousand years old. The most famous early calculation of the earth's age was by the Irish archbishop Ussher in the seventeenth century. By a careful study of the chronology implied by the "begats" and the ages of the patriarchs given in the Book of Genesis, he inferred that the creation had taken place in 4004 B.C. Because the earth's surface is geologically active, we have no surface rock that dates back to the earth's formation. The oldest rock presently on the earth's surface is slightly less than four billion years old. Dating of moon rocks and meteorites, which have changed little since their first formation (the moon is geologically inactive and has no weather), indicates an age for the solar system of about four and a half billion years. So Archbishop Ussher's estimate was off by a factor of about a million.

Compared to the age of the earth, dinosaurs are relative latecomers. The first dinosaurs appeared about 225 million years ago, after 95 percent of the earth's present history had passed. Still, the dinosaurs lived, flourished, and went extinct in the depths of time. To get some idea of how long ago *T. rex* became extinct, consider this: *T. rex* was one of the last of the dinosaurs. These last ones died about 65 million years ago. Let 1 inch represent 100 years—a century, close to the maximum age that humans live. One foot therefore represents the time since the Dark Ages. One foot ago, Viking hordes were plundering the British Isles. Two feet stand for the time since the Athenians put Socrates on trial, and 3 feet would take us far back into the Bronze Age. On this scale, where 1 inch stands for 100 years, we have 100 (years per inch) × 12 (inches per foot) × 5,280 (feet per mile) =

6,336,000 years per mile. So if 1 inch stands for 100 years, it would take over 10 miles to represent the amount of time that has passed since the extinction of the dinosaurs. Think about that the next time you have to walk a mile!

The dinosaurs lived so long ago that the earth itself has changed since that time. During the 1960s geologists learned that continents are not permanent. They rest on massive crustal plates that abut other plates along massive faults. The San Andreas Fault in California is where the Pacific and North American plates come into contact. Plates move with respect to one another, sometimes colliding, sometimes moving apart, sometimes grinding past each other in opposite directions, and sometimes with one plate boundary forced underneath another ("subduction"). The Himalayas are still rising today as the plate carrying the Indian subcontinent continues its collision with the Asian plate. Such massive crustal movements have vastly changed the face of the earth over geological time. In the Permian, the geological era just before the time of the dinosaurs, all of the earth's landmass was joined into one massive supercontinent called Pangaea ("all-earth"). Naturally, there was only one great ocean, Panthalassa ("all-sea"). Early in the time of the dinosaurs, this supercontinent broke up into a northern continent, Laurasia, and a southern one, Gondwana. By the end of the dinosaur age, the continents had started to assume something like their present shapes and relative positions.

Because of the great tectonic forces that have shaped the earth, particular areas of the earth's surface have changed greatly in location and in topography over geological time. For instance, the region around Pittsburgh, Pennsylvania, which definitely did not have a tropical climate when I lived there, was on the equator at one time. Areas that once were seafloor now are the peaks of high mountains. Marine fossils are found high in the Himalayas. Arid and semiarid lands, such as the great plains of North America, were once the floors of shallow epicontinental seas. Naturally, the climate in particular places, and of the earth as a whole, has changed radically over time. During much of the dinosaurs' time the earth's climate was much warmer than now. Dinosaurs lived comfortably at very high latitudes in both the Arctic and the Antarctic. There were no polar ice caps then. One hundred fifty million years ago what is now the badlands of Wyoming and Montana, where winter wind chills of 60 below are not at all uncommon, had a topography and climate like those of Houston or New Orleans.

The geological timescale can be somewhat intimidating and confusing. What do paleontologists mean when they speak of the "upper Cretaceous," say? The earth's history is encoded in sedimentary rock. Sedimentary rock is formed by the deposition of mineral sediments. As the deposition process continues over vast amounts of time, the sediment solidifies into rock, forming the various strata that you can often see when road builders cut through a hillside. Naturally, unless the original position has been disturbed by geological forces, the lower strata will be older, having been deposited earlier, than the higher strata. When they first began to study sedimentary strata, geologists noticed that fossils were completely absent from the lowest, and therefore oldest, layers. At a certain level, however, fossils suddenly become abundant. Rock layers in different locales can be grouped together because characteristic types of fossils are predominant in different strata. For instance, the name "Cretaceous" comes from the Latin for "chalky" because of the enormous deposits of chalk—the skeletons of innumerable tiny organisms—that were laid down during this time. By noticing such correlations between strata in different locales, geologists devised schemes for classifying them. Thus, "upper Cretaceous" refers to those layers of rock deposited in Cretaceous time that are higher up in the rock column, and therefore younger, than other Cretaceous rock. At first the classification scheme allowed geologists only to chart relative ages of strata. For example, Jurassic rocks were found below Cretaceous ones and were therefore known to be older, but just how old such rocks were was unknown. With the development of radiological dating in the twentieth century, however, it became possible to determine absolute ages. The Age of Dinosaurs, the time when dinosaurs lived, is the Mesozoic era. The Mesozoic era encompasses three shorter "periods": the Triassic, from 245–213 million years ago (ma), the Jurassic, 213–144 ma, and the Cretaceous, 144–65 ma. The Mesozoic era therefore extended from 245 to 65 ma.

Perhaps we should pause to reflect on these numbers. Dinosaurs existed for about 160 million years. We humans have recognized their existence for only about one-millionth of that time. The entire sweep of human history, from the first pharaohs of Egypt to the present day, is only around 5,000 years. That is 1/32,000 of the time the dinosaurs existed. As a dinosaur-crazy kid I had a book with a timeline of the Dinosaur Age and the following Age of Mammals on which each inch represented a million years. The timeline ran through the book with a tiny, barely visible sliver of the last inch on the last page standing for human history.

The earliest dinosaurs evolved in the late Triassic. Remains of some of the earliest dinosaurs were found in the badlands of Argentina. Among these early dinosaurs were *Herrerasaurus* and *Eoraptor*. These were bipedal, carnivorous dinosaurs that had the basic body plan seen in more developed form in much later and better-known "theropod" dinosaurs such as *Allosaurus* and *Tyrannosaurus*. This basic plan, which underwent many modifications in the Mesozoic, was one of the most successful animal designs in the history of life. Apparently all other dinosaurs, including the huge, four-footed herbivores of the Jurassic and Cretaceous, evolved from such small, two-footed forms. Actually, paleontologists have found this out only fairly recently. For a long time it was thought that the term "dinosaur" did not name a "clade"—a particular branch of the evolutionary tree of life. Rather the term seemed to encompass two distinct branches, the saurischia, or "lizard-hipped," dinosaurs, and the ornithischia, or "bird-hipped," dinosaurs. These names refer to the different pelvic anatomy of the two groups. Some, such as *T. rex* and *Apatosaursus,* had lizardlike hipbones; others, such as *Triceratops* and *Iguanodon,* had birdlike hipbones. Now paleontologists recognize so many anatomical similarities among all the dinosaurs that they agree that Richard Owen was right in 1842 to create a single group called "Dinosauria."

By the end of the Triassic, dinosaurs were the dominant type of land animal on Earth, having replaced "therapsid," or mammal-like, reptiles that had dominated earlier (*Dimetrodon,* which lived far back in the Permian, is the most famous of the mammal-like reptiles). How did the dinosaurs triumph? Paleontologists are divided on the question. Some give the traditional Darwinian explanation that dinosaurs were upright, smarter, faster, and meaner, and so simply outcompeted other groups. Other paleontologists point to the mass-extinction events that occurred in the late Triassic and see these as opening new ecological niches for dinosaurs by killing off the previous occupants.

At any rate, once established, the dinosaurs continued to flourish, diversifying into an amazing assortment of creatures. The late Jurassic, about 150 million years ago, featured many of the most familiar dinosaurs. It was then that the great sauropods lived. Giants among giants, the sauropods were the largest land animals ever, some up to 40 meters (130 feet) in length and others weighing more than 50 tonnes (55 tons) (a tonne is a metric ton, i.e., 1,000 kilograms). Consider that the largest African elephants weigh about 5 tonnes. The sauropods had long tails and necks, with relatively tiny heads and massive bodies supported by columnar legs. The sauropods were herbivores, and it must have taken a prodigious amount of foliage to maintain their bulk. *Apatosaurus* (formerly known as *Brontosaurus;* see Chapter 4), *Diplodocus, Camarasaurus,* and *Brachiosaurus* are among the best-known sauropods.

The sauropods shared their world with a number of other familiar creatures. The bizarre stegosaurs had rows of plates or spikes running down their backs, and often spiked tails, which they could no doubt wield with great effect if threatened. While the largest species were as big as elephants, they all had tiny brains, usually described as walnut-sized. The most curious thing about stegosaurs is how they managed to mate with all the fearsome plates and spikes running down their backs. As famous as the stegosaurs and sauropods was the predator *Allosaurus. Allosaurus* was a big (up to 12 meters [40 feet] long), carnivore with a massive head on a very strong neck and with large claws on both the legs and forelimbs.

Many illustrators have depicted these Jurassic creatures, and it is easy to see why dinosaurs appeal to the imagination so vividly. First, of course, is their sheer size. Even a moderate-sized carnivorous dinosaur, such as *Ceratosaurus nasicornis,* was at least as big as a Kodiak brown bear, the largest land-dwelling carnivore now alive (500 kilograms, or 1,100 pounds). As indicated earlier, the biggest sauropods would have dwarfed the largest African elephants. The ground must literally have shaken when herds of sauropods trudged along. Also, dinosaurs were decorated with an ostentatious variety of horns, spikes, plates, spines, clubs, sails, claws, and teeth. Their colors are conjectural, though it seems likely that many were more brightly colored than the dull green, brown, or olive drab of most illustrations. It is pleasing to picture a vast herd of *Apatosaurus* crossing an alluvial flat with a stately but inexorable gait while small groups of hungry allosaurs skirt around the edges, waiting for a weaker animal to lag behind.

Even more interesting than dinosaurs' appearance is how they acted. Here the evidence is weaker since behavior does not fossilize. Still, there are many lines of evidence that give us hints about dinosaur lifestyles. For instance, fossilized footprints contain a great deal of information about how dinosaurs walked or ran, whether they were solitary or herd animals, and the sorts of environments they frequented. R. McNeill Alexander ingeniously derived a formula that lets us tell approximately how fast an animal moved from studying its fossilized footprints. Paleontologists have found that dinosaurs generally moved fairly slowly, as would be expected of creatures that size.

However, footprints of carnivorous dinosaurs indicate that some of them could move at least as fast as a human sprinter. Movies depict dinosaurs as always fighting, but, like creatures living today, they probably spent most of their time peacefully. Unquestionably, though, dinosaurs did live in a violent world; many of their bones show evidence of massive injuries. When dinosaurs did fight, these titanic clashes were certainly spectacles of noise and violence that nature has seldom equaled.

Sauropods seem to have been especially misrepresented in the popular media. Traditionally, they were shown shoulder-deep in a swamp on the assumption that such massive creatures would need the buoyancy of water to stand up for long, and that their poor dental apparatus could handle only soft water plants. Even movies such as *Jurassic Park* assume that they were docile because they were herbivores. Yet fossil bone and footprint evidence shows that sauropods walked upright on dry land with their tails held off the ground. As for their temperament, just because an animal is a herbivore does not necessarily mean that it is docile. The hippopotamus, rhinoceros, and cape buffalo are among the most dangerous animals in Africa. To survive in their environment, sauropods probably had to have a real mean streak. Being swatted with a sauropod tail would have been like being beaten with a steel girder wielded by Paul Bunyan.

Dinosaurs shared their world with other creatures as fascinating as they were. In the skies were the flying pterosaurs, many of bizarre form and some much larger than any bird. One, *Quetzalcoatlus,* had a wingspan of 12 meters (40 feet; appropriately enough, the skeleton of this largest pterosaur was found in Texas). Also in the air, but mostly on the ground and in the water, were the first birds. The earliest recognized bird was the famous *Archaeopteryx lithographica,* which comes from the late Jurassic. During the Cretaceous, birds diversified, but many retained "primitive" reptilian features such as teeth.

The oceans held some creatures that would look familiar to us. Sharks have not changed all that much since the Mesozoic. However, sharks were not the top predators in the Mesozoic seas. The gigantic marine reptile *Kronosaurus* had jaws more massive and powerful than even *T. rex*'s. Long-necked plesiosaurs and the ichthyosaurs, which looked like enormous dolphins, were other common denizens of the seas. In the Cretaceous, mosasaurs, which were related to the varanid lizards—monitors such as the famous Komodo dragon are varanids—were also huge, ferocious predators. Even fresh waters were not safe.

Giant crocodiles, up to 15 meters (50 feet) long, preyed upon dinosaurs. Some invertebrates were also impressive, such as the very numerous ammonoids, cephalopods with beautiful, tightly coiled shells. Ammonoid shells are among the most common fossils of the Mesozoic.

The Jurassic produced the largest dinosaurs, but dinosaurs became most diverse during the Cretaceous. In the early Cretaceous iguanodontids became the most common herbivorous dinosaurs. They are described in Chapter 2. The villains of the movie *Jurassic Park,* the deadly velociraptors, actually evolved in the Cretaceous. Velociraptors were part of a group, the dromaeosaurs, which included perhaps the most lethal land predators ever seen on Earth (before humans). The dromaeosaurs were fast and agile, with skeletons well adapted for speed and maneuverability. They had powerful jaws and grasping forearms with sharp claws, but their deadliest weapon no doubt was the sicklelike claw on each foot. These claws were retracted while the creature was running but could be extended while the leg raked down with a powerful downward kick. To make matters worse, they may have hunted in packs, allowing them to kill prey much larger than themselves.

Perhaps the most common dinosaurs of the late Cretaceous were the hadrosaurs, or "duck-billed" dinosaurs. Their ducklike "bills" contained a battery of teeth well adapted to the grinding of plant matter. Some hadrosaurs, such as *Corythosaurus* and *Parasaurolophus,* developed ornate headgears, which were hollow and may have served as resonating chambers to deepen and amplify their calls. If this hypothesis is true, these were probably the loudest dinosaurs. When we imagine scenes of the late Cretaceous, we also have to imagine the calls—perhaps sonorous, perhaps earsplitting—of the hadrosaurs resounding through forest and plain.

The late Cretaceous also saw the evolution of the ceratopsians, the horned dinosaurs. The most famous of these was *Triceratops,* which means "three-horn-face." *Triceratops* had three long horns projecting from a massive skull supported by a broad, tanklike body. The largest specimens were about 9 meters (30 feet) long. Near the turn of the twentieth century Charles Knight painted one of the most famous dinosaur illustrations in Chicago's Field Museum. It shows a *Tyrannosaurus* squaring off against a *Triceratops* for what is sure to be a battle to the death. Actually, though dinosaurs were not notable for their intelligence, probably no *Tyrannosaurus* would have been stupid enough to make a head-on attack against a healthy, full-grown *Triceratops.* Any predator foolish enough to do so would very probably become a dinosaur shish kebab.

Finally, and most famously of all, the Cretaceous gave birth to the tyrannosaurids. The best known of these, in fact probably the best-known dinosaur of all, was *Tyrannosaurus rex*. *T. rex* flourished at the very end of the dinosaur age. Long depicted as the most ferocious and fearsome predator of all time, *T. rex* has recently had an image problem because some paleontologists, such as Jack Horner, now argue that *T. rex* was probably exclusively a scavenger rather than an active hunter. Carrion feeders, unfairly no doubt, are not esteemed as highly as predators. Many nations have had the eagle as a symbol; none has had the buzzard. Perhaps Calvin of the *Calvin and Hobbes* comic strip said it best: "It would be so bogus if *T. rex* only ate things already dead!" Yet Horner's evidence is intriguing. *T. rex* was a very large animal and does not seem to have been built for speed. If an animal weighing five tons or more were to trip and fall—a distinct possibility while pursuing prey—it would probably suffer so many broken bones and internal injuries that it would never get up again. It might be replied that though it may not have been terribly fast compared to a lion or tiger, it was faster than its presumed prey, which consisted of similarly huge and ponderous creatures. As for the dangers of pursuit, it might have been an ambush hunter, lunging from cover to grab the unwary. In this case, it would not have engaged in long, dangerous chases, only short bursts of activity. On the other hand, perhaps young tyrannosaurs were active hunters, but full-grown ones were scavengers. As is usually the case in paleontology, the evidence permits various interpretations.

The dinosaurs, and many other creatures such as the pterosaurs, mosasaurs, and ammonoids, went extinct at the end of the Cretaceous. There is considerable evidence that dinosaurs, after reaching a peak of diversity about seventy-four million years ago, began a gradual decline. Unquestionably, though, something delivered the coup de grâce. In Chapter 8 I tell the story of the bitter, vicious controversy that raged over the reasons for the mass extinction that ended the Mesozoic. Many regard the dinosaurs' demise with sadness. Many a kid has fantasized that somehow, somewhere, there is a lost forest where *T. rex* continues to prowl. But if the dinosaurs had not gone extinct, it would be most unlikely that we would be here. During the Mesozoic, our mammalian ancestors were small, nondescript, probably nocturnal creatures. While dinosaurs were dominant, there simply was no free ecological space for large mammals to occupy. Only after the extinction of the dinosaurs did the remarkable evolutionary radiation of mammals occur.

The dinosaurs were the lords of the Mesozoic. Their demise offers a deep lesson to us: The earth does not grant tenure. On the contrary, 99 percent of all species that have ever lived are now extinct. Even more sobering is the realization that several different hominid species have lived on Earth. Only one survives—*Homo sapiens.* We tend to think that the "sapiens" part of that name, our intelligence, gives us an edge that will allow us to escape the fate of other species. Well, we certainly have been successful in the short term—there are six billion of us on the planet, making us by far the most numerous species of large animal. By contrast, only about two thousand tigers survive in the wild. But the deepest lesson of the dinosaurs is the awful, incomprehensible depth of time. Sixty-five million years from now, will any of our descendants be alive? If so, will they be in any sense human? Will they remember any of our literature, our art, our science, our religions? After all, geological time shows that the only thing permanent is the fact of change. Evolution on Earth began with the first self-replicating molecules four billion years ago, and it will continue until the sun dies.

Further Reading

Several books mentioned in the Further Reading sections following the preceding chapters are the best general books on dinosaurs. These include the anthology *The Complete Dinosaur,* edited by J. O. Farlow and M. K. Brett-Surman; David Norman's *Illustrated Dinosaur Encyclopedia;* John Noble Wilford's *The Riddle of the Dinosaur;* and Don Lessem's *Dinosaurs Rediscovered* (originally titled *The Kings of Creation*). Almost as valuable as the Farlow and Brett-Surman anthology is the *Encyclopedia of Dinosaurs* (San Diego: Academic Press, 1997), edited by P. J. Currie and K. Padian. Good books on particular types of dinosaurs include John Horner and Don Lessem's *The Complete T. rex* (New York: Simon and Schuster, 1993) and Peter Dodson's *The Horned Dinosaurs* (Princeton, NJ: Princeton University Press, 1996). Gregory S. Paul's *Predatory Dinosaurs of the World* (New York: Simon and Schuster, 1988) is interesting and contains the author's own excellent illustrations but makes many very controversial claims about how fast these creatures ran and how they lived.

Some of the most interesting recent books on dinosaurs attempt to reconstruct their lives and habits. John Horner's books can be particularly recommended as works that try to tell us about dinosaur behavior and lifestyles. His book *Digging Dinosaurs: The Search That Unraveled the Mystery of*

Baby Dinosaurs (New York: Workman, 1988), cowritten with J. Gorman, tells about his discovery of dinosaur eggs and nests. These discoveries prompted Horner to name one dinosaur *Maiasaura,* which means "good mother lizard," because of evidence that it kept nests and nurtured its young. A more recent book coauthored with E. Dobb, *Dinosaur Lives: Unearthing an Evolutionary Saga* (San Diego: Harcourt Brace, 1997), gives evidence for other aspects of dinosaur behavior.

There are two college-level textbooks on dinosaurs, *Dinosaurs: The Textbook* by Spencer G. Lucas (Dubuque, IA: William C. Brown Publishers, 1994) and *The Evolution and Extinction of the Dinosaurs* by David E. Fastovsky and David B. Weishampel (Cambridge: Cambridge University Press, 1996). Both are very good, well illustrated, and considerably livelier than the usual cut-and-dried textbooks.

A word of caution: There are many good books on dinosaurs. I have been able to list only a few of them here. However, there are also many very bad books on dinosaurs, especially books for children. These books are often written by people who do not know much about dinosaurs and who don't mind making a few bucks by passing on their ignorance to others. Some years ago Don Lessem and other paleontologists formed the Dinosaur Society to safeguard against misinformation. If a book bears the seal of the Dinosaur Society, it will contain accurate information—otherwise, caveat emptor.

Appendix II
Evolution

What Is Evolution?

Evolution is about populations of organisms and how they come to be genetically different from their ancestral populations. Individuals do not evolve; they either succeed in producing offspring or they do not, and that is their importance for evolution. A sexually reproducing organism passes on one copy of each of its genes to each of its offspring. The other parent also contributes one copy of its genes, so each offspring gets two copies of each gene. A gene is just a segment of DNA, the molecule that encodes genetic information—the information needed to make an organism. Genes come in different varieties, and these varieties are called "alleles." Different alleles can have different effects on an organism's "phenotype"—the collective anatomical, physiological, or behavioral features of an organism. Whether we have brown eyes or blue, or whether our blood type is O, A, B, or AB, depends on the alleles we have inherited from out parents.

Now, each parent of a sexually reproducing species will have two copies of each of its genes—one from each of *its* parents. As noted earlier, when it reproduces it can pass along only one of those copies to each offspring. Therefore, an allele that a parent has might not be passed on to its offspring. But the loss of a particular allele in a particular act of reproduction is not important for evolution. What matters is the change of "gene frequency" across time in successive generations of whole populations of organisms. When we consider a population as a whole, we have to talk about the "gene pool," the totality of genetic information in a given population at a given time. The gene pool will therefore include *all* of the alleles found in *all* members of a given population at a given time. A famous equation, which (mathphobes rejoice!) need not be repeated here, predicts that under ideal conditions the proportions of different alleles found in the gene pool of a generation of offspring will be the same as that found in the gene pool of the parental generation. In short, under ideal conditions, the totality of genetic information will remain in equilibrium in succeeding generations.

But the real world is not an ideal world. In the real world all sorts of factors can upset the predicted equilibrium. In the real world, the genetic composition of succeeding generations changes over time. Rare alleles can become common, and common ones can become rare or even disappear entirely. New genes can be produced, which then spread through succeeding generations. Genes that are turned off in one generation can be turned on in another. Further, these genetic changes can accumulate so that a descendant generation can be genetically very different from an ancestor one.

Sometimes the descendants of a particular ancestral population will split into two different lines that are physically separated from one another so that there is no gene flow between the two. It then sometimes happens that one such line will not diverge very far from the ancestral population, but the other one will. If this genetic divergence becomes great enough, "reproductive isolation" can occur. That is, if the two descendant populations are reunited in the same physical space, they either cannot or will not interbreed, so they remain two separate populations. This means that a new species has been born, since reproductive isolation is the most widely used criterion for demarcating separate species. "Speciation," the production of new species, is the most important evolutionary event. This development of new species from the physical isolation and subsequent genetic divergence of different populations, called "allopatric speciation," may not be the only way speciation occurs, but it is probably the most common.

We may therefore define "evolution" as follows: Evolution is the gradual process of the accumulation of genetic change in succeeding generations of organisms. Further, such accumulating genetic change accounts for the origin of new species and therefore explains the diversity of life on Earth. A theory of evolution does two things: (1) it postulates that evolution in the sense just defined actually occurs, and (2) it offers an account of *how* such evolution occurs. For instance, the leading current theory, neo-Darwinian evolutionary theory, asserts that evolution occurred and explains that occurrence in terms of a certain set of mechanisms and processes. The fact that an evolutionary theory such as neo-Darwinism has two components, the assertion of the *fact* of evolution and an account of its *cause,* leads to confusion, which has been mischievously exploited by creationists. A noted scientist might criticize neo-Darwinism (leading the scientifically ignorant popular media to shout "Darwin Wrong!" in headlines), thereby meaning to question aspects of the neo-Darwinian explanation of the *cause* of evolution, *not* the fact of evolution itself. The *fact* of evolution is as well established as any fact can be in science. The continuing debates are over the *causes* of evolution.

How Do We Know That Evolution Occurred?

Since the only persons these days who deny that evolution has occurred are the various sorts of creationists (or "intelligent-design" theorists), I shall state the case for evolution with reference to creationist claims. Small-scale evolution, "microevolution," has been observed hundreds of times in nature. Studies by the noted Princeton University biologists Peter and Rosemary Grant have traced the evolution of successive generations of birds, the famous "Darwin's finches," in the Galápagos Islands. These studies document the effects of natural selection, showing that, just as Darwin predicted, changing environmental factors, such as the vagaries of drought and flood, have evolutionary consequences for populations. Since we can *see* such evolution as it takes place in real time, not even creationists can deny its reality. Their tactic is to belittle the importance of such small-scale evolution and then deny that its occurrence is evidence for large-scale evolutionary changes—such as the descent of birds from reptiles. That is, they deny that the gradual accumulation of genetic change in successive generations of organisms can account for evolution above the species level. In their view, *no* natural process can account for major changes, such as the emergence of one class (birds, say) from another (reptiles). Only direct divine intervention into natural processes can bring this off. Therefore we have to ask what evidence there is for "macroevolution," the claim that such large-scale changes can be explained by the gradual accumulation of genetic change in succeeding generations of organisms.

Because it takes place over geological time, macroevolution cannot be directly observed. We cannot watch reptiles evolve into birds, for instance. But isn't science all about observation? How can something be scientific if we cannot observe it? In fact, science is full of processes that never have and never will be directly observed. No one has ever observed an electron stimulated by a passing photon to emit a parallel photon of precisely the same energy, yet the entire technology of lasers depends on just this process. Very likely no one will ever observe the center of the sun, but astrophysicists are quite confident about what nuclear processes occur there. Scientists constantly refer to innumerable processes that are too big, too small, too fast, too slow, too inaccessible, or too far away in time or space to be observed.

Why do scientists believe in processes that they cannot observe? Because the postulation of such processes provides the *best theoretical explanation* for large bodies of data. Precisely the same grounds support the theory of evolution. Innumerable data points, which are otherwise unexplained, are powerfully and economically explained by descent with modification. Indeed, when one considers the enormous simplifying and unifying power of evolutionary explanations, one must say with the great geneticist Theodosius Dobzhansky, "Nothing in biology makes sense except in the light of evolution."

The obvious place to look for evidence of macroevolution is the fossil record. First, we need to be clear about what kind of evidence the fossil record can be expected to provide. Critics of evolution have long clamored for "missing links" and have loudly asserted that such links are indeed missing. For such critics, the ideal "missing link" would be a direct ancestor of a descendant group that is exactly halfway between the descendant group and a postulated ancestor. For humans, since humans are hypothesized to have evolved from apelike ancestors, the ideal missing link would be a direct human ancestor whose features were exactly halfway between an ape's and a human's. Its teeth would be exactly intermediate between a modern human's and an ape's, its brain size would be exactly between the current human and ape averages, its gait would be just between an ape's shamble and the human stride, and so on. But the expectation that there ever was such a creature, much less that we ought to find it in the fossil record, is based on several erroneous assumptions.

First, evolution simply does not work like that. There almost certainly never was a creature exactly half ape and half human. The expectation that there was such a creature comes not from evolutionary biology but from the hierarchical thinking of the Great Chain of Being that ranks all creatures from "highest" to "lowest" with all intermediate stages in place. What evolutionary theory predicts is that the intermediates between an ancestral and an evolved group will be mosaics with some features almost unchanged from the "primitive" ancestral condition and others evolved in different degrees toward the "advanced," or derived, condition. And this is exactly what we find. When we look at fossil hominids, such as *Homo erectus, Homo habilis,* or *Australopithecus,* we find that they have some features much like those of modern humans and others that are more apelike. For instance, *Australopithecus* seems to have had a predominately bipedal gait, more like humans than apes. However, it had the proportionately longer arms and smaller brain size of apes. As I noted in Chapter 7, when we look at the first known bird, *Archaeopteryx lithographica,* we find that it has a few distinctly avian features, such as wings, feathers, larger eyes and brain, and a furcula (wishbone). Otherwise, its anatomy is almost totally reptilian. In fact, the collectors of some specimens of *Archaeopteryx* initially identified their finds as *Compsognathus,* a small dinosaur.

A second erroneous assumption of those who demand missing links is that such links, even if they existed, must be found in the fossil record. Fossilization, especially of large vertebrates, is not a dependable process, nor is finding the fossils. Very likely only a small fraction of the vertebrate species that have lived have left a fossil record. Further, even if we do find something intermediate between two major groups, we cannot be sure that the creature was a direct ancestor of the later group. *Archaeopteryx* may have been the ancestor of modern birds, or it may have been a side branch that came to a dead end after splitting off from the ancestral line. There is no way to be sure.

So evolutionists do not expect perfect missing links to have existed. Even if they did exist, they do not expect that we must find them in the fossil record. Nevertheless, though they are relatively rare, we occasionally find clear examples of transitional fossils. These are fossil organisms that have characteristics intermediate between an ancestor and a descendant group—whether that fossil organism was on the direct line between these groups or on a side branch. The most famous examples of such transitional creatures are the fossil horses. Unfortunately, as originally presented by O. C. Marsh, these creatures were depicted as evolving in a straight line from the earliest "dawn horse," *Eohippus* (now called *Hyracotherium*), to the present-day *Equus.* No present-day evolutionary biologist holds that the fossil horses represent a straight line leading to the modern horse. Rather, they see these creatures as representing some of the many ramifying branches of the evolution of the horse family. However, they recognize that many of the fossil horses had features intermediate between the earliest and living horses.

Here, then, we have a clear-cut difference in what creationists and evolutionists expect about the fossil record. Evolutionists expect that there will be some clear-cut instances of transitional fossils. Creationists, whether the cruder "young-earth" variety or the more sophisticated "intelligent-design" sort, agree in denying that macroevolution, evolution at higher than the species level, has taken place. That is, they deny that the natural accumulation of genetic change in succeeding generations of organisms is sufficient to account for major transformations (e.g., reptile to bird). Therefore, they predict that there will be *no* genuine intermediates between different classes of organisms, between reptiles and mammals or reptiles and birds, say. There will not even be transitional fossils between orders within a class, say between cetaceans (whales and dolphins) and any other order of mammals. It is therefore unsurprising that they are so desperate to deny the many clear instances of such transitional fossils. The existence of such fossils clearly falsifies creationist expectations about the fossil record and supports the claims of macroevolution.

One of the most remarkable series of transitional fossils documents the amazing story of whale evolution. From about sixty-five to thirty-seven million years ago in the Paleocene and Eocene Periods, there lived a group of land-dwelling mammals, which, though they were ungulates (hoofed animals), were carnivorous and may have behaved like hyenas. Called "mesonychids," these creatures would not be anyone's first guess as a likely ancestor of the sperm whale. Yet in 1979, while prospecting for fossils in Pakistan, the paleontologist Phillip Gingerich made a remarkable find: an animal that, though only coyote-sized, had the distinctive anatomical traits of a whale, and so was named *"Pakicetus."* Yet its teeth closely resemble those of land-dwelling mesonychids—so closely that paleontologists (who know far more about teeth than any dentist) had always regarded such teeth as belonging to mesonychids until they found the jaws those teeth came from.

Pakicetus is therefore the oldest and most primitive cetacean yet found, dating from the early Eocene of about fifty million years ago. It lacked the cranial adaptations of later whales for efficient underwater hearing and probably led an amphibious lifestyle. *Ambulocetus,* found in Pakistan by Gingerich's colleague Hans Thewissen, also dated from the early Eocene. It had robust forelimbs and very large hindfeet for swimming, but its feet terminated in a convex hoof like that of mesonychids. *Ambulocetus* also retained the mesonychian tail structure. Like *Pakicetus,* it probably was amphibious. Later in the Eocene lived *Indocetus,* which had a more "advanced," that is, whalelike, skull but a postcranial anatomy that retained a number of primitive features such as a long neck and long hindlimbs. It could still support its weight on land and so was probably amphibious. Later still lived *Rhodocetus,* which retained some primitive features, such as a pelvis that articulated directly with the sacrum as in land animals and unlike modern whales. However, it also had many modern features, such as unfused sacral vertebrae for greater flexibility in swimming. Its hind limbs were also reduced compared to earlier cetaceans, but not nearly so much as later ones. Finally, about forty million years ago lived *Basilosaurus,* which was fully aquatic and could not have supported its weight on land. Its hind limbs were greatly reduced and could not have assisted in swimming.

We see then that over a ten-million-year period whales evolved from small, amphibious creatures that closely resembled land-dwelling mesonychids to fully aquatic creatures with vestigial hindlimbs that were no longer able to support themselves on land. Allow me to stress that this is precisely what we expect to see if macroevolution has occurred, and not at all what we expect if God had to bring about whales by an act of special creation. If God specially created the first whale, then it must have been *Pakicetus* or something very similar. But if an all-powerful God wanted to create a whale, why not go directly to a blue or sperm whale rather than make, sequentially over time, a series of creatures that look just like transitional forms? Long ago the philosopher René Descartes said that God is no deceiver, but he would have to be if creationism were true.

There is an even simpler way to argue for evolution from the fossil record. The most obvious thing about the fossil record is that the further you go back in time, the less the organisms alive at that time resemble those alive today. If we look at the fossils from one million years ago, in general the creatures look very familiar. There are some that have gone extinct since then, such as saber-toothed cats, mastodons, and woolly rhinoceroses, but we find very many of the same species, or at least the same genera, that we have today. Go back ten million years and things are quite different. There is nothing even remotely human. However, though they differed from modern types, there are lions, horses, camels, elephants, giraffes, whales, bats, and other familiar forms. Go back one hundred million years and there are no longer any lions, horses, whales, or indeed any mammals much larger than an opossum. Instead, the mighty dinosaurs dominated the landscape, and huge marine reptiles ruled the seas. Birds had teeth, and flying reptiles filled the air. Go back five hundred million years and there are no mammals, birds, reptiles, amphibians, or any vertebrates at all except a few jawless fishes. The creatures that lived then looked like, and indeed were, beings from another world. Go back one billion years, and there is nothing more advanced than colonies of single-celled animals.

How do we explain these obvious, undeniable facts of the fossil record? How do we explain the emergence through geological time of increasingly familiar creatures? One way is to postulate evolution—a historical process of the gradual emergence of new forms by a transformation of old ones. Looked at from the end of this process, where we currently are, we would expect to see the features of the fossil record described in the preceding paragraph. But if the only alternative is that the major groups of organisms were specially created by the deity, many apparently unanswerable questions arise. Why did the creator string out his creation over billions of years rather than create everything at once (maybe in six literal days, as the "young-earth" fundamentalist creationists preposterously assert)? If humanity is the pinnacle of creation, as creationists devoutly believe, then why go through interminable ages of useless trilobites and dinosaurs before humans very belatedly come onto the scene?

In fact, just what *is* the creationist explanation of, say, the origin of birds? Since they reject macroevolution, and so deny the gradual development of birds from nonavian ancestors, creationists *must* say that it was something like this: God said, "Let there be birds!" and (pause for drum roll) *poof! Archaeopteryx* emerges from a blaze of light! Naturally, I'm being flippant here, but if they do not claim that *something* like this really happened, just what are they claiming? Creationists often complain that evolutionists arbitrarily reject hypotheses of intelligent design. But it does not seem intolerant to view a "hypothesis" of the origin of birds in terms of the inscrutable exercise of occult powers by an invisible being in unknowable ways and for unknowable purposes as a less than satisfactory explanation.

There are very many other lines of evidence supporting evolution. One piece of evidence is still striking, though it was recognized before Darwin published *The Origin of Species.* Richard Owen, the brilliant antievolutionary comparative anatomist discussed in Chapters 2 and 3, noted what he called "homologies" in the skeletal anatomy of different animals. He noted that despite their very different external appearances, many such features shared an underlying structure. To take just one of many examples, consider the wing of a bat, the flipper of a dolphin, the paw of a bear, and the hand of a human. Although these appendages look very different and are used for very different purposes, the underlying anatomy is a set of variations on the same theme. For Owen, these homologies were evidence that organisms were realizations of a metaphysical archetype (see Chapter 3), but this hypothesis seemed to create far more problems than it solved. Darwin pointed out that such homologies are very naturally explained by a process whereby ancestral structures are retained but modified to be used for different purposes by descendants.

Creationists might try to explain such homologies by saying that God employed a common plan to adapt organisms to their environments. However, many of these homologies make no adaptive sense at all. That is, they do not make any sense from the viewpoint of someone trying to design an organism to perform certain functions in a given environment. An almighty Designer could design a wing or an eye from scratch; evolution has to work with what is already there.

These are just a very few of the very many pieces of evidence that conclusively establish that evolution as defined here has actually occurred. Consult the Further Reading section for sources that give more of this evidence.

Aren't There Good Arguments against Evolution?

No. Evolution has always been a controversial topic, and many arguments have been offered against it, but each has been decisively refuted. Nevertheless, some of the arguments against evolution themselves have a long evolutionary history. Sometimes such an argument is offered and then refuted, but instead of suffering extinction, it retreats, mutates into new form, and issues forth again. Again defeated, it may hibernate for a long time until it changes form once more.

Consider, for instance, an argument that Darwin himself examined in the *Origin,* the existence of "organs of extreme perfection and complication." (By the way, Darwin was scrupulously honest in conceding the difficulties faced by his theory, thereby displaying an integrity and modesty never shown by his creationist critics.) It is hard to see how evolution by natural selection (Darwin's theory) could account for the existence of organs of extreme perfection, such as the human eye. According to Darwin, the eye had to evolve from something simpler, ultimately from a noneye. However, natural selection supposedly works by preserving those useful features that an organism has and perfecting them over succeeding generations. But what possible use is half an eye? The human eye is a delicate, finely tuned mechanism. Each part is intricately interconnected and interrelated with other parts so that they all work together harmoniously to achieve sight. Take away one part and the whole thing ceases to function. Bits and pieces of an eye would have no use at all and would not be preserved by natural selection should they come into being. It seems then that the eye either functions as a completed whole or not at all. Therefore, it appears absurd to view it as having been assembled out of parts. It had to come into existence as a complex, fully functional whole, a feat a creator could accomplish, but not natural selection.

Darwin's reply was, in effect, that half an eye definitely is better than none. He pointed out that many gradations in complexity exist between the simplest light-detecting organs found in some one-celled creatures and the elaborate mammalian eye. Each of these organs, of whatever degree of complexity, serves its possessor well and would give it an advantage over a competing creature with poorer light-detecting abilities. Natural selection does not have to assemble an eye out of disparate, functionless bits. Rather, it takes structures that are already serving some valuable function and finds ways to incrementally change or improve that functioning. The many gradations between the simplest and most complex light-detecting organs show that there is no great, unbridgeable gap in eye evolution that natural selection could not cross.

So, to the objection that organs of extreme perfection are just too complex to have come about by an evolutionary process, Darwin convincingly showed how they could have done so. Yet basically the same sort of objection was revived in Michael Behe's 1996 book *Darwin's Black Box*. Behe, a biochemist, argues that some of the biochemical processes necessary for life exhibit "irreducible complexity." These very intricate processes, such as the Krebs cycle, the final link in the chain of biochemical reactions whereby aerobic respiration occurs, function only when all the pieces are present and all the subprocesses are doing their jobs. Take away any piece of this complex whole and the entire process collapses. The subprocesses do not do anything valuable by themselves; their function matters only when integrated with the whole. Such complex processes seemingly could not have developed from something simpler; nothing simpler would have worked at all. Therefore, Behe argues that the Darwinian explanation that such complex processes gradually developed from molecular tinkering with simpler processes just cannot hold. All such irreducibly complex processes must have come into existence all at once as a fully functional whole. Only creation by a designing intelligence seems to be capable of such a feat.

Behe's arguments are thoroughly dissected in *Finding Darwin's God* (1999) by Kenneth R. Miller, a leading cell biologist at Brown University (and also, by the way, a devout Christian). He takes one of Behe's most plausible examples of "irreducible complexity," the structure of the cilia found in many eukaryotic cells, and shows that such complexity is not irreducible at all. In fact, like Darwin pointing to simpler eyes, Miller points to many cells that have cilia or flagella that operate well with a much simpler structure than those Behe considers. Further, Miller cites clear evidence, both from the fossil record and from laboratory experiments, showing that the neo-Darwinian mechanisms of mutation and natural selection are indeed adequate to produce apparently irreducibly complex systems. He shows how even biochemical processes such as the Krebs cycle could have been built up by "molecular tinkering" from simpler processes that served other functions. In short, Brown shows that Behe's version of the appeal to design is no more successful than those Darwin faced.

Numerous books have addressed creationist critiques of evolution; I have more than a dozen on my shelves. See the Further Reading section, where I recommend my favorites of these.

Further Reading

George Bernard Shaw once quipped, "Those who are down on something usually are not up on it." This is more true for evolution than for practically any other subject. Persons who venomously attack evolution often have only the vaguest knowledge of the subject. More insidiously, other critics do seem to know a good bit about evolutionary theory but intentionally misrepresent its claims. Knowledge, of course, is the best defense against aggressive ignorance and straw-man attacks. Two excellent textbooks on evolution are *Evolution,* 2nd ed., by Mark Ridley (Cambridge, MA: Blackwell Science, 1996), and *Evolution* by Monroe V. Strickberger (Boston: Jones and Bartlett, 1990). There are many popular books on evolution. The two best, in my view, are David Young's *The Discovery of Evolution* (Cambridge: Cambridge University Press, 1992) and Colin Patterson's *Evolution,* 2nd ed. (Ithaca, NY: Cornell University Press, 1999).

Also of great interest is *What Evolution Is* by Ernst Mayr, perhaps the leading evolutionary biologist of the twentieth century. On a personal note, I had the privilege of meeting Mayr ten years ago when he was only eighty-nine years old. *What Evolution Is* is the latest of several books that he has published in the tenth decade of his astonishingly productive life. Clearly, Ernst Mayr not only possesses great knowledge of evolutionary theory; he is the living exemplification of the principle of survival of the fittest.

For those who want their evolutionary theory spiced with strong opinion, Stephen Jay Gould and Richard Dawkins have written many books on evolutionary topics—from *very* different perspectives. In fact, a fun way to learn about recent debates about evolutionary theory is to read what Gould and Dawkins have to say about each other. The philosopher Kim Sterelny has written an amusing book on their disputes, *Dawkins vs. Gould: Survival of the Fittest* (Cambridge: Icon Books, 2001). For solid, well-written information on Darwin and evolution, written with philosophical insight, any of the numerous writings of Michael Ruse may be recommended, especially *The Evolution Wars* (Santa Barbara, CA: ABC-CLIO, 2001).

A very good book dealing specifically with the topic of macroevolution is Carl Zimmer's *At the Water's Edge: Fish with Fingers, Whales with Legs, and How Life Came Ashore but Then Went Back to Sea* (New York: Free Press, 1998), which tells the story of the marvelous fossils documenting whale evolution. Zimmer's book completely refutes the creationist canard that there are no true transitional fossils and shows how strong the evidence really is for major transformations in the history of life. Zimmer also wrote the beautifully illustrated *Evolution: The Triumph of an Idea* (New York: HarperCollins, 2001), which is the companion volume to the PBS series of the same name.

Books critical of creationism have come in two waves. The earlier wave addressed the claims of the "young-earth" creationists, which received much exposure in the early 1980s. The second wave came in the late 1990s and early 2000s in response to the rise of the latest incarnation of creationism, "intelligent-design" theory. The two best of those first-wave books were Douglas J. Futuyma's *Science on Trial: The Case for Evolution* (New York: Pantheon Books, 1982) and Philip Kitcher's *Abusing Science: The Case against Creationism* (Cambridge, MA: MIT Press, 1982). Futuyma is an evolutionary biologist, and Kitcher is a philosopher of science; both are among the best in their fields. Between them, Kitcher and Futuyma deftly dissected the logical fallacies, factual errors, and sleazy tactics of the young-earthers. Futuyma's book also contains the clearest, most cogent statement of the evidence for evolution that I have ever read.

Rob Pennock's *Tower of Babel: The Evidence against the New Creationism* (Cambridge, MA: MIT Press, 1999) spearheaded the second wave of creationist literature. Pennock is a professional philosopher of science, and he is particularly effective against the claims of Phillip Johnson, an activist for the "new" creationism. Johnson claims that science in general, and evolutionary theory in particular, are dominated by a dogmatic commitment to metaphysical naturalism—the philosophy that denies the supernatural (including God) and asserts that only the natural is real. Pennock very effectively argues that the naturalism of science is a methodological, not a metaphysical, assumption that is completely justified by the requirements of scientific practice. Pennock also edited a massive volume, *Intelligent Design*

Creationism and Its Critics: Philosophical, Theological, and Scientific Perspectives (Cambridge, MA: MIT Press, 2001). This work brings together the chief representatives of the neocreationist view and allows them a full and fair statement of their position. Pennock also gives the critical response to those views by leading scientists, philosophers, and theologians.

Kenneth R. Miller's *Finding Darwin's God: A Scientist's Search for Common Ground between God and Evolution* (New York: HarperCollins, 1999) is a remarkable book. Miller is a leading cell biologist and a professor of biology at Brown University. He is also a devout Christian who argues vigorously for the compatibility of Darwinism and religious belief. His book contains clear and cogent refutations of some of the recent creationist claims, such as Behe's "irreducible complexity." Another notable book is Niles Eldredge's *The Triumph of Evolution and the Failure of Creationism* (New York: W. H. Freeman, 2000). Eldredge is a noted paleontologist and prolific author who developed, with Stephen Jay Gould, the theory of "punctuated equilibrium." His book debunks both the old and the new forms of creationism. Together, the books of Pennock, Miller, and Eldredge should serve as tombstones for the neocreationist view.

APATOSAURUS: A huge sauropod of the late Jurassic, sometimes referred to as "Brontosaurus." A specimen of *Apatosaurus louisae* was mounted in the Carnegie Museum of Natural History with the wrong head from 1934 to 1979.

ARCHAEOPTERYX: Archaeopteryx lithographica, the earliest known bird, found in the Solnhofen limestone in Bavaria. It has feathers and a furcula (wishbone) but also has teeth, claws on its wings, and a reptilian skeleton.

ARCHETYPE: For Richard Owen the archetype was a generalized ideal form or building plan, conceived by God, which is embodied in all actual creatures of that type.

CHICXULUB STRUCTURE: Claimed to be the remnant of the crater formed when the Cretaceous-ending object struck the earth sixty-five million years ago. Considered a volcanic structure by opponents of that hypothesis.

CLADE: Also called a "monophyletic group." A taxonomic group that consists of all and only those groups that descended from a most recent common ancestor.

COMPSOGNATHUS: A small theropod with many birdlike features.

CONVERGENT EVOLUTION: The independent evolutionary development of similar characteristics by different lineages of organisms.

CRETACEOUS: The final period of the Mesozoic era. Dinosaurs became most diverse in this period before declining and becoming extinct at its close.

DARWINISM: The theory of evolution developed from the work of Charles Darwin (1809–1882) that emphasized natural selection as the dominant cause of evolution.

DECCAN TRAPS: Vast areas of flood basalt in the Indian subcontinent. This basalt was deposited in a series of massive eruptions that occurred at the end of the Cretaceous.

DEINONYCHUS: Deinonychus antirrhopus, a small but lethal predator of the early Cretaceous. *Deinonychus* is one of the dromaeosaurs, which have numerous birdlike features. Discovered and described by John Ostrom.

DINOSAUR: From the taxonomic category *Dinosauria* named by Richard Owen in 1842, derived from the Greek for "terrible lizard." Dinosaurs were not lizards, however; they were more closely related to crocodiles and birds.

"DINOSAUR ORTHODOXY": A term used derisively by the paleontologist Robert Bakker to denote a view of dinosaurs as sluggish, stupid, and obsolescent.

DIPLODOCUS: A massive sauropod of the Late Jurassic. An articulated skeleton of *Diplodocus carnegii* was discovered in 1899 and mounted in the Carnegie Museum of Natural History.

ECTOTHERMY: The "cold-blooded" pattern of thermoregulation whereby animals cannot generate sufficient body heat by metabolic processes and must therefore gain heat from the environment. An ectothermic animal is called an "ectotherm."

ENDOTHERMY: The "warm-blooded" pattern of thermoregulation whereby animals maintain high body-core temperature by the maintenance of a high level of basal metabolism. An endothermic animal is called an "endotherm."

EVOLUTION: The historical process of descent with modification whereby descendant populations tend to depart indefinitely from the forms of ancestor populations.

FOSSIL: Any remains of an ancient organism, including petrified bones or wood, tracks, burrows, casts, excrement, etc.

"GROUND-UP" THEORY: The theory that avian flight originated when creatures that ran and leaped after prey evolved powered flight. The "ground-up" theory is favored by those who advocate a theropod ancestry for dinosaurs.

HAVERSIAN BONE: Bone tissue that is highly vascularized and is associated with animals with active, warm-blooded lifestyles.

IGUANODON: A large herbivorous dinosaur of the early to mid-Cretaceous, first named and described by Gideon Mantell.

IMPACT THEORY: The theory that the K/T extinctions were caused by the impact of a large (10 kilometers) extraterrestrial body with the earth. According to this theory, the direct effects of the impact, and the lingering environmental damage, would have been lethal to many groups of organisms.

JURASSIC: The second period of the Mesozoic era. The largest dinosaurs lived late in this period, as did *Archaeopteryx.*

K/T: The dividing line between the Cretaceous period, the last period of the Mesozoic Era, and the succeeding Tertiary, or Cenozoic, Era. The "K" comes from the German word for Cretaceous, "Kreide."

K/T EXTINCTION: The mass extinction that occurred at the end of the Cretaceous period. The dinosaurs disappeared in this extinction along with the pterosaurs, large marine reptiles such as the mosasaurs, and the ammonites.

LAMARCKISM: A theory of evolution based on the work of Jean Baptiste de Lamarck, which emphasized evolutionary progress through the inheritance of acquired characteristics.

MACROEVOLUTION: Large-scale evolutionary change that occurs over geological time.

MESOZOIC: The geological era comprising the Triassic—248–213 million years ago (ma), Jurassic—213–214 ma, and Cretaceous—144–65 ma periods. Dinosaurs evolved in the Triassic and flourished in the Jurassic and Cretaceous, then declined and went extinct at the end of the Cretaceous.

METABOLISM: All of the internal chemical processes of an organism, such as digestion, necessary for maintaining life.

MONOPHYLETIC GROUP: See CLADE.

NATURALISM (METAPHYSICAL): The philosophical position that asserts that nothing exists beyond nature.

NATURALISM (METHODOLOGICAL): The attempt to explain natural phenomena purely in terms of natural causes. Scientists generally assume such naturalism as a heuristic (method involving experimental or trial-and-error methods to aid in problem-solving) guiding their search for explanatory theories.

ORNITHISCHIA: The "bird-hipped" dinosaurs, which included the ceratopsian horned dinosaurs and the "duck-billed" hadrosaurs.

PALEONTOLOGY: The science that studies the life of the past.

PHYLOGENY: The evolutionary history of a species or other taxon.

PLATE TECTONICS: The theory that the crust of the earth consists of several massive plates that move over geological time, carrying the continents with them.

PTEROSAUR: Flying reptiles of the Mesozoic.

ROYAL SOCIETY: The preeminent society of British scientists.

SAURISCHIA: The "lizard-hipped" dinosaurs, which included the massive sauropods such as *Apatosaurus* and the theropods such as *Tyrannosaurus.*

SAUROPOD: Enormous four-footed herbivorous dinosaurs with long necks, long tails, and columnar legs supporting a massive body—such as *Apatosaurus, Brachiosaurus,* and *Diplodocus.* These were the largest dinosaurs and the largest land creatures that ever lived.

SHOCKED QUARTZ: Quartz crystals that have been deformed in particular ways by intense pressure. Certain types of shocked quartz seem to be associated only with sites of meteorite impact or nuclear explosions.

SIDEROPHILE ELEMENTS: "Iron-loving" elements such as platinum and iridium that concentrate where iron is present. Since most of the earth's iron is in the core, almost all of the platinum or iridium is there also, leaving trace amounts in the earth's crust. Extraterrestrial bodies such as asteroids and meteoroids have much higher amounts of these elements.

SOCIAL CONSTRUCTIVISM: The view that the theories and factual claims of science are purely cultural artifacts fabricated by scientists to serve local social and political agendas. In this view, scientific claims are not rigorously constrained by objective empirical reality; rather, "reality" is whatever scientists agree to call "real."

SPECIES: Among living organisms, "species" is defined as a naturally interbreeding group that produces fertile offspring. Among fossil creatures, "species" is defined by body structure.

STANDING CROP: The total living biomass of a given type of organism that a particular environment can sustain at any given time.

TAXON: Any term, such as "species" or "genus," used in the scientific classification of organisms. For example, for classifying humans, *Homo* is the genus and *sapiens* is the species.

TAXONOMY: The practice of systematically classifying organisms into types.

THEROPOD: Bipedal carnivorous dinosaurs, including carnosaurs such as *Allosaurus,* and coelurosaurs, which included tyrannosaurs and dromaeosaurs such as *Deinonychus.*

TRANSITIONAL FOSSIL: A fossil creature that has features intermediate between two major groups of organisms. For instance, *Archaeopteryx* is a mosaic with some avian and mostly reptilian features.

"TREES-DOWN" THEORY: The theory that avian flight originated when gliding creatures evolved the capacity for powered flight. The "trees-down" theory is favored by those who oppose a theropod origin for birds.

TRIASSIC: The first period of the Mesozoic era. Dinosaurs evolved in this period.

TSUNAMI: An enormous wave, often called a "tidal wave" (though it has nothing to do with tides), caused by an earthquake, eruption, landslide, or impact, that can cause great devastation to surrounding coastlines.

UNDERDETERMINATION: The fact that indefinitely many theories are logically compatible with any set of evidence or data.

UNIFORMITARIANISM: A term used in various ways but generally associated with the view of the earth sciences put forward by Charles Lyell in *Principles of Geology*. Since Lyell, many earth scientists have held it methodologically improper to explain phenomena in terms of processes or events of a kind or degree not previously observed.

VICTORIAN AGE: The age defined by the reign (1837–1901) of Victoria, queen of Great Britain and Ireland and empress of India.

VOLCANIC THEORY: The theory that massive volcanism so disrupted the environment that the K/T extinction resulted.

WHIG HISTORY: History written by the winners. Whig history is poor history because it removes historical figures and episodes from their proper context and evaluates them by how well they anticipated current science.

1677	Robert Plot, curator of the Oxford Ashmolean Museum, describes and illustrates the end of a large dinosaur thighbone.
1763	The Oxford researcher Richard Brookes gives the descriptive name *Scrotum humanum* to Plot's fragment of dinosaur bone.
1776	American Independence declared.
1809	The French scientist Jean Baptiste de Lamarck proposes his evolutionary theory in his book *Philosophie zoologique*.
Ca. 1815	Reverend William Buckland of Oxford University discovers some giant teeth and a portion of a lower jaw with a tooth. He later names the creature *Megalosaurus*.
1815	Battle of Waterloo.
1824	Buckland reads his paper on *Megalosaurus* to the Geological Society in London. This was the first scientific description of a creature now recognized as a dinosaur.
1825	Gideon Mantell reports his discovery of *Iguanodon* to the Royal Society.
	Richard Owen begins his career with the Hunterian Museum of the Royal College of Surgeons.

1830	The first volume of Charles Lyell's *Principles of Geology* is published. The uniformitarian principles, and concomitant gradualism, inculcated by Lyell have a tremendous impact on the practice of the earth sciences.
1841	Richard Owen reports on the terrestrial fossil reptiles to the British Association for Advancement of Science.
1842	Owen coins the term *"Dinosauria"* in the published version of his report to the BAAS.
1844	The anonymous publication of the book *Vestiges of the Natural History of Creation* (actually written by Robert Chambers) precipitates a furious response to its evolutionary thesis.
1854	The Crystal Palace opens at Sydenham near London. Benjamin Waterhouse Hawkins's full-sized sculptures of Owen's dinosaurs are a prime attraction.
1855	Ferdinand Vandiveer Hayden explores the Montana territories and discovers dinosaur bones.
1859	Charles Darwin publishes *The Origin of Species*.
1860	Owen publishes his *Paleontology,* which concludes that the

	ultimate cause of natural phenomena cannot be material.
1861	A nearly complete specimen of *Archaeopteryx lithographica* is found.
1861–1865	American Civil War.
1868	T. H. Huxley publishes his essay "On the Animals Which are Most Nearly Intermediate between Birds and Reptiles," in which he argues for the similarity of birds and dinosaurs.
1870	O. C. Marsh leads a group of Yale students on a fossil-hunting expedition to the Wild West.
1876	The Peabody Museum of Natural History opens at Yale University.
1877	The American Museum of Natural History opens in New York.
	Marsh begins to describe dinosaurs from the incredibly rich bone fields of Como Bluff, Wyoming, including *Stegosaurus, Apatosaurus,* and *Allosaurus.*
1882	The United States Geological Survey appoints Marsh as its official vertebrate paleontologist.
1890	The Cope-Marsh feud reaches a crescendo when Cope's charges against Marsh and John Wesley Powell are published in the New York *Herald* newspaper. Marsh fires back at length, accusing Cope of professional incompetence and suggesting that Cope's anger is fueled by envy.
1896	The Carnegie Museum of Natural History opens in Pittsburgh.
1898	Andrew Carnegie reads about the discovery of the "world's

	most colossal creature" and determines that it will be housed in his museum. When the creature is not forthcoming, an expedition is sent to the west to find a dinosaur for Carnegie.
1899	A Carnegie Museum expedition led by W. H. Reed discovers the complete, articulated skeleton of *Diplodocus carnegii* in Wyoming at a site they call Sheep Creek.
1901	Queen Victoria dies.
	Teddy Roosevelt becomes president of the United States.
1905	At the request of various emperors, kings, and other heads of state, the Carnegie Museum makes casts of *Diplodocus carnegii* and donates them to museums around the world.
1908–1909	American O. P. Hay and German Gustav Tornier publish articles claiming that the giant sauropods such as *Diplodocus* did not stand erect but crawled on splayed-out limbs like lizards.
1909	Earl Douglass begins work at the Carnegie Quarry in Utah. He discovers an articulated skeleton of *Apatosaurus,* but the skull is missing.
1910	W. J. Holland, director of the Carnegie Museum of Natural History, blasts Tornier in an article in *The American Naturalist* and vigorously defends the upright pose of the sauropods.
1914–1918	World War I.
1938	Dinosaur footprints examined by R. T. Bird in the Paluxy River in Texas show unequivocally that sauropods did have an erect stance.

1939–1945	World War II.
1964	John Ostrom of Yale University discovers the skeleton of the ferocious predator *Deinonychus*.
1968	Robert Bakker, while an undergraduate at Yale, publishes his first article on dinosaurs.
1969	Landing on moon.
1971	Bakker publishes article in the journal *Evolution* arguing that dinosaurs were warm-blooded, precipitating a lively controversy.
1975	Bakker's article in *Scientific American* proclaims a "dinosaur renaissance," alleging that dinosaurs were warm-blooded and that the old "dinosaur orthodoxy" is dead.
1978	An American Association for the Advancement of Science symposium debates the warm-blooded dinosaur hypothesis. The papers presented at the conference are published two years later under the title *A Cold Look at the Warm-Blooded Dinosaurs*.
1980	Ronald Reagan elected president of the United States. Luis Alvarez, Walter Alvarez, Helen Michel, and Frank Asaro publish an article in the journal *Science* claiming an extraterrestrial cause of the mass extinction at the end of the Cretaceous. A furious controversy erupts over their thesis.
1991	Alan Hildebrand publishes an article in the journal *Geology* claiming that the Chicxulub structure in the Yucatán is the remnant of the impact crater. Charles Officer and other opponents of the impact hypothesis remain deeply skeptical.
1994	Fragmented comet Shoemaker/Levy 9 crashes into Jupiter, demonstrating that massive impacts do still occur in the solar system.
1996	Alan Feduccia publishes his book *The Origin and Evolution of Birds,* which rejects the theory of dinosaur ancestry for birds and argues for a trees-down origin of avian flight.
1998–2003	Feathered dinosaurs such as *Caudipteryx* and *Microraptor* are discovered in China and described in the scientific literature.

Alexander, R. M. 1989. *Dynamics of Dinosaurs and Other Extinct Giants.* New York: Columbia University Press.

———. 1997. Engineering a Dinosaur. In J. O. Farlow and M. K. Brett-Surman, eds., *The Complete Dinosaur.* Bloomington: Indiana University Press, pp. 414–425.

Allaby, A., and M. Allaby. 1991. *The Concise Oxford Dictionary of the Earth Sciences.* Oxford: Oxford University Press.

Alvarez, L. W. 1987. *Alvarez: Adventures of a Physicist.* New York: Basic Books.

Alvarez, L.W., W. Alvarez, F. Asaro, and H. V. Michel. 1980. Extraterrestrial Cause for the Cretaceous-Tertiary Mass Extinction. *Science* 208: 1095–1108.

Alvarez, W. 1997. *T. rex and the Crater of Doom.* Princeton, NJ: Princeton University Press.

Alvarez, W., and F. Asaro. 1990. An Extraterrestrial Impact. *Scientific American* 263, 4: 78–84.

Archibald, J. D. 1996. *Dinosaur Extinction and the End of an Era.* New York: Columbia University Press.

Bakker, R. T. 1968. The Superiority of Dinosaurs. *Discovery* 3: 11–22.

———. 1971. Dinosaur Physiology and the Origin of Mammals. *Evolution* 25: 636–658.

———. 1972. Anatomical and Ecological Evidence of Endothermy in Dinosaurs. *Nature* 238: 81–85.

———. 1975. Dinosaur Renaissance. *Scientific American* 232: 58–78.

———. 1980. Dinosaur Heresy—Dinosaur Renaissance. In R. D. K. Thomas and E. C. Olson, eds., *A Cold Look at the Warm-Blooded Dinosaurs.* Boulder: Westview Press, pp. 351–472.

———. 1986. *The Dinosaur Heresies.* New York: William Morrow and Company.

———. 1987. The Return of the Dancing Dinosaur. In S. J. Czerkas and E. C. Olson, eds., *Dinosaurs Past and Present,* Vol. I. Los Angeles: Natural History Museum of Los Angeles County, pp. 38–64.

———. 1994. The Bite of the Bronto. *Earth* 3, 6: 26–35.

Barrick, R. E., M. K. Stoskopf, and W. J. Showers. 1997. Oxygen Isotopes in Dinosaur Bone. In J. O. Farlow and M. K. Brett-Surman, eds., *The Complete Dinosaur.* Bloomington: Indiana University Press, pp. 474–490.

Behe, M. 1996. *Darwin's Black Box.* New York: The Free Press.

Bennett, A. F., and B. Dalzell. 1973. Dinosaur Physiology: A Critique. *Evolution* 27: 170–174.

Berman, D. S., and J. S. McIntosh. 1978. Skull and Relationships of the Upper Jurassic Sauropod Apatosaurus (Reptilia, Saurischia). *Bulletin of the Carnegie Museum of Natural History* 8: 1–35.

Bernstein, R. 1983. *Beyond Objectivism and Relativism: Science, Hermeneutics, and Praxis.* Philadelphia: University of Pennsylvania Press.

Brett-Surman, M. K. 1997. Ornithopods. In J. O. Farlow and M. K. Brett-Surman, eds., *The Complete Dinosaur.* Bloomington: Indiana University Press, pp. 330–346.

Bowler, P. J. 1976. *Fossils and Progress.* New York: Science History Publications.

————. 1989. *Evolution: The History of an Idea,* revised ed. Berkeley: University of California Press.

Cadbury, D. 2000. *Terrible Lizard: The First Dinosaur Hunters and the Birth of a New Science.* New York: Henry Holt and Company.

Chapman, R. E. 1997. Technology and the Study of Dinosaurs. In J. O. Farlow and M. K. Brett-Surman, eds., *The Complete Dinosaur.* Bloomington: Indiana University Press, pp. 112–135.

Chinsamy, A. 1993. Bone Histology and Growth Trajectory of the Prosauropod Dinosaur *Massopondylus carinatus. Modern Geology* 18: 319–329.

Cope, E. D. 1868. On the Origin of Genera. *Proceedings of the Academy of Natural Sciences, Philadelphia* 20: 242–300.

Courtillot, V. E. 1990. A Volcanic Eruption. *Scientific American* 263, 4: 85–92.

Crompton, A. W., and S. M. Gatesy. 1989. Review of *Predatory Dinosaurs of the World* by Gregory Paul. *Scientific American* 260: 110–113.

Croswell, K. 1997. *Planet Quest—The Epic Discovery of Alien Solar Systems.* New York: Free Press.

Currie, P. J., and K. Padian. 1997. *Encyclopedia of Dinosaurs.* San Diego: Academic Press.

Darwin, C. 1860a. Letter to Charles Lyell. *The Life and Letters of Charles Darwin,* Vol. II. Ed. by F. Darwin. New York: Basic Books, 1959, p. 94.

————. 1860b. Letter to J. S. Henslow. *The Life and Letters of Charles Darwin,* Vol. II. Ed. by F. Darwin. New York: Basic Books, 1959, p. 96.

————. 1979. *The Origin of Species.* New York: Anvil Books.

Dawkins, R. 1992. Progress. In E. A. Keller and E. F. Lloyd, eds., *Key Words in Evolutionary Biology.* Cambridge, MA: Harvard University Press, pp. 263–272.

————. 1987. *The Blind Watchmaker.* New York: W. W. Norton.

Dennett, D. C. 1995. *Darwin's Dangerous Idea: Evolution and the Meanings of Life.* New York: Simon and Schuster.

de Ricqlès, A. J. 1980. Tissue Structures of Dinosaur Bone. In R. D. K. Thomas and E. C. Olson, eds., *A Cold Look at the Warm-Blooded Dinosaurs.* Boulder: Westview Press, pp. 103–139.

Desmond, A. J. 1976. *The Hot-Blooded Dinosaurs: A Revolution in Palaeontology.* New York: The Dial Press.

————. 1982. *Ancestors and Archetypes: Paleontology in Victorian London 1850–1875.* Chicago: University of Chicago Press.

————. 1989. *The Politics of Evolution.* Chicago: University of Chicago Press.

————. 1997. *Huxley: From Devil's Disciple to Evolution's High Priest.* Reading, MA: Addison-Wesley.

Desmond, A. J., and J. Moore. 1991. *Darwin: The Life of a Tormented Evolutionist.* New York: Warner Books.

Di Gregorio, M. A. 1984. *T. H. Huxley's Place in Natural Science*. New Haven: Yale University Press.

Dingus, L., and T. Rowe. 1998. *The Mistaken Extinction: Dinosaur Evolution and the Origin of Birds*. New York: W. H. Freeman.

Dodson, P. 1996. *The Horned Dinosaurs*. Princeton, NJ: Princeton University Press.

Douglass, E. Correspondence. Transcribed and ed. by Elizabeth Hill. Archives of the Department of Vertebrate Paleontology, Carnegie Museum of Natural History, Pittsburgh, PA.

Eldredge, N. 2000. *The Triumph of Evolution and the Failure of Creationism*. New York: W. H. Freeman.

Farlow, J. O. 1980. Predator/Prey Biomass Ratios, Community Food Webs and Dinosaur Physiology. In R. D. K. Thomas and E. C. Olson, eds., *A Cold Look at the Warm-Blooded Dinosaurs*. Boulder: Westview Press, pp. 55–83.

————. 1990. Dinosaur Energetics and Thermal Biology. In D. B. Weishampel, P. Dodson, and H. Osmolska, eds., *The Dinosauria*. Berkeley: University of California Press, pp. 43–55.

Farlow, J. O., and M. K. Brett-Surman, eds. 1997. *The Complete Dinosaur*. Bloomington: Indiana University Press.

Farlow, J. O. and R. E. Chapman. 1997. In J. O. Farlow and M. K. Brett-Surman, eds., *The Complete Dinosaur*. Bloomington: Indiana University Press, pp. 519–553.

Fastovsky, D. E., and D. B. Weishampel. 1996. *The Evolution and Extinction of the Dinosaurs*. Cambridge: Cambridge University Press.

Feduccia, A. 1973. Dinosaurs as Reptiles. *Evolution* 27: 166–169.

————. 1996. *The Origin and Evolution of Birds*. New Haven: Yale University Press.

Feyerabend, P. 1975. *Against Method*. London: NLB.

Flannery, T. 2002. Dinosaur Crazy. *The New York Review of Books* 49, 1 (January 17): 33–36.

Frankel, C. 1999. *The End of the Dinosaurs: Chicxulub Crater and Mass Extinctions*. Cambridge: Cambridge University Press.

Futuyma, D. J. 1982. *Science on Trial: The Case for Evolution*. New York: Pantheon Books.

Galileo Galilei. 1989. *Siderius Nuncius*. Translated by A. Van Helden. Chicago: University of Chicago Press.

Gillespie, N. C. 1979. *Charles Darwin and the Problem of Creation*. Chicago: University of Chicago Press.

Gilmore, C. W. 1936. Osteology of *Apatosaurus,* with Special Reference to Specimens in the Carnegie Museum. *Memoirs of the Carnegie Museum* 11: 176–300.

Glen, W. 1994. Introduction; What the Impact/Volcanism/Mass Extinction Debates Are About; and How Science Works in the Mass Extinction Debates. In W. Glen, ed., *The Mass Extinction Debates: How Science Works in a Crisis*. Stanford: Stanford University Press, pp. 7–91.

Glen, W., ed. 1994. *The Mass Extinction Debates: How Science Works in a Crisis*. Stanford: Stanford University Press.

Glut, D. F., and M. K. Brett-Surman. 1997. Dinosaurs and the Media. In J. O. Farlow and M. K. Brett-Surman, eds., *The Complete Dinosaur*. Bloomington: Indiana University Press, pp. 675–706.

Gould, S. J. 1981. *The Mismeasure of Man.* New York: W. W. Norton.

————. 1991. *Bully for Brontosaurus.* New York: W. W. Norton.

Gross, P. R., and N. Levitt. 1994. *Higher Superstition: The Academic Left and Its Quarrels with Science.* Baltimore: The Johns Hopkins University Press.

Hanson, V. D. 2001. *Carnage and Culture.* New York: Doubleday.

Hay, O. P. 1908. On the Habits and the Pose of the Sauropodous Dinosaurs, Especially of Diplodocus. *The American Naturalist* 43: 672–681.

————. 1910. On the Manner of Locomotion of the Dinosaurs, Especially Diplodocus, with Remarks on the Origin of the Birds. *Proceedings of the Washington Academy of Sciences* 12: 1–25.

Heilmann, G. 1927. *The Origin of Birds.* New York: D. Appleton.

Hellman, H. 1998. *Great Feuds in Science: Ten of the Liveliest Disputes Ever.* New York: John Wiley and Sons.

Holland, W. J. 1910. A Review of Some Recent Criticisms of the Restorations of Sauropod Dinosaurs in the Museums of the United States with Special Reference to that of Diplodocus Carnegiei in the Carnegie Museum. *The American Naturalist* 44: 259–283.

Horner, J., and E. Dobb. 1997. *Dinosaur Lives: Unearthing an Evolutionary Saga.* San Diego: Harcourt Brace.

Horner, J., and J. Gorman. 1988. *Digging Dinosaurs: The Search That Unraveled the Mystery of Baby Dinosaurs.* New York: Workman.

Horner, J., and D. Lessem. 1993. *The Complete T. rex.* New York: Simon and Schuster.

Huggett, R. 1997. *Catastrophism: Asteroids, Comets, and Other Dynamic Events in Earth History.* London: Verso.

Hull, D. 1973. *Darwin and His Critics: The Reception of Darwin's Theory of Evolution by the Scientific Community.* Chicago: University of Chicago Press.

Huxley, T. H. 1868a. On the Animals Which Are Most Nearly Intermediate between Birds and Reptiles. *The Annals and Magazine of Natural History; Zoology, Botany, and Geology* (series 4) 2: 66–75.

————. 1868b. On the Physical Basis of Life. *Selected Works of Thomas H. Huxley: Methods and Results.* New York: D. Appleton (no date), pp. 130–165.

Jaffe, M. 2000. *The Gilded Dinosaur.* New York: Three Rivers Press.

Ji Q., P. J. Currie, M. A. Norrell, and Ji S. 1998. Two Feathered Dinosaurs from Northeastern China. *Nature* 393: 753–761.

Kitcher, P. 1982. *Abusing Science: The Case against Creationism.* Cambridge, MA: MIT Press.

Klee, R. 1997. *Introduction to the Philosophy of Science: Cutting Nature at Its Seems.* Oxford: Oxford University Press.

Kohl, M. F., and J. S. McIntosh. 1997. *Discovering Dinosaurs in the Old West.* Washington, DC: Smithsonian Institution Press.

Lanham, U. 1973. *The Bone Hunters: The Heroic Age of Paleontology in the American West.* New York: Dover Publications.

Latour, B. 1987. *Science in Action: How to Follow Scientists and Engineers through Society.* Cambridge, MA: Harvard University Press.

————. 1988. *The Pasteurization of France.* Cambridge, MA: Harvard University Press.

Latour, B. and Woolgar, S. 1986. *Laboratory Life: The Construction of Scientific Facts*, 2nd ed. Princeton: Princeton University Press.

Lessem, D. 1992. *Dinosaurs Rediscovered: New Findings Which Are Revolutionizing Dinosaur Science.* New York: Simon and Schuster.

Lovejoy, A. O. 1936. *The Great Chain of Being.* Cambridge, MA: Harvard University Press.

Lucas, S. G. 1994. *Dinosaurs: The Textbook.* Dubuque, IA: William C. Brown Publishers.

Machamer, P., M. Pera, and A. Baltas. 2000. *Scientific Controversies.* Oxford: Oxford University Press.

Mantell, G. 1825. On the Teeth of the *Iguanodon. Philosophical Transactions of the Royal Society* 115: 179–186.

————. 1833 *Geology of South-East England.* London: Thomas Longman.

Marsh, O. C. 1876. Recent Discoveries of Extinct Mammals. *American Journal of Science* 12: 59–61.

————. 1877. Notice of a New and Gigantic Dinosaur. *American Journal of Science,* (series 3) 14: 87–88.

————. 1879. Notice of New Jurassic Reptiles. *American Journal of Science* (series 3) 18: 501–505.

_____. 1883. Principal Characters of American Jurassic Dinosaurs. *American Journal of Science* (series 3) 26: 81–85.

Martin, L. D. 1983. The Origin of Birds and of Avian Flight. In R. F. Johnston, ed., *Current Ornithology,* Vol. 1. New York: Plenum Press, pp. 105–129.

————. 1998. The Big Flap. *The Sciences* 38, 2: 39–44.

Matthew, W. D. 1910. The Pose of the Sauropodous Dinosaurs. *The American Naturalist* 54: 547–560.

Mayr, E. 1982. *The Growth of Biological Thought.* Cambridge, MA: Belknap Harvard.

————. 2002. *What Evolution Is.* New York: Basic Books.

McGinnis, H. 1982. *Carnegie's Dinosaurs.* Pittsburgh: Board of Trustees, Carnegie Institute.

McGowan, C. 1983. *In the Beginning: A Scientist Shows Why the Creationists Are Wrong.* Toronto: Macmillan.

————. 2001. *The Dragon Seekers.* Cambridge, MA: Perseus Publishers.

McIntosh, J. S., M. K. Brett-Surman, and J. O. Farlow. 1997. Sauropods. In J. O. Farlow and M. K. Brett-Surman, eds., *The Complete Dinosaur.* Bloomington: Indiana University Press, pp. 264–290.

McPhee, J. 1998. *Annals of the Former World.* New York: Farrar, Straus, and Giroux.

Miller, K. R. 1999. *Finding Darwin's God: A Scientist's Search for Common Ground between God and Evolution.* New York: HarperCollins.

Monastersky, R. 1998. *Science News* 153, 26: 404.

Norman, D. 1985. *The Illustrated Encyclopedia of Dinosaurs.* New York: Crescent Books.

Officer, C., and J. Page. 1996. *The Great Dinosaur Extinction Controversy.* Reading, MA: Addison-Wesley.

Ostrom, J. H. 1969. Terrible Claw. *Discovery* 5, 1: 1–9.

————. 1976. *Archaeopteryx* and the Origin of Birds. *Biological Journal of the Linnean Society* 8: 91–182.

Owen, R. 1842. Report on British Fossil Reptiles: Part II. *Report of the British Association for Advancement of Science for 1841:* 60–204.

————. 1860. *Paleontology.* Edinburgh: Adam and Charles Black.

Padian, K. 1987. The Case of the Bat-Winged Pterosaur. In S. J. Czerkas and E. C. Olson, eds., *Dinosaurs Past and Present,* Vol. II. Los Angeles: Natural History Museum of Los Angeles County, pp. 64–81.

Padian, K., and L. M. Chiappe. 1997. Bird Origins. *Encyclopedia of Dinosaurs.* In P. J. Currie and K. Padian, eds., *Encyclopedia of Dinosaurs.* San Diego: Academic Press, pp. 71–79.

Parsons, K. 2001. *Drawing Out Leviathan: Dinosaurs and the Science Wars.* Bloomington: Indiana University Press.

————, ed. 2003. *The Science Wars: Debating Scientific Knowledge and Technology.* Amherst, NY: Prometheus Books.

Patterson, C. 1999. *Evolution,* 2nd ed. Ithaca, NY: Cornell University Press.

Paul, G. S. 1988. *Predatory Dinosaurs of the World.* New York: Simon and Schuster.

Pennock, R. T. 1999. *Tower of Babel: The Evidence against the New Creationism.* Cambridge, MA: MIT Press.

————, ed. 2001. *Intelligent Design Creationism and Its Critics: Philosophical, Theological, and Scientific Perspectives.* Cambridge, MA: MIT Press.

Pera, M. 1994. *The Discourses of Science.* Chicago: University of Chicago Press.

Powell, J. L. 1998. *Night Comes to the Cretaceous: Dinosaur Extinction and the Transformation of Modern Geology.* New York: W. H. Freeman.

Raup, D. M. 1996. *The Nemesis Affair: A Story of the Death of Dinosaurs and the Ways of Science.* New York: W. W. Norton.

Reid, R. E. H. 1997. Dinosaur Physiology: The Case for "Intermediate" Dinosaurs. In J. O. Farlow and M. K. Brett-Surman, eds., *The Complete Dinosaur.* Bloomington: Indiana University Press, pp. 449–473.

Rescher, N. 1997. *Objectivity: The Obligations of Impersonal Reason.* Notre Dame, IN: University of Notre Dame Press.

Richards, R. J. 1992. *The Meaning of Evolution.* Chicago: University of Chicago Press.

Ridley, M. *Evolution,* 2nd ed. 1996. Cambridge, MA: Blackwell Science.

Riggs, E. S. 1903. Structure and Relationships of Opisthocoelian Dinosaurs, Part I: *Apatosaurus,* Marsh. *Publications of the Field Columbian Museum, Geology* 2: 165–196.

Roberts, J. H. 1988. *Darwinism and the Divine in America.* Madison: University of Wisconsin Press.

Romer, A. S. 1956. *Osteology of the Reptiles.* Chicago: University of Chicago Press.

Rothbart, D. 1990. Demarcating Genuine Science from Pseudoscience. In P. Grim, ed., *Philosophy of Science and the Occult,* 2nd ed. Albany: State University of New York Press, pp. 111–122.

Rothschild, B. M. 1997. Dinosaur Paleopathology. In J. O. Farlow and M. K. Brett-Surman, eds., *The Complete Dinosaur.* Bloomington: Indiana University Press, pp. 420–448.

Ruben, J., Andrew Leitch, Willem Hillenius, Nicholas Geist, and Terry Jones. 1997. New Insights into the Metabolic Physiology of Dinosaurs. In J. O. Farlow and M. K. Brett-Surman, eds., *The Complete Dinosaur.* Bloomington: Indiana University Press. pp. 505–518.

Rudwick, M. 1976. *The Meaning of Fossils,* 2nd ed. Chicago: University of Chicago Press.

————. 1985. *The Great Devonian Controversy.* Chicago: University of Chicago Press.

Ruse, M. 2000. *The Evolution Wars: A Guide to the Debates.* Santa Barbara, CA: ABC-CLIO.

————. 2001. *Can a Darwinian Be a Christian?* Cambridge: Cambridge University Press.

Russell, D. A., and P. Dodson. 1997. The Extinction of the Dinosaurs: A Dialogue between a Catastrophist and a Gradualist. In J. O. Farlow and M. K. Brett-Surman, eds., *The Complete Dinosaur.* Bloomington: Indiana University Press. pp. 662–672.

Sampson, S. 1997. Dinosaur Combat and Courtship. In J. O. Farlow and M. K. Brett-Surman, eds., *The Complete Dinosaur.* Bloomington: Indiana University Press, pp. 383–393.

Shapin, S., and S. Schaffer. 1986. *Leviathan and the Air-Pump.* Princeton, NJ: Princeton University Press.

Shipman, P. 1998. *Taking Wing:* Archaeopteryx *and the Evolution of Bird Flight.* New York: Simon and Schuster.

Sterelny, K. 2001. *Dawkins vs. Gould: Survival of the Fittest.* Cambridge: Icon Books.

Strickberger, M. V. 1990. *Evolution.* Boston: Jones and Bartlett.

Thomas, R. D. K., and E. C. Olson, eds. 1980. *A Cold Look at the Warm-Blooded Dinosaurs.* Boulder: Westview Press.

Tornier, G. 1909. Wie war der Diplodocus carnegii Wirklich Gebaut? *Sitzungsbericht der Gesellschaft Naturforschender Freunde zu Berlin* (April 20): 193–209.

Torrens, H. 1997. Politics and Paleontology: Richard Owen and the Invention of Dinosaurs. In J. O. Farlow and M. K. Brett-Surman, eds., *The Complete Dinosaur.* Bloomington: Indiana University Press, pp. 175–190.

Tuchman, B. W. 1994. *The Guns of August,* reprint ed. New York: Ballantine.

————. 1996. *The Proud Tower: A Portrait of the World before the War, 1890–1914.* New York: Ballantine.

von Däniken, E. 1987. *Chariots of the Gods?* reissue ed. New York: Berkley Pub Group.

Wall, J. F. 1970. *Andrew Carnegie.* New York: Oxford University Press.

Ward, P. D. 1998. *Time Machines: Scientific Explorations in Deep Time.* New York: Springer.

Wilford, J. N. 1985. *The Riddle of the Dinosaur.* New York: Vintage Books.

Woolgar, S. 1988. *Science: The Very Idea.* London: Tavistock Publications.

Young, D. 1992. *The Discovery of Evolution.* Cambridge: Cambridge University Press.

Xu, X, Zhonge Zhou, Xiaolin Wang, Xuewem Kuang, Fucheng Zhang, and Xiangke Du. 2003. Four-Winged Dinosaurs from China. *Nature* 421: 335–340.

Zimmer, C. 1998. *At the Water's Edge: Fish with Fingers, Whales with Legs, and How Life Came Ashore but Then Went Back to Sea.* New York: Free Press.

————. 2001. *Evolution: The Triumph of an Idea.* New York: HarperCollins.

Documents

Chapter 2

The documents for Chapter 2 include Gideon Mantell's "Notice on the Iguanodon"(1825) and an extract from Richard Owen's "Report on British Fossil Reptiles, Part II" (1842). Both are technical pieces written in spare scientific prose. Nevertheless, anyone who cares about dinosaurs will find them most interesting. Much can be read between the lines of dry description of teeth and bones. Mantell clearly knew he was on to something big—speaking both literally and figuratively. Of course, his estimates for the size of the Iguanodon as upward of sixty feet were soon shown to be exaggerated. Still, we can imagine the excitement and awe felt by Mantell and others when they first realized that England—the center of everything orderly and civilized in their minds—had once been a primeval wilderness populated by exotic reptilian monsters. The thought had to be both thrilling and disturbing.

The selection from Owen's "Report" contains the very passage in which Owen gave us the term "Dinosauria." It also includes Owen's description of Iguanodon. Owen's account is far more detailed and thorough than Mantell's and shows that very precise observations and measurements (down to 1/25,000 of an inch!) were possible at that time. However, some of Owen's own feelings are readily discernible. This selection concludes with a quote from the Reverend William Buckland's contribution to the Bridgewater Treatises. These were works by some of the most reputable scientists attempting to show evidence of divine design in the natural world. Owen agreed with Buckland that the teeth of Iguanodon showed a marvelous degree of adaptation, thereby giving evidence of providential design. It is interesting to contrast this attitude with the methodological materialism recommended by Huxley in one of the readings for the next chapter.

Notice on the Iguanodon, a Newly Discovered Fossil Reptile, from the Sandstone of Tilgate Forest, in Sussex

Gideon Mantell

F. L. S. and M. G. S. Fellow of the College of Surgeons, &c. In a Letter to Davies Gilbert, Esq. *M.P.V.P.R.S. &c. &c. &c. Communicated by* D. Gilbert, Esq.

Read February 10, 1825.

Sir,

I avail myself of your obliging offer to lay before the Royal Society, a notice of the discovery of the teeth and bones of a fossil herbivorous reptile, in the sandstone of Tilgate forest; in the hope that, imperfect as are the materials at present collected, they will be found to possess sufficient interest to excite further and more successful investigation, that may supply the deficiencies which exist in our knowledge of the osteology of this extraordinary animal.

The sandstone of Tilgate forest is a portion of that extensive series of arenaceous strata, which constitutes the iron-sand formation, and in Sussex forms a chain of hills that stretches through the county in a W. N. W. direction, extending from Hastings to Horsham. In various parts of its course, but more particularly in the country around Tilgate and St. Leonard's forests, the sandstone contains the remains of saurian animals, turtles, birds, fishes, shells, and vegetables. Of the former, three if not four species belonging to as many genera are known to occur, viz. the crocodile, megalosaurus, plesiosaurus, and the iguanodon, the animal whose teeth form the subject of this communication. The existence of a gigantic species of crocodile in the waters which deposited the sandstone, is satisfactorily proved by the occurrence of numerous conical striated teeth, and of bones possessing the osteological characters peculiar to the animals of that genus; of the megalosaurus, by the presence of teeth and bones resembling those discovered by Professor BUCKLAND in the Stonesfield slate; and of the plesiosaurus, by the vertebrae and teeth analogous to those of that animal.

The teeth of the crocodile, megalosaurus and plesiosaurus, differ so materially from each other, and from those of the other lacertae, as be to identified without difficulty; but in the summer of 1822, others were discovered in the same strata, which although evidently referable to some herbivorous reptile, possessed characters so remarkable, that the most superficial observer would have been struck with their appearance, as indicating something novel and interesting. As these teeth were distinct from any that had previously come under my notice, I felt anxious to submit them to the examination of persons whose knowledge and means of observation were more extensive than my own; I therefore transmitted specimens to some of the most eminent naturalists in this country, and on the continent. But although my communications were acknowledged with that candour and liberality which constantly characterises the intercourse of scientific men, yet no light was thrown upon the subject, except by the illustrious BARON CUVIER, whose . . . remarks induced me to pursue my investigations with increased assiduity, but hitherto they have not been attended with the desired success, no connected portion of the skeleton having been discovered. Among the specimens lately collected, some however were so perfect, that I resolved to avail myself of the obliging offer of Mr. CLIFT, (to whose kindness and liberality I hold myself particularly indebted) to assist me in comparing the fossil teeth with those of the recent lacertae in the Museum of the Royal College of Surgeons. The result of this examination proved highly satisfactory, for in an Iguana which Mr. STUTCHBURY had prepared to present to the college, we discovered teeth possessing the form and structure of the fossil specimens. . . .

The teeth . . . although varying from each other in some particulars, do not present greater dissimilarity than the differences arising from age, and the situation they respectively occupied in the jaw, would be liable to produce. Like the teeth of the recent iguana, the crown of the tooth is accuminated; the edges are strongly serrated or dentated; the outer surface is ridged, and the inner smooth and convex; and as in that animal the secondary teeth appear to have been formed in a hollow in the base of the primary ones, which they expelled as they increased in size. From the appearance of the fangs in such fossil teeth as are in a good state of preservation, it seems probable that they adhered to the inner side of the maxillae, as in the iguana, and were not placed in separate alveoli, as in the crocodile. The teeth appear to have been hollow in the young animals, and to have become solid in the adult. The curved teeth probably occupied the front of the jaw, and those which are nearly straight, the posterior part.

It appears unnecessary to dwell longer on the resemblance existing between the recent and fossil teeth. Whether the animal to which the latter belonged should be considered as referable to existing genera, differing in its specific characters only, or should be placed in the division of the enalio-sauri of Mr. CONYBEARE, which includes marine genera only, cannot at present be determined. If however any inference may be drawn from the nature of the fossils with which its remains associated, we may conclude, that if amphibious, it was not of marine origin, but inhabited rivers or fresh-water lakes; in either use the term IGUANODON derived from the form of the teeth, (and which I have adopted at the suggestion of the Rev. W. CONYBEARE) will not, it is presumed, be deemed objectionable.

It has already been mentioned, that of the bones of oviparous quadrupeds found in the sandstone of Tilgate forest, some are decidedly referable to the crocodile, and others to the megalosaurus and iguanodon; but our knowledge of the osteology of the latter is at present so limited that until some connected portion of the skeleton shall be discovered, it is impossible to distinguish the bones of the one from those of the other. Since, however, the teeth of the iguanodon are not known to occur in the Stonesfield slate, perhaps such of the bones from Tilgate forest as resemble those figured and descried by Professor BUCKLAND . . . may be attributed to the megalosaurus; while others not less gigantic may be assigned to the iguanodon. That the latter equalled, if not exceeded the former in magnitude, seems highly probable; for if the recent and fossil animal bore the same relative proportions, the tooth must have belonged [sic] an individual upwards of sixty feet long; a conclusion in perfect accordance with that deduced by Professor BUCKLAND from a femur, and other bones in my possession. . . .

From: Report on British Fossil Reptiles, Part II

Richard Owen

Dinosaurians

This group, which includes at least three well-established genera of Saurians, is characterized by a large sacrum composed of five anchylosed vertebrae of unusual construction, by the height and breadth and outward sculpturing of the neural arch of the dorsal vertebrae, by the twofold articulation of the ribs to the vertebrae, viz. at the anterior part of the spine by a head and tubercle, and along the rest of the trunk by a tubercle attached to the transverse process only; by broad and sometimes complicated coracoids and long and slender clavicles, whereby Crocodilian character of the vertebral column are combined with a Lacertian type of the pectoral arch; the dental organs also exhibit the same transitional or annectent characters in a greater or less degree. The bones of the extremities are of large proportional size, for Saurians; they are provided with large medullary cavities, and with well developed and unusual processes, and are terminated by metacarpal, metatarsal and phalangeal bones, which, with the exception of the ungual phalanges, more or less resemble those of the heavy pachydermal Mammals, and attest, with the hollow long-bones, the terrestrial habits of the species.

The combination of such characters, some, as the sacral ones, altogether peculiar among Reptiles, others borrowed, as it were, from groups now distinct from each other, and all manifested by creatures far surpassing in size the largest of existing reptiles, will, it is presumed, be deemed sufficient ground for establishing a distinct tribe or sub-order of Saurian Reptiles, for which I would propose the name of *Dinosauria.*

Of this tribe the principal and best established genera are the *Megalosaurus,* the *Hylaeosaurus,* and the *Iguanodon;* the gigantic Crocodile-lizards of the dry land, the peculiarities of the osteological structure of which distinguish them as clearly from the modern terrestrial and amphibious *Sauria,* as the opposite modifications for an aquatic life characterize the extinct *Enaliosauria,* or Marine Lizards.

Iguanodon Mantelli, Cuv.

The bones of an enormous reptile, successively discovered in the Wealden strata by Dr. Mantell, interpreted by their discoverer with the aid of Cuvier and Clift, named *Iguanodon* by Conybeare, lastly found in juxtaposition to the extent of nearly half the skeleton, in the green-sand quarries of Mr. Benstead, offer not the least marvellous or significant evidences of the inhabitants of the now temperate latitudes during the earlier oolitic periods of the formation of the earth's crust.

With vertebra subconcave at both articular extremities, having, in the dorsal region, lofty and expanded neural arches, and doubly articulated ribs, and characterized in the sacral region by their unusual number and complication of structure; with a Lacertian pectoral arch and unusually large bones of the extremities excavated by large medullary cavities and adapted for terrestrial progression;—the *Iguanodon* was also distinguished by teeth, resembling in shape those of the Iguana, but in structure differing from the teeth of every other known Reptile, and unequivocally indicating the former existence, in the Dinosaurian Order, of a gigantic representative of the small group of living lizards which subsist on vegetable substances.

Of this remarkable Reptile, the results of personal examination of almost all the recognisable remains that have hitherto been collected in public or private museums, are here given.

Teeth:—The value of the ordinary external characters of the teeth of the oviparous Vertebrata has never perhaps been placed in so striking a point of view as in the leading steps to the discovery of the *Iguanodon,* which cannot be better recounted than in the words of Dr. Mantell.

After noticing the ordinary organic remains which characterize the sandstone of the Tilgate Forest, and his discovery, in the summer of 1822, of other teeth distinguished by novel and remarkable characters, the indefatigable explorer of the Wealden proceeds to state:—". . . in an Iguana which Mr. Stutchbury had prepared to present to the College, we discovered teeth possessing the form and structure of the fossil specimens."

The important difference which the fossil teeth presented in the form of their grinding surface was afterwards pointed out by Cuvier, and recognised by Dr. Mantell, and the combination of this dental distinction with the vertebral and costal characters, which prove the *Iguanodon* not to have belonged to the same group of Saurians as that which includes the Iguana and other modern lizards, rendered it highly desirable to ascertain, by the improved modes of investigating dental structure, the actual amount of correspondence between the *Iguanodon* and Iguana in this respect. This I have endeavoured to do in my general description of the Teeth of Reptiles, from which the following account is abridged.

The teeth of the *Iguanodon,* though resembling most closely those of the Iguana, do not present an exact magnified image of them, but differ in the greater relative thickness of the crown, its more complicated external surface, and, still more essentially, in a modification of the internal structure, by which the *Iguanodon* equally deviates from every other known reptile.

As in the Iguana, the base of the tooth is elongated and contracted, while the crown is expanded, and smoothly convex on the inner side; when first formed it is acuminated, compressed, its sloping sides serrated, and its external surface traversed by a median longitudinal ridge, and coated by a layer of enamel, but beyond this point the description of the tooth of the *Iguanodon* indicates characters peculiar to that genus. In most of the teeth that have hitherto been found, three longitudinal ridges traverse the outer surface of the crown, one on each side of the median primitive ridge; these are separated from each other and from the serrated margins of the crown by four wide and smooth longitudinal grooves. The relative width of these grooves varies in different teeth; sometimes a fourth small longitudinal ridge is developed on the outer side of the crown. The marginal serrations, which, at first sight, appear to be simple notches, as in the Iguana, present under a low magnifying power the form of transverse ridges, themselves notched, so as to resemble the mammillated margins of the unworn plates of the elephant's grinder: slight grooves lead from the interspaces of these notches upon the sides of the marginal ridges. These ridges or dentations do not extend beyond the expanded part of the crown: the longitudinal ridges are continued further down, especially the median ones, which do not subside till the fang of the tooth begins to assume its subcylindrical form. The tooth at first increases both in breadth and thickness; it then diminishes in breadth, but its thickness goes on increasing; in the larger and fully formed teeth, the fang decreases in every diameter, and sometimes tapers almost to a point. The smooth unbroken surface of such fangs indicates that they did not adhere to the inner side of the maxilla, as in the Iguana, but were placed in separate alveoli, as in the Crocodile and Megalosaur: such support would appear, indeed, to be indispensable to teeth so worn by mastication as those of the *Iguanodon.*

The apex of the tooth soon begins to be worn away, and it would appear, by many specimens, that the teeth were retained until nearly the whole of the crown had yielded to the daily abrasion. In these teeth, however, the deep excavation of the remaining fang plainly bespeaks the progress of the successional tooth prepared to supply the place of the worn out grinder. At the earlier stages of abrasion a sharp edge is maintained at the external part of the tooth by means of the enamel which covers that surface of the crown; the prominent ridges upon that surface give a sinuous contour to the middle of the cutting edge, whilst its sides are jagged by the lateral serrations: the adaptation of this admirable dental instrument to the cropping and comminution of such tough vegetable food as the *Clathrariae* and similar plants, which are found buried with the *Iguanodon,* is pointed out by Dr.

Buckland, with his usual felicity of illustration, in his ' Bridgewater Treatise,' vol. i. p. 246. . . .

The microscopical examination of the structure of the Iguanodon's teeth [also] contributes additional evidence of the perfection of their adaptation to the offices to which their more obvious characters had indicated them to have been destined. . . .

And if the following reflections were natural and just after a review of the external characters of the dental organs of the *Iguanodon,* their truth and beauty become still more manifest as our knowledge of their subject becomes more particular and exact:

"In this curious piece of animal mechanism we find a varied adjustment of all parts and proportion of the tooth, to the exercise of peculiar functions, attended by compensations adapted to shifting conditions of the instrument, during different stages of its consumption. And we must estimate the works of nature by a different standard from that which we apply to the productions of human art, if we can view such examples of mechanical contrivance, united with so much economy of expenditure, and with such anticipated adaptations to varying conditions in their application, with feeling a profound conviction that all this adjustment has resulted from design and high intelligence."—(Buckland's *Bridgewater Treatise,* vol. i p. 249).

Chapter 3

The two selections from the writings of T. H. Huxley included here reveal two sides of this remarkable man. Huxley displays his scientific talents in the article "On the Animals Which Are Most Nearly Intermediate between Birds and Reptiles." Good science requires bold hypotheses and rigorous empirical support. Huxley's thesis is bold—that the facts of paleontology support evolution by showing that in the past there were reptiles much more birdlike and birds much more reptilelike than any living today. He supports this thesis with detailed and careful analysis of the anatomical evidence. Though the piece is written in sober, plain scientific style, Huxley's rhetorical skills can be discerned. He admits that like a property owner who cannot produce a proper title deed, the defender of a scientific theory cannot feel secure when vital evidence is missing. For the defender of evolution, the apparent lack of transitional forms between major groups was such an embarrassment (this is no longer so—see Appendix 2). However, the fossil record fills in these gaps in some cases. As Huxley observes: "If I cannot produce the complete title-deeds of the doctrine of animal evolution, I am able to show a considerable piece of parchment evidently belonging to them."

The next selection contains the closing pages of Huxley's essay "On the Physical Basis of Life." Here we encounter Huxley the philosopher of science. Huxley raises the question— still a major issue in his day—of whether biological inquiry should proceed on completely materialist grounds. He argues that it certainly should, but he is very careful to state that this is not (pace some recent creationist critics of science) because scientists are dogmatically committed to a materialist metaphysics. On the contrary, he very pointedly rejects any such metaphysical claim as completely unknowable—indeed just as imponderable as the politics of extraterrestrials. Instead, materialism is a pragmatic assumption. That is, because materialistic explanations have so far worked so well, scientists are completely justified in pursuing further inquiries along materialist lines.

Royal Institution of Great Britain

"On the Animals which are most nearly intermediate between Birds and Reptiles"

February 7, 1868

Professor Huxley, LL.D., F.R.S.

Those who hold the doctrine of evolution (and I am one of them) conceive that there are grounds for holding that the world, with all that is in it and on it, did not come into existence in the condition in which we now see it, nor in anything approaching that condition.

On the contrary, they hold that the present conformation and composition of the earth's crust, the distribution of land and water, and the infinitely diversified forms of animals and plants which constitute its present population, are merely the final terms in an immense series of changes which have been brought about, in the course of immeasurable time, by the operation of causes which are more or less similar to those which are at work in the present day.

Perhaps this doctrine of evolution is not maintained consciously and in its logical integrity by a very great number of persons. But many hold particular applications of it without committing themselves to the whole; and many, on the other hand, favour the general doctrine without giving an absolute assent to its particular applications.

Thus, one who adopts the nebular hypothesis in astronomy, or is a uniformitarian in geology, or a Darwinian in biology, is so far an adherent of the doctrine of evolution.

And, as I can testify from personal experience, it is possible to have a complete faith in the general doctrine of evolution, and yet to hesitate in accepting the nebular, or the uniformitarian, or the Darwinian hypotheses in all their integrity and fulness; for many of the objections that are brought against these hypotheses affect them only, and, even if they be valid, leave the general doctrine of evolution untouched.

On the other hand, it must be admitted that some arguments which are adduced against particular forms of the doctrine of evolution would very seriously affect the whole doctrine if they were proof against refutation.

For example, there is an objection which I see constantly and confidently urged against Mr. Darwin's views, but which really strikes at the heart of the whole doctrine of evolution, so far as it is applied to the organic world.

It is admitted on all sides that existing animals and plants are marked out by natural intervals into sundry very distinct groups: insects are widely different from fish, fish from reptiles, reptiles from mammals, and so on. And out of this fact arises the very pertinent objection, How is it, if all animals have proceeded by gradual modification from a common stock, that these great gaps exist?

We, who believe in evolution, reply that these gaps once were non-existent; that the connecting forms existed in previous epochs of the world's history, but that they have since died out.

Naturally enough then, we are asked to produce these extinct forms of life. Among the innumerable fossils of all ages which exist, we are asked to point to those which constitute the connecting forms.

Our reply to this request is, in most cases, an admission that such forms are not forthcoming; and we account for this failure of the needful evidence by the known imperfection of the geological record. We say that the series of formations with which we are acquainted is but a small fraction of those which have existed, and that between those which we know are great breaks and gaps.

I believe that these excuses have great force; but I cannot smother the uncomfortable feeling that they are excuses.

If a landed proprietor is asked to produce the title-deeds of his estate, and is obliged to reply that some of them were destroyed in a fire a century ago, that some were carried off by a dishonest attorney, and that the rest are in a safe somewhere, but that he really cannot lay his hands on them, he cannot, I think, feel pleasantly secure, though all his allegations may be correct and his ownership indisputable. But a doctrine is a scientific estate, and the holder must always be able to produce the title-deeds, in the way of direct evidence, or take the penalty of that peculiar discomfort to which I have referred.

You will not be surprised, therefore, if I take this opportunity of pointing out that the objection to the doctrine of evolution, drawn from the supposed absence of intermediate forms in the fossil state, certainly does not hold good in all cases. In short, if I cannot produce the complete title-deeds of the doctrine of animal evolution, I am able to show a considerable piece of parchment evidently belonging to them.

To superficial observation no two groups of beings can appear to be more entirely dissimilar than reptiles and birds. Placed side by side, a Humming-bird and a Tortoise, an Ostrich and a Crocodile offer the strongest contrast, and a Stork seems to have little but animality in common with the Snake it swallows.

Careful observation has shown, indeed, that these obvious differences are of a much more superficial character than might have been suspected, and that reptiles and birds do really agree much more closely than birds with mammals, or reptiles with amphibians. But still, "though not as wide as a church-door or as deep as a well," the gap between the two groups, in the present world, is considerable enough.

Without attempting to plunge you into the depths of anatomy, and confining myself to that osseous system to which those who desire to compare extinct with living animals and are almost entirely restricted, I may mention the following as the most important differences between all the birds and the reptiles which at present exist.

1. The pinion of a bird, which answers to the hand of a man or to the fore paw of a reptile, contains neither more nor fewer than three fingers. These answer to the thumb and the two succeeding fingers in man, and have their metacarpals connected together by firm bony union, or ankylosed. Claws are developed upon the ends of at most two of the three fingers (that answering to the thumb and the next), and are sometimes entirely absent.

   No reptile with well-developed fore limbs has so few as three fingers; nor are the metacarpal bones in these ever united together; nor do they present fewer than three claws at their terminations.

2. The breast-bone of a bird becomes converted into a membrane bone, and the ossification commences in it from at least two centres. The breast-bone of no reptile becomes converted into a membrane bone, nor does it ever ossify from several distinct centres.

3. A considerable number of caudal and lumbar, or dorsal, vertebrae unite together with the proper sacral vertebrae of a bird to form its "sacrum." In reptiles the same region of the spine is constituted by the one or two sacral vertebrae.

4. In birds the haunch-bone (ilium) far in front of, as well as behind, the acetabulum; the ischia and pubes are directed backwards, almost parallel with it and with one another; the ischia do not unite in the ventral middle line of the body.

   In reptiles, on the contrary, the haunch-bone is not produced in front of the acetabulum; and the axes of the ischia and pubes diverge and lie more or less at right angles to that of the illium. The ischia always unite in the middle ventral line of the body.

5. In all birds the axis of the thigh-bone lies nearly parallel with the median plane of the body (as in ordinary mammals) in the natural position of the leg. In reptiles it stands out at a more or less open angle with the median plane.

6. In birds, one half of the tarsus is inseparably united with the tibia, the other half with the metatarsal bone of the foot. This is not the case in reptiles.

7. Birds have never more than four toes, the fifth being always absent. The metatarsal of the hallux, or great toe, is always short and in-complete above. The other metatarsals are ankylosed together and unite with one half of the tarsus, so as to form a single bone, which is called the tarso-metatarsus.

   Reptiles with completely developed hind limbs have at fewest four toes, the metatarsals of which are all complete and distinct from one another.

How far can this gap be filled by a reference to the records of the life of past ages?

This question resolves itself into two:—

1. Are there any fossil birds more reptilian than any of those now living?
2. Are there any fossil reptiles more bird-like than living reptiles?

And I shall endeavor to show that both these questions must be answered in the affirmative.

It is very instructive to note by how mere a chance it is we happen to know that a fossil bird, more reptilian in some respects than any now living, once existed.

Bones of birds have been obtained from rocks of very various dates in the Tertiary series without revealing any forms but such as would range themselves among existing families.

A few years ago the great Mesozoic formations had yielded only a few fragmentary ornitholites which have been discovered in the Cambridge greensand, and which are insufficient for the complete determination of the affinities of the bird to which they belonged.

However, the very fine calcareous mud of the ancient Oolitic sea-bottom which has now hardened into the famous lithographic slate of Solenhofen, and has preserved innumerable delicate organisms of the existence of which we would otherwise have been, in all probability, totally ignorant, in 1861 revealed the impression of a feather to the famous palaeontologist Hermann von Meyer. Von Meyer named the unknown bird to which this feather belonged *Archaeopteryx lithographica;* and in the same year the independent discovery by Dr. Haberlein of the precious skeleton of *Archaeopteryx* itself, which now adorns the British Museum, demonstrated the chief characters of this very early bird. But it must be remembered that this feather and this imperfect skeleton are the sole remains of birds which have yet been obtained in all that great series of formations known as Wealden and Oolite, which partly lie above, and partly correspond with, the Solenhofen slates.

Though some palaeontologists may be forced, by a sense of consistency, to declare that the class of birds was created in the sole person of *Archaeopteryx* during the deposition of the Solenhofen slates and disappeared during the Wealden, to be recreated in the Greensand, to vanish once more during the Cretaceous epoch and reappear in the Tertiaries, I incline to the hypothesis that many birds besides *Archaeopteryx* existed throughout all this period of time, and that we know nothing about them, simply because we do not happen to have hit upon those deposits in which their remains are preserved.

Now, what is this *Archaeopteryx* like? Unfortunately the skull is lost; but the leg and foot, the pelvis, the shoulder-girdle, and the feathers, so far as their structure can be made out, are completely like those of existing ordinary birds.

On the other hand, the tail is very long, and more like that of a reptile than that of a bird in this respect. Two digits of the manus have curved claws, much stronger than those of any existing bird; and, to all appearance, the metacarpal bones are quite free and disunited.

Thus it is a matter of fact that, in certain particulars, the oldest known bird does exhibit a closer approximation to reptilian structure than any modern bird.

Are any fossil reptiles more bird-like than those which now exist?

. . . The *Dinosauria,* a group of extinct reptiles . . . which occur throughout the whole series of the Mesozoic rocks, and are, for the most part, of gigantic size, appear to me to furnish the required conditions. . . .

But a single specimen, obtained from those Solenhofen slates to the accident of whose existence and usefulness in the arts palaeontology is so much indebted, affords a still nearer approximation to the "missing link" between reptiles and birds. This is the singular reptile which has been described and named *Compsognathus longipes* by the late Andreas Wagner, and some of the more recondite ornithic affinities of which have since been pointed out by Gegenbaur. Notwithstanding its small size (it was not much more than 2 feet in length), this reptile must, I think, be placed among, or close to, the *Dinosauria;* but it is still more bird-like than any of the animals which are ordinarily included in that group.

Compsongnathus longipes has a light head, with toothed jaws, supported upon a very long and slender neck. The ilia are prolonged in front of and behind the acetabulum. The pubes seem to have been remarkably long and slender. . . . The fore limb is very small. The bones of the manus are unfortunately shattered; but only four claws are to be found, so that possibly each manus may have had but two clawed digits.

The hind limb is very large, and disposed as in birds. As in the latter class, the femur is shorter than the tibia—a circumstance in which *Compsognathus* is more ornithic than the ordinary *Dinosauria.*

The proximal division of the tarsus is ankylosed with the tibia, as in birds. In the foot the distal tarsals are not united with the three long and slender metatarsals, which answer to the second, third, and fourth toes. Of the fifth toe there is only a rudimentary metatarsal. The hallux is short, and the metatarsal appears to be deficient at its proximal end.

It is impossible to look at the conformation of this strange reptile and to doubt that it hopped or walked, in an erect or semierect position, after the manner of a bird, to which its long neck, slight head, and small anterior limbs must have given it an extraordinary resemblance.

I have now, I hope, redeemed my promise to show that, in past times, birds more like reptiles than any now living, and reptiles more like birds than any now living, did really exist.

But, on the mere doctrine of chances, it would be the height of improbability that the couple of skeletons, each unique of its kind, which have been preserved in those comparatively small beds of Solenhofen slate, which record the life of a fraction of Mesozoic time, should be the relics, the one of the most reptilian of birds, and the other of the most ornithic of reptiles.

And this conclusion acquires a far greater force when we reflect upon that wonderful evidence of the life of the Triassic age which is afforded us by the sandstones of Connecticut. It is true that these have yielded neither feathers nor bones; but the creatures which traversed them when they were the sandy beaches of a quiet sea have left innumerable tracts which are full of instructive suggestion. Many of these tracks are wholly indistinguishable from those of modern birds in form and size; others are gigantic three-toed impressions, like those of the Weald in our own country; others are more like the marks left by existing reptiles or Amphibia.

The important truth which these tracks reveal is, that at the commencement of the Mesozoic epoch bipedal animals existed which had the feet of birds, and walked in the same erect or semierect fashion. These bipeds were either birds or reptiles, or more probably both; and it can hardly be doubted that a lithographic slate of Triassic age would yield birds so much more reptilian than *Archaeopteryx,* and reptiles so much more ornithic than *Compsognathus,* as to obliterate completely the gap which they still leave between reptiles and birds. . . .

In conclusion, I think I have shown cause for the assertion that the facts of palaeontology, so far as birds and reptiles are concerned, are not opposed to the doctrine of evolution, but, on the contrary, are quite such as that doctrine would lead us to expect; for they enable us to form a conception of the manner in which birds may have been evolved from reptiles, and thereby justify us in maintaining the superiority of the hypothesis that have been so originated to all hypotheses which are devoid of an equivalent basis of fact.

From: On the Physical Basis of Life

T. H. Huxley

I have endeavoured, in the first part of this discourse, to give you a conception of the direction towards which modern physiology is tending; and I ask you, what is the difference between the conception of life as the product of a certain disposition of material molecules, and the old notion of an Archaeus governing and directing blind matter within each living body, except this—that here, as elsewhere, matter and law have devoured spirit and spontaneity? And as surely as every future grows out of past and present, so will the physiology of the future gradually extend the realm of matter and law until it is co-extensive with knowledge, with feeling, and with action.

The consciousness of this great truth weighs like a nightmare, I believe, upon many of the best minds of these days. They watch what they conceive to be the progress of materialism, in such fear and powerless anger as a savage feels, when, during an eclipse, the great shadow creeps over the face of the sun. The advancing tide of matter threatens to drown their souls; the tightening grasp of law impedes their freedom; they are alarmed lest man's moral nature be debased by the increase of his wisdom.

If the "New Philosophy" be worthy of the reprobation with which it is visited, I confess their fears seem to me to be well founded. While, on the contrary, could David Hume be consulted, I think he would smile at their perplexities, and chide them for doing even as the heathen, and falling down in terror before the hideous idols their own hands have raised.

For, after all, what do we know of this terrible "matter," except as a name for the unknown and hypothetical cause of states of our own consciousness? And what do we know of that "spirit" over whose threatened extinction by matter a great lamentation is arising, like that which was heard at the death of Pan, except that it is also a name for an unknown and hypothetical cause, or condition, of states of consciousness? In other words, matter and spirit are but names for the imaginary substrata of groups of natural phaenomena.

And what is the dire necessity and "iron" law under which men groan? Truly, most gratuitously invented bugbears. I suppose if there be an "iron" law, it is that of gravitation; and if there be a physical necessity, it is that a stone, unsupported, must fall to the ground. But what is all we really know, and can know, about the latter phaenomena? Simply, that, in all human experience, stones have fallen to the ground under these conditions; that we have not the smallest reason for believing that any stone so circumstanced will not fall to the ground; and that we have, on the contrary, every reason to believe that it will so fall. It is very convenient to indicate that all the conditions of belief have been fulfilled in this case, by calling the statement that unsupported stones will fall to the ground, "a law of Nature." But when, as commonly happens, we change *will* into *must,* we introduce an idea of ne-

cessity which most assuredly does not lie in the observed facts, and has no war-
ranty that I can discover elsewhere. For my part, I utterly repudiate and anathema-
tise the intruder. Fact I know; and Law I know; but what is this Necessity, save an
empty shadow of my own mind's throwing?

But, if it is certain that we can have no knowledge of the nature of either
matter or spirit, and that the notion of necessity is something illegitimately thrust
into the perfectly legitimate conception of law, the materialistic position that there
is nothing in the world but matter, force, and necessity, is as utterly devoid of justi-
fication as the most baseless of theological dogmas. The fundamental doctrines of
materialism, like those of spiritualism, and most other "isms," lie outside "the limits of philosophical inquiry," and David Hume's great service to humanity is his ir-
refragable demonstration of what these limits are. Hume called himself a sceptic,
and therefore others cannot be blamed if they apply the same title to him; but that
does not alter the fact that the name, with its existing implications, does him gross
injustice.

If a man asks me what the politics of the inhabitants of the moon are, and I
reply that I do not know; that neither I, nor any one else, has any means of knowing; and that, under these circumstances, I decline to trouble myself about the subject at all, I do not think he has any right to call me a sceptic. On the contrary, in
replying thus, I conceive that I am simply honest and truthful, and show a proper
regard for the economy of time. So Hume's strong and subtle intellect takes up a
great many problems about which we are naturally curious, and shows us that they
are essentially questions of lunar politics, in their essence incapable of being answered, and therefore not worth the attention of men who have work to do in the
world. And he thus ends one of his essays—

"If we take in hand any volume of Divinity, or school metaphysics, for instance, let us ask, *Does it contain any abstract reasoning concerning quantity or number? No. Does it contain any experimental reasoning concerning matter of fact and existence? No.* Commit it then to the flames; for it can contain nothing but sophistry
and illusion."

Permit me to enforce this most wise advice. Why trouble ourselves about
matters of which, however important they may be, we do know nothing, and can
know nothing? We live in a world which is full of misery and ignorance, and the
plain duty of each and all of us is to try to make the little corner he can influence
somewhat less miserable and somewhat less ignorant than it was before he entered
it. To do this effectually it is necessary to be fully possessed of only two beliefs: the
first, that the order of Nature is ascertainable by our faculties to an extent which is
practically unlimited; the second, that our volition counts for something as a condition of the course of events.

Each of these beliefs can be verified experimentally, as often as we like to
try. Each, therefore, stands upon the strongest foundation upon which any belief
can rest, and forms one of our highest truths. If we find that the ascertainment
of the order of nature is facilitated by using one terminology, or one set of symbols, rather than another, it is our clear duty to use the former; and no harm can
accrue, so long as we bear in mind, that we are dealing merely with terms and
symbols.

In itself it is of little moment whether we express the phaenomena of matter in terms of spirit; or the phaenomena of spirit in terms of matter: matter may be regarded as a form of thought, thought may be regarded as a property of matter—each statement has a certain relative truth. But with a view to the progress of science, the materialistic terminology is in every way to be preferred. For it connects thought with the other phaenomena of the universe, and suggests inquiry into the nature of those physical conditions, or concomitants of thought, which are more or less accessible to us, and a knowledge of which may, in future, help us to exercise the same kind of control over the world of thought, as we already possess in respect of the material world; whereas, the alternative, or spiritualistic, terminology is utterly barren, and leads to nothing but obscurity and confusion of ideas.

Thus there can be little doubt, that the further science advances, the more extensively and consistently will all the phaenomena of Nature be represented by materialistic formulae and symbols.

But the man of science, who, forgetting the limits of philosophical inquiry, slides from these formulae and symbols into what is commonly understood by materialism, seems to me to place himself on a level with the mathematician, who should mistake the x's and y's with which he works his problems, for real entities—and with this further disadvantage, as compared with the mathematician, that the blunders of the latter are of no practical consequence, while the errors of systematic materialism may paralyse the energies and destroy the beauty of a life.

Chapter 4

To get an appreciation of just how nasty the Cope-Marsh feud got to be, I include a chapter from Url Lanham's book The Bone Hunters. *This chapter shows how Cope, bitter and nearly broke after his investments had failed disastrously, and burning with resentment at Marsh's success, took his feud into the public arena. With the help of a sympathetic newspaper reporter, he blasted Marsh and the U.S. Geological Survey. Marsh retaliated, and the reading public was titillated by a sordid display of battling scientists. Cope got his revenge, but at the cost of besmirching his own reputation and diminishing his place in history. Far worse than the damage done to the reputation of either man was the bad effect on science. As I note in Chapter 4, the malicious rivalry led Marsh to rush into print without taking the proper care and exercising the caution absolutely necessary in paleontology. One result was that, on Marsh's authority,* Apatosaurus *was given the wrong head for many years.*

The patient and careful work of people like Earl Douglass, people more dedicated to science than to stroking their distended egos, was hampered by Marsh's poor judgment. Included here is a letter from Douglass to W. J. Holland telling of his great disappointment when, after vast labors of excavation, no skull was found at the end of the neck of Apatosaurus. *This is a moving testimonial to the fact that science can be a harsh taskmaster, dispensing disappointment along with reward. Douglass did find a skull associated with the skeleton, which he suspected was actually the skull of* Apatosaurus, *but he could not be sure because it was not found at the end of the neck. Marsh, unwilling to wait to get proper evidence, just stuck on an unassociated skull . . . which happened to be dead wrong.*

[Cope's] Revenge

(*From* The Bone Hunters: The Heroic Age of
Paleontology in the American West)

Url Lanham

After his failure as a mining promoter, Cope tried everywhere, and without success, to find a job. In a letter to his wife he says, "I wrote to Prof. Baird about securing a place made vacant by Mr. Stejneger in the Smithsonian, but I found the place did not exist," and in 1888, "The application I made for a position at the N.Y. American Museum of Natl. History has also been declined." He even considered a post as president of the new Stanford University, and tried to get a position as paleontologist for the Geological Survey, even though Marsh was already a member of the Survey. He gave lectures of all sorts for hire, including religious and philosophical discourses, and wrote popular articles for magazines.

Cope made plans and negotiated, without success, to set up a new science institution which would give employment to himself. He propagandized for one such establishment in the pages of the *American Naturalist:* "We have colleges and universities enough in most of the States, but there has not yet been established a single school where knowledge is produced, which corresponds in scope with the numerous institutions where it is taught." He thought that an endowment of a million dollars would be enough to support the six departments needed to cover all fields of knowledge. To keep out administrators, he said that the charter of this precursor of the modern graduate school should require "that the position of director should be forfeited by that one who should not produce some original work of merit every year or two."

Year after year Cope was in Washington to lobby Congressmen for an appropriation that would pay him to write a concluding volume to his "Bible." At times he seemed close to success, but he failed. In 1886, he wrote to his wife, "*My bill was thrown out* on a technicality, no doubt in the interest of Powell. . . . We will have to live very cheaply for a year." But that poverty was to some extent relative for Cope is shown by this comment, in a letter of 1887, "If I get the appropriation I will not have to sell the Labradorite [a beautiful and valuable stone] table this year at least."

During the latter part of the 1880s the Geological Survey was periodically under fire from Congress, mainly because of the socialistic leanings of Powell. Since farming in the arid region, where precipitation amounted to only fifteen inches or less per year, was impossible without irrigation, Powell said that settlers should band together in small groups to develop irrigation facilities for common use, somewhat in the manner that the Mormons had settled the irrigable lands of Utah. In his classic "Report on the Lands of the Arid region of the United States," Powell says of the lands suited for grazing:

> The great areas over which stock must roam to obtain subsistence usually prevents the practicability of fencing the lands. It will not pay to fence the pasturage fields, hence in many cases the lands must be occupied by herds roaming in common; for poor men cooperative pasturage

is necessary, or communal regulations for the occupancy of the ground and for the division of the increase of the herds. Such communal regulations have already been devised in many parts of the country.

Powell was an admirer of the pioneer American anthropologist Lewis Morgan, who furnished some of the underpinnings for the work of Marx and Engels, and said in a lecture before the Anthropological Society, "individualism is transmitted into socialism, egoism into altruism, and man is lifted above the brute to an immeasurable height."

Although Cope on occasion made halfhearted attempts to get assistance from Powell, in general he thought that his best chance for getting Congress to give him money was to inflame prejudice against Powell, and to discredit the Survey. In 1885, a year when the situation was especially critical for Powell, Cope, with the help of his mining engineer F. M. Endlich, and a friend, Persifor Frazer, got together a lengthy attack on Powell and Marsh which was circulated privately among the members of Congress. Cope, however, did not succeed in bringing down Powell, who successfully defended his position until 1892. The sapping operations conducted by Cope may have weakened his defense, but the proximate causes of the downfall of Powell and Marsh involved factors much more fundamental than the squabble between Cope and the federal geological establishment.

In 1885 Cope had begun seriously to gather ammunition to destroy Marsh. This he did by cultivating Marsh's laboratory assistants in the hope that he could turn them against their employer.

The large staff that Marsh had assembled included (besides the field collectors, technicians who prepared the bones, and artists who drew them) a number of assistants who in fact, if not in practice, had the status of academic colleagues entitled to publish their own papers in vertebrate paleontology and to rank as co-author with Marsh on some of the papers that came from the Yale laboratory. The first of these hired intellectuals, and the most indispensable to Marsh, was Oscar Harger, an undersized, scholarly, semi-invalid with an enlarged heart. Harger had worked his way through Yale College to a bachelor's degree in mathematics in 1868 by doing statistics for an insurance company. Poor, and lacking in drive, he could not progress to independent standing, and fell under the domination of Marsh, who owned him from 1870 until his death, at forty-four, in 1887.

Harger was Marsh's eyes and brain. Marsh, after his student days, read little in vertebrate paleontology; Harger read everything. Harger could relate the facts discovered in the laboratory to the contemporary world of paleontological research, and he provided sound judgment that helped Marsh to convert the work of the laboratory into the standard scientific commodity, publication. In his gentle way, Harger yearned to be a part of the scientific community by writing and publishing on his own or with Marsh as co-author, but Marsh was adamant in keeping him an invisible subordinate. Two of the Yale professors begged Marsh to give Harger scientific freedom, and even the president of Yale, Timothy Dwight, knew the case of Harger, whom he called a "kindly and gentle spirit," but Marsh was unmoved.

Samuel Williston came to the Yale laboratory from Kansas in 1876. A strong and independent spirit who achieved careers in medicine, entomology, and administration, as well as paleontology, he got what he could from Marsh by way of training and money, then, about the time of his friend Harger's death, criticized Marsh in language that fairly smoked with rage, and left Marsh's laboratory to become a staff member of the Yale Medical School. He was nine years on the Marsh payroll, during which time he got both an M.D. and a Ph.D. He eventually became professor of geology at the universities of Kansas and Chicago.

When Marsh became vertebrate paleontologist for the federal government in 1882, he got funds to expand his laboratory. Remembering the Prussian laboratories of his student days in Europe, where a single professor ruled as undisputed autocrat over assistants who loved to be subservient to capricious and arbitrary authority, Marsh imported three laboratory assistants from Germany—Max Schlosser, Otto Meyer, and Georg Baur. Marsh bungled the job. Two of the three left him soon after their arrival, and the other, Georg Baur, was a monumentally bad choice for an assistant, since he came from a long line of aristocratic German professors.

Baur negotiated an agreement with Marsh to be allowed to publish under his own name, and during the six years he was at Yale published seventy-five papers. The only hold that Marsh had over him was to keep him in a state of carefully modulated poverty. Like a coal miner in a company town, Baur was continually in debt to his employer, and it took a good deal of maneuvering for him to get clear of Marsh and find another job. One of his final publications based on his experiences in the Yale laboratory was an exposé of Marsh published in the *American Naturalist*. Convinced of his own superiority over Marsh, he tore the reputation of his one-time employer to shreds in a paper that is perhaps unique in American scientific literature. Later, while paleontologist at the University of Chicago, Baur's health collapsed. Always unstable, he became insane and died in a German asylum at the age of forty-four.

To join this love feast there came, in the mid-1880s, Edward Drinker Cope. He lurked about the laboratory at New Haven, taking the assistants aside for confidential chats, and trying to get a look at the fossils that Marsh was working on, to the point where Marsh had to forbid visitors access to the laboratory. Cope wrote his wife that "I want to stop over a day in New Haven and see Dr. Baur"; to Osborn that "I spent a day at New Haven, recently, with Baur and Williston"; and he consorted with Baur at scientific meetings.

By 1885 Cope had made good progress in bringing disaffection in the Marsh laboratory to a head. He was particularly concerned about the problem of getting Baur out of hock, and wrote to Osborn:

> The 4 men at Yale are anxious to publish in both Europe and America their statement, and I naturally am willing that they should. There is however one difficulty. Dr. Baur who is as important as any of the four & furnishes a good deal of the backbone, is in Marsh's debt some $650 and he can do nothing till that is paid as he is a married man. I want to find someone to lend him the money or part of it, because if it is paid out of his wages, he cannot pay it till next April, which will be too late for the paper to be of any service in the investigation. Although I am entirely impecunious, I have taken upon me to try & raise $200 of this amount.

Cope asked Osborn to contribute some money, pointing out that "The publication of the paper will be a blessing to American Science generally, for the demoralizing effect of Marsh's success is incalculable. . . . He has completely suppressed my work." Cope sums up the results of his researches into the operation of Marsh's laboratory by saying that Marsh is "more of a pretender than even I had supposed him to be. . . . It is now clear to me that Marsh is simply a scientific-political adventurer who has succeeded, in ways other than those proceeding from scientific merit, in placing himself in the leading scientific position in the country." The modern scientist will recognize this as a rather judicious portrait of some of his own scientific competitors.

Probably Cope made use of the material from his scientific espionage when he circularized Congress with the scandal sheet on Marsh and Powell in 1887, but it did not come to the surface until the winter of 1889–90. That fall Cope got a small foothold in the domain of security with an appointment as professor of geology and mineralogy at the University of Pennsylvania. However, this gave him little more than a bare living—there was little for field work, research, and publication in the style to which he was accustomed.

The only large financial asset remaining, after he had sold his mining stocks for what they would bring, was his collection of fossils, into which he had poured nearly $80,000 of his own money, besides increasing their value with years of work getting them into the literature, and making types of many of them. As he wrote his wife, "My entire future in a financial sense . . . depends on that collection so far as I can see now."

Cope was stunned to receive, in December of 1889, a demand from the Secretary of the Interior, John W. Noble, that he turn over to the United States National Museum (and thus to Marsh) his entire collection of vertebrate fossils from the Tertiary and Cretaceous formations of the West, on the grounds that they were government property, since Cope had been a member of the Hayden Survey. The Secretary acted on the recommendation of Powell, who doubtless was egged on by Marsh. Powell was fully aware of the way in which Cope built up his collection, and of the fact that Cope had been an unpaid "volunteer" for the Survey. The move was pure malice, perhaps made with the knowledge that Cope would be able to defend himself against the manifestly unfair claim, but only after providing a good deal of fun for his tormentors.

Cope reacted promptly, although the revenge he exacted may have been something short of what he first had in mind, since he wrote to Osborn, "When a wrong is to be righted, the press is the best and most Christian medium of doing it. It replaces the old time shot gun & bludgeon & is a great improvement." Cope decided to give the scandalous information he had accumulated about Marsh and about Powell's Survey over the past twenty-eight years to the press. He had for all this time kept a drawer of what he called "Marshiana"; and since Marsh immediately produced a return barrage of equally slanderous tales about Cope, Marsh must have kept ready for use a store of "Copeiana," most likely kept locked in a steel safe.

The instrument for Cope's revenge was to be a young journalist for the New York *Herald* named William Hosea Ballou, whom he had known for several years. When Cope laid out his project before Ballou, the journalist must have thought the squabble had great possibilities for publicity, since Marsh was well known, and criticizing the Powell Survey was always good copy. When he saw Cope's article, he must also have seen that the attack was very ill advised, since Cope took on too many people at once: Marsh, a majority of the National Academy of Sciences, and the entire United States Geological Survey. But he apparently did nothing to dissuade Cope, and before publication took the article to both Marsh and Powell in the hopes of setting a lively dispute going from the very beginning.

Cope's article appeared January 12, 1890, prefaced by a fake interview scene contrived by Ballou.

> I saw Professor Cope at his residence, No. 2102 Pine Street, Philadelphia, a few days ago. He was engaged in making new species of new fossil animals but he cheerfully put aside his work when the nature of my call was made known to him. He is a man not over forty-five years of age and of distinguished appearance.
>
> "What is the origin, Professor, of this war against the Geological Survey?" I asked.
>
> "It may be found in the outrageous order I have received from the Secretary of the Interior to turn over my collections to the National Museum at Washington. I have not more than a bushel of specimens belonging to the government and to those it is welcome. The fact is, I sent out my own collecting parties and secured my collection at an expense of about $80,000 of my own money, to say nothing of the value of all the time I have expended upon them."
>
> "Who is the author of the order?"
>
> "Why to be sure, who but Major Powell. The object of this absurd order to place my collections in the National Museum is to gain control of them, so that my work may be postponed until it has been done by Professor Marsh of Yale College, and this in spite of the fact that the preliminary work has been already published by me, and that the truth is sure to come out at some future time."

In the article Cope recited an extract from his long list of Marsh's technical errors, such as calling a dinosaur bone a buffalo horn, along with more serious but more esoteric errors which could not have much impressed the lay reader. He pointed out mistakes made by the amateur and self-taught geologist Powell. He accused Powell of packing the Academy with supporters of the Survey, and of buying influence by hiring Congressmen's sons and college professors in the Survey. He claimed that Marsh's scientific writings were stolen from other authors without acknowledgment, and those not stolen were written by his hired assistants. Cope then presented some of the testimony gathered from Marsh's employees. Among them was a letter written to him by Williston in 1886, which he had then predicted would cause "great execution" when used at the proper time:

> I wait with patience the light that will surely be shed over Professor Marsh and his work. Is it possible for a man whom all his colleagues call a liar to retain a general reputation for veracity! . . . I do not worry about his ultimate position in science. He will find his level, possibly fall below it. There is one thing I have always felt was a burning disgrace—that such a man should be chosen to the

highest position in science as the president of the National Academy of Science while men of the deepest erudition and unspotted reputation are passed by unnoticed. Professor Marsh did once indirectly request me to destroy Kansas fossils rather than let them fall into your hands. It is necessary for me to say that I only despised him for it.

The assertion of Professor Marsh that he devotes his entire time to the preparation of his reports is so supremely absurd, or rather so supremely untrue, that it can only produce an audible smile from his most devoted admirers. I have known him intimately for ten years. During most of the time while in his employ I never knew him to do two consecutive, honest days' work in science, nor am I exaggerating when I say that he has not averaged more than one hour's work per day. He is absent from the Museum fully half of the time, and when in New Haven he rarely appears at the museum till two o'clock or later and stays but an hour or two, devoting his time chiefly to the most absurd details and old maid crotchets. The larger part of the papers published since my connection with him in 1878 have been either the work or the actual language of his assistants. At least I can positively assert that papers have been published on Dinosaurs which were chiefly written by me. . . .

Professor Marsh's reputation for veracity among his colleagues is very slight. In fact he has none. . . . Those who know him best say—and I concur in the opinion—that he has never been known to tell the truth when a falsehood would serve the purpose as well. Those are strong statements to make of one holding such a position as he does, but I state them the more freely from the fact that everybody here [Yale College] concurs in them. He has no friends here save those who do not know him well.

When Marsh saw the advance copy of the newspaper article, he confronted Williston, and extracted from him a modified disclaimer, which only allowed that it was published without Williston's permission, not that the statement was false. As the newspaper said, "Next Professor Williston was mysteriously moved and wrote to the *Herald* that his letters concerning Professor Marsh were mostly written some years ago, under exasperating circumstances, before he had become connected with Yale College" [as a professor of anatomy].

Georg Baur also took part in the onslaught. Cope published a series of articles by Marsh's assistants in the *American Naturalist,* which were then bound into a pamphlet designed to serve as a companion volume to the newspaper articles. Baur's contribution was published in March, and gave an overview of the whole dispute, being entitled "A review of the charges against the paleontological department of the U.S. Geological Survey, and of the defense made by Prof O. C. Marsh." Following is an excerpt.

I will now give a short review of the charges made against Professor Marsh, and of his defence, based on an experience of nearly six years, during which I was an assistant of Prof. Marsh, paid by the U.S. Geological Survey.

    1.   In the New York *Herald* of January 12th, Prof. E. D. Cope, of the University of Pennsylvania, stated, "The collections made by Prof. Marsh, as the vertebrate paleontologist of the Geological Survey . . . are all stored at Yale College, with no assured record as to what belongs to the Government and what to the college."

To this Professor Marsh replied that "every specimen belonging to the government is kept by itself, and no mixing with the Yale Museum collections is possible." Prof. H. F. Osborn and Dr. O. Meyer have sustained this fully, and I am glad to say that great care is taken at the Yale Museum in this regard. But this is irrelevant to the question raised by Prof. Cope, for, of course, the labeling is entirely in the hands of Prof. Marsh, without any control from the Geological Survey. In this connection there is one thing that I can not quite understand; how it is that the splendid specimens of horned dinosaurs became the property of Prof. Marsh, and not of the government. Can Prof. Marsh pay his collectors this month out of his own pocket, and the following out of the pocket of the government?

    2.   The next statement made in the *Herald* is, that these collections "are locked away from the people, and no one is allowed to see them, not even visiting scientists." This Prof. Marsh admits is in part true.

He says, that "visiting scientists of good moral character are always welcome." Now I may mention, that a scientist of "good moral character," well known in this country and in Europe, wanted to see the material of the Dinocerata shortly after the volume on this order had been published. When he arrived at New Haven he was told by Prof Marsh that he was sorry not to be able to show him the material, since it had been boxed up lately and was inaccessible. The fact is, that the whole material was spread on a large table in the room where the conversation took place. By the order of the professor the fossils had been covered up with cloth the day before.

To the dispute as to how much Marsh relied on the work of his assistants, Baur addressed these comments:

The fact is that a great part of the descriptive and general part of most of Prof. Marsh's papers is the work of his assistants. Prof. Marsh asks them questions, the answers of which he either immediately puts down in black and white, or he makes out a list of questions to be worked out by his assistants, for instance: "What are the principal characters of the skull of the Sauropoda?" or, "What are the relations between the different groups of Dinosaurs?" and so on. The assistant, if not yet fully familiar with these questions, begins to work; he goes over the whole literature, a thing rarely done by the Professor, and studies the specimens in the collection. After this is done, the Professor receives the notes of the assistant, or he asks questions, writing down the answers he receives. In this way he accumulates a great quantity of notes, written in his own handwriting, or in that of the assistant. By comparing and using these notes it is easy for him to dictate a paper to any person who can write. This person, of course, when asked, can testify that the work was dictated by Prof. Marsh, without telling a falsehood.

Baur gave a vivid picture of the master at work on really important questions. Marsh had finished the main body of the work on the Dinocerata monograph, and wanted to write a conclusion that placed the whole subject in perspective. He invited Baur to his home on a Sunday for a conference. After various pointed questions, which settled some important matters, Marsh asked him how the ungulates, the group to which the Dinocerata belonged, were to be classified. Baur gave him a classification that had been proposed by Cope. It became clear that Marsh had never taken the trouble to read Cope's paper, but was happy enough to use the classification, without credit, although he changed Cope's names to those of his own liking.

The primary function of a scientific leader like Marsh is to exercise sound judgment as to where to borrow from others (and whether or not to acknowledge it); that is, he has to know how to recognize a good idea, rather than create it. The creation of ideas is generally left to the solitary thinker. Scholarly brooding is utterly foreign to the scientific entrepreneur, who has to move quickly and aggressively in order to prosper in the scientific market place. Marsh's sound judgment on scientific matters was his most valuable trait.

Baur described a pathetic scene in the Marsh laboratory. Marsh had written a laudatory review of his own book on the Dinocerata and asked Harger, then Williston, to sign it. Both refused, and Marsh had to be content with the initials of the "lady typewriter."

Marsh and Powell were given space in the January 12 issue of the *Herald* to make a preliminary counterattack, then Marsh let loose the full barrage a week later, taking up an entire page of the newspaper. He began:

> The author of the recent attack upon me and my work is Professor E. D. Cope, and he has at last placed publicly on record the slanders he has secretly been repeating for years. Whether he makes the statements directly or conceals them in the form of an interview with himself or others they are his own. He has devoted some of his best years to its preparation and to the preparation of the public for it, and it may thus be regarded as the crowning work of his life.

Now was the chance for Marsh to publish what must have been one of his favorite stories, here given a headline by Ballou:

Professor Cope had described an extinct reptile from the cretaceous of Kansas, under the name *Elasmosaurus,* and his published description made this the most remarkable animal of ancient or modern times. Besides his original description, he had read a paper on the subject before the American Association, in which he explained the marvelous creature. In a communication published by the Boston Society of Natural History in 1869, he placed it in a new order, *Streptosauria.* In the *American Naturalist* he gave a restoration of the animal, represented as alive. This was afterward copied in *Appleton's Cyclopedia.* Finally, an extensive description was published in the *Transactions of the American Philosophical Society,* Vol. XIV, in which a full restoration of this wonderful creation was given.

The skeleton itself was arranged in the Museum of the Philadelphia Academy of Natural Sciences, according to this restoration, and when Professor Cope showed it to me and explained its peculiarities I noticed that the articulations of the vertebrae were reversed and suggested to him gently that he had the whole thing wrong end foremost. His indignation was very great, and he asserted in strong language that he had studied the animal for many months and ought at least to know one end from the other.

It seems he did not, for Professor Leidy in his quiet way took the last vertebrae of the tail, as Cope had placed it, and found it to be the atlas and axis, with the occipital condyle of the skull in position. This single observation of America's most distinguished comparative anatomist, whom Cope has wronged grievously in name and fame, was a demonstration that could not be questioned, and when I informed Professor Cope of it his wounded vanity received a shock from which it has never recovered, and he has since been my bitter enemy. Professor Cope had actually placed the head on the end of the tail in all his restorations, but now his new order was not only extinct, but extinguished.

Actually, Cope had turned *Elasmosaurus* right end to twenty years before, in the *Transactions of the American Philosophical Society,* where he wrote,

The determination of the extremities of this species was rendered difficult from the fact that Leidy in his descriptions of Cimoliasaurus, reverses the relations of the vertebrae, viewing the cervicals as caudals and lumbars, and describing the caudals as belonging to another genus. Not suspecting this error, I arranged the skeleton of *Elasmosaurus* with the same relation of the extremities, and the more willingly as the distal cervicals present an extraordinary attenuation, even for this type, and also as the discoverer assured me that the fragments of the cranium were found at the extremity which is properly caudal. Viewed in this light many details of the structure were the reverse of those ordinarily observed among reptiles, whence I was induced to consider it as the type of a peculiar group of high rank. This view is, of course, abandoned on a correct interpretation of the extremities. Leidy detected the error in this arrangement, and the correction extends to Cimoliasaurus as well.

Cope had charged that Marsh's fame as the discoverer of the true evolution of the horse was won by plagiarizing from himself and Kovalevsky, the brilliant Russian paleontologist. Marsh countered with the observation that Kovalevsky was best known for being as great a thief of other people's specimens as Cope, and that "Kovalevsky was at last stricken with remorse and ended his unfortunate career by blowing out his own brains. Cope still lives, unrepentant."

Powell, like Marsh replied with carefully measured prose, and offered this free analysis of Cope's character:

> I am not willing to be betrayed into any statement which will do injustice to Professor Cope. He is the only one of the coterie who has scientific standing. The others are simply his tools and act on his inspiration. The Professor himself has done much valuable work for science. He has made great collections in the field and has described these collections with skill. Altogether he is a fair systematist. If his infirmities of character could be corrected by advancing age, if he could be made to realize that the enemy which he sees forever haunting him as a ghost is himself . . . he could yet do great work for science.

The controversy went on through several editions of the *Herald,* with most of the gunfire coming from the Marsh/Powell camp. Cope tried to wind up the affair with

> The recklessness of assertion, the erroneousness of statement and the incapacity of comprehending our relative positions on the part of Professor Marsh render further discussion of the trivial matters upon which we disagree unnecessary, and my time is too fully occupied on more important subjects to permit me to waste it upon personal affairs which are already sufficiently before the public. Professor Marsh has recorded his views *aere perennius,* and may continue to do so without personal notice by E. D. Cope.

Marsh had the final say, with a comment to the effect that like the boy who twisted the mule's tail, Cope now looked worse but knew more. However, the public image that Marsh had fashioned for himself also was considerably less handsome than when it had emerged from the hands of its creator.

While the newspaper war was going on, Marsh tried to get at Cope through the provost of the University of Pennsylvania, where Cope held his new professorship. (Cope said that the administration tried to fire him.) The provost was at the time involved in a case of blackmail that had been successfully kept from the public. The paleontologist W. B. Scott in his memoirs writes that "Marsh wrote to the Provost, demanding that he silence Cope, on pain of having his own scandals aired. The *Herald,* getting wind of this, let the Provost know that he would have cause to regret any attempt to interfere with Cope's freedom of speech."

The year 1890 was a drought year, and Congress brought pressure on the Geological Survey to produce immediately plans for reservoirs and irrigation projects. Powell, however, thought that more basic research had to be done, and in spite of the unfavorable publicity brought the Survey by Cope's newspaper campaign, successfully warded off attacks that were designed to fragment the Survey and get the scientific jurisdiction of the public domain out of his hands. Then in 1892 came a financial recession, and the public clamor for the mythical homestead sites that were supposed to exist in the arid West brought renewed pressure to open these lands, and this time Marsh proved to be the weak link in Powell's armor.

In 1892 a Congressman from Alabama, Hilary A. Herbert, who had a perfect record for voting against every bill designed to protect the public domain, proposed in Congress that the funds for paleontology be cut from the appropriation to the United States Geological Survey. He had been influenced by conversations with Professor Alexander Agassiz of Harvard (son of Louis Agassiz, founder of the Harvard Museum of Comparative Zoology), who besides being a scholar was a wealthy mining engineer with holdings in railways and copper mines. Agassiz thought that the federal government should have nothing to do with basic research in such scholarly subjects as paleontology, that this could be left to private funding and to college professors who would do research for nothing to enhance their own prestige.

Agassiz gave Herbert plenty of ammunition for his attack on Powell, and also in some way Herbert became aware of Marsh's huge, elegant, tinted-paper, gilt-edged monograph on Odontornithes, the toothed birds, which had been published as a report of the King Survey of the 40th parallel, and brought it before Congress. Ignoring the fact (which had been made known to Herbert) that Marsh had paid for printing the deluxe volume with his own money, and had paid for the collecting and research that went into it, Herbert made "birds with teeth" into a catchword for a war on what was called the waste of government funds in research.

The strange alliance of Cope, reactionary Congressmen, and Professor Agassiz made Congress again go through the gamut of charges against Marsh and Powell that Cope had been making for the past ten years. This time Congress was able to move quickly and with effect. It was Marsh's turn to be stunned with bad news from Washington. On July 20, 1892, he got a telegram from Powell which read "Appropriation cut off. Please send your resignation at once."

Letter from Earl Douglass to Dr. Holland

Jensen, Utah, Nov. 10, 1909

My dear Dr. Holland,
I know you will be anxious to learn all you can about the Dinosaurs here, on your return to America, so I will try to have a letter there on your arrival.

Though it takes constant effort to get some things that are needed here the work has been progressing rather more rapidly than I expected. There has been an almost unaccountable scarcity of men here yet I was fortunate in getting all the men I needed and just the men I want—men who are resourceful and ever ready to do their part.

Plaster is very expensive here when we have had to send away for it. It costs 6 [cents] or 7 [cents] per lb. A man here however undertook to make some for us. He made nearly a half a ton and it works splendidly. Am sorry that I did not get him to make about two tons. Shall get him to make a lot more if he has not gone away and will make it. He charges me 3 1/2 [cents]. You will conclude, before I am through that it will take lots of material as well as much labor.

I have used up all the alcohol in Vernal. Have ordered 25 gal from Grand Junction which ought to have been here long ago. Am making every effort to hurry it up. I had to send two men and teams over 50 miles for lumber. They got nearly 2000 feet but begin to see that we probably will not have nearly enough as the Dinosaur skeletons are increasing in numbers. I could not get a hundred feet of lumber in Vernal.

The bones of the best skeletons are in a gray sandstone and, a little distance from the surface are splendidly preserved. In some places the sandstone is very hard but back of the rock that contains the bones—between that and the hard sandstone beds—there is usually, so far, a softer stratum that enables us to get back of (geologically below) the bones.

The beds dip at an angle of nearly 60° to the south so excavating to uncover the bones is more nearly vertical than horizontal. This dip to the south with the high sandstone ledge on the north makes almost the best possible conditions for excavating at this time of the year. We haven't suffered with the cold a day, and even now its pretty hot in the middle of the day when the sun shines. The skeletons lie in such a position too that they could easily be enclosed and one could work in the roughest weather. We went within less than a half mile of the place when you and I went from the dredge to Murray's ranch. You remember the fine exposures of rock on our right Green River being on our left.

Now as to the skeletons themselves. Sometimes conditions are such that only one or two can work at the principal diggings and I send the extra men to two other places in the same beds. Only a few rods away is the skeleton of a comparatively small Dinosaur, apparently with short laterally compressed teeth with finely serrate edges. Some of the teeth seem to be long curved and conical. The bones of this skeleton are not scattered, but those of the neck are considerably broken and flattened.

About 20 rods from here I found a foot. Apparently with nearly all the metatarsals and phalanges including three ungual phalanges. One bone which I thought was a metacarpal is about 30 in. long and 15 inches across the upper end but flattened and concave-convex. But in uncovering the foot we came to the femur, and, apparently, the tibia and fibula lying parallel so we have them all in one block. But the femur is only four feet long so the supposed metacarpal would hardly be that bone of that animal. In uncovering the hind limb we struck a pelvis, two vertebrae &c. so the omnicient [sic] only knows how much there is of that animal.

But I've just got to the good part of the story and your [sic] tired, and so am I for I've been working to day. But the nature of the remainder of the story is such as to have a tendency to cost a fellow who is interested in Dinosaurs.

Well we went down on the big Brontosaurus until we came to four sternal ribs and true ribs pointing in a direction which indicated that the back bone was bent like a bow, so that the spines of the anterior portion of the back might point nearly vertically downward. We found that we must not go down any farther until we got the upper portions out. We then began to work in back to the left of the pelvis to get below the thing and discovered a series of vertebrae running to the left (west) and upward. They looked like cervicals and we concluded that the beast had thrown his neck and head back over his back so instead of going down any-where from ten to thirty feet for the neck and digging for months in a state of sus-pense for the skull, here it was right before us and if the skull had been buried there and hadn't come to the surface and been weathered out we would soon have it. Well we followed it with beating hearts—the neck I mean, and it turned down-ward a little. I could almost see the skull I was so sure of it for was there not a se-ries of 8 cervicals undisturbed and in natural position, but when I had come to what I thought was the third of fourth cervical I dug ahead and there were no more. Apparently it had gone the way of all Brontosaur skulls. How disappointing and sickening. We thought as one of my men said, "If there isn't a skull there there aint a dam bit of use of looking for one anywhere."

Afterward I dug down and discovered that there was another series of verte-brae turned bottom side up and going right nearly parallel with the first series. We uncovered a series of 7 of the cervical ribs with their posterior processes pointing westward showing that the neck was going toward the head to the right and to-ward the pelvis that had been uncovered.

In the mean time we had found a little Dinosaur to the left of the pelvis. When we had secured the tail, pelvis, femur, tibia, sacrum &c of this (in one block) we saw the rib of another cervical making 8 in all and the lower portions all large. So we must have been mistaken about the parallel series. It is yet a puzzle, but I do not intend to let my curiosity lead to endangering the safety of the speci-men. At any rate there is a series of 8 cervicals going right toward and running under the pelvis of Dinosaur no. 1 and it cant be far to the skull if it is in place with reference to the neck, but it may be a long time before we see it as the pelvis, ver-tebrae &c have to be removed.

I am not certain yet, but everything indicates that this neck belongs to an-other individual than No. 1 (the first discovered). It is difficult if not impossible to see how a Dinosaur could have gotten his neck detached from his body &c &c. Then too it lies at a little lower level. It looks as if one lay top of the other. The back line of the little dinosaur is running on perfectly connected as far as we have gone at least 8 or 10 vert. anterior to the sacrum.

But I'm tired of Dinosaurs and I fain would sleep. So good night.

Nov. 11
Earl Douglass.

Chapter 5

The conflict between W. J. Holland and Gustav Tornier over the posture of Diplodocus *is the focus of this chapter. However, Tornier's article is in German and, except for the two brief pas-sages I translated, has to the best of my knowledge never been rendered into English. So in-*

On the Habits and the Pose of the Sauropodous Dinosaurs . . .

Dr. Oliver P. Hay

To most persons the habits of living animals are more interesting than is their anatomy. The same is probably even more true with respect to the extinct animals. However, when it comes to determining the habits of extinct animals, their aquatic or terrestrial habitat, their modes of progression, their bearing on their limbs, their food and their ways of procuring it, their modes of attack and defense against their enemies, their manner of reproduction, etc., we meet with many difficulties.

The Sauropoda, and especially the species of Diplodocus, offer a fine illustration of the difficulties mentioned. Were they aquatic, or terrestrial, or amphibious? Did they affect dry lands, or swamps, or rivers and lakes? Did they eat vegetable food or did they prey on other animals? Did they chew their food or did they bolt it? Did they bring forth living young or did they lay immense eggs? Did they make bold attacks on their enemies or were they timid and cowardly creatures? Did they walk only, or swim only, or did they employ both methods of transporting their huge bodies? If they walked, was it on all four legs or on the hinder ones only? If on all four, did they carry their bodies high above the ground, after the manner of the ox and the horse, or did they carry them low down, like the crocodiles, perhaps dragging their bellies on the ground?

To some of these questions more or less definite answers have been made and accepted; others remain unanswered. It is pretty well agreed that a part of their time was passed in the water; that they could swim readily; that they walked mostly on all fours; that to some extent at least they went about on land; that their food was mainly, if not wholly, vegetable; and that they had imperfect or no means of chewing it.

We are assisted in understanding the habits of those creatures by a knowledge of the nature of their environment. And this we must determine from the character of the deposits in which their bones are discovered and from the kinds of animals and plants accompanying them. Investigation has shown that their remains occur in sandstones and clays which were certainly laid down in fresh waters having no great amount of motion. The accompanying animals are other dinosaurs, some herbivorous, others carnivorous; besides crocodiles, turtles, freshwater fishes and freshwater shells. Some of the plants that occur in the deposits certainly lived in fresh water.

Hatcher has discussed at length the nature of the region in which the species of Diplodocus and their allies lived, as well as the habits of the Sauropoda in general; and the present writer agrees with him on most points. Hatcher believed that the Atlantosaurus beds were deposited, not in an immense freshwater lake, as held by some geologists, but over a comparatively low and level plane, which was occupied by perhaps small lakes connected by an interlacing system of river channels. The climate was warm and the region was overspread by luxuriant forests and broad savannas. The area thus occupied included large parts of the present states of Colorado, New Mexico, Utah, Montana, and the Dakotas. In his memoir on Diplodocus, Hatcher compares the conditions prevailing in that region during the Upper Jurassic to those now found about the mouth of the Amazon and over some of the more elevated plains of western Brazil. In such regions the rivers, fed from distant elevated lands, must have been subject to frequent inundations.The beds of the streams were continually shifting, and there existed numerous abandoned channels that were filled with stagnant water. An animal that lived in such a region would be compelled to adapt itself to a more or less aquatic life, and this adaptation would be reflected to a greater or less extent in the structure of the animal. Marsh had concluded from the position of the eternal nares of *Diplodocus* that it was addicted in some measure to an aquatic existence. The feet too are of rather peculiar structure, the inner toes being strongly clawed, the outer toes greatly reduced; but the meaning of this is differently interpreted.

The Food of Diplodocus

The particular sort of food eaten by the species of Diplodocus is unknown, but nobody doubts that it was vegetable. The teeth were pencil-like in form and they were entirely confined to the front of the jaws. By general consent, they could have been employed only for prehension of food, not at all for its mastication. Hatcher suggested that the teeth might have been useful in detaching from the bottoms and shores the tender and succulent aquatic and semi-aquatic plants that must have grown there in abundance. Osborn says that "the food probably consisted of some very large and nutritious species of water plant. The anterior claws may have been used in uprooting such plants. . . . The plants may have been drawn down the throat in large quantities without mastication." In a restoration of Diplodocus by Mr. Charles W. Knight, the animal is represented as standing on its hind legs and preparing to bite off the terminal bud of a towering cycad. Holland thinks that the teeth were better adapted for raking and tearing off from the rocks soft masses of clinging algae than for securing any other forms of vegetable food now represented in the waters of the world.

To the present writer the suggestion of Dr. Holland has in it more of probability than any of the others presented. If the food-plants sought by Diplodocus had been large and such as required uprooting by the great claws of the reptile the prehension and manipulation of the masses would have been liable to break the slender teeth and would certainly have produced on them perceptible wear. The upper teeth of the original of Marsh's [illustrations of Diplodocus teeth] show no wear, so far as the writer can determine. The mandibular teeth are not well exposed to view.

With respect to Osborn's theory, it is well to take into consideration also the probable ability of the reptile to digest great masses of undivided and unmasticated vegetation. Against the theory suggested by restoration it may be urged that the teeth, pointed or slightly chisel-shaped, are poorly adapted for cropping leaves and great buds; most of all, the teeth leave spaces between them, like the teeth of a great comb, an arrangement not favorable to their functioning as cutting instruments. The teeth could hardly have been used for scraping algae from rocks, either, for that usage would have produced evident and rapid wear. It is more probable that the food consisted of floating algae and of plants that were loosely attached to the bottoms of stagnant bayous and ponds. Hatcher has reported the finding of the seeds and the stems of a species of Chara near the Marsh quarry, where many Sauropoda have been found. This alga, it seems to the writer, would have been admirably adapted to the needs of Diplodocus. It could be easily gathered into the mouth as the reptile swam or crawled lazily about or rested itself and retracted and extended its long neck. The long and highly vaulted palate would have permitted a considerable mass to be collected, out of which, by pressure of the tongue, superfluous water might have been squeezed between the spaced teeth. In addition to various algae there were probably other floating plants.

The Posture of Diplodocus

Marsh presented no restoration of Diplodocus, but he did furnish restorations of Brontosaurus; and he stated that he regarded it as representing the general form and proportions of the Sauropoda. In this figure Brontosaurus is shown as walking with the body high above the ground and with the limbs, especially the hinder ones, about as straight as they are in the elephant.

So far as the bearing of Brontosaurus and Diplodocus on their limbs is concerned, Marsh's example has been almost slavishly followed ever since. No one, so far as the writer knows, has ventured to defend in print a more crocodilian posture. Osborn grants that there is room for wide differences of opinion as regards the habits and means of locomotion of these gigantic animals and states that some hold the opinion that on land at least these reptiles had rather the attitude of the alligator. The same writer says in Nature, vol. 73, 1906, p. 283 that Dr. Matthew and Mr. Gidley have maintained the latter view. However, the trend of opinion seems to have been in the opposite direction. Osborn suggested that Diplodocus might lift its fore limbs from the ground and support itself on the hinder legs and the tail. This idea has found expression in Knight's restoration referred to above. Osborn's general notion of Diplodocus seems, however, to be that it was essentially an aquatic animal, long, light-limbed, and agile, and capable of swimming rapidly by means of its great tail, provided, as he thought, with a vertical fin; yet occasionally going about on land. Hatcher opposed the view that Diplodocus was aquatic; and he showed that there is no evidence of the presence of a vertical fin. The compression of the centra where the fin is supposed to have been situated seems to have been slight, and the neural spines are not higher than elsewhere. The present writer finds neither in the feet nor in the tail any special arrangements for swimming. For navigation in its restricted waters no fin was needed. Almost any colubrid snake makes fair progress in the water, notwithstanding the absence both of a compressed tail and of a vertical fin.

Hatcher's final view does not, after all, appear to have been greatly different from that of Osborn. He held that Diplodocus, as well as most of the Sauropoda, were essentially terrestrial animals; but that they passed much, perhaps most, of their time in shallow water, where they could wade about and search for food. He believed that they were ambulatory, but quite capable of swimming. Hatcher's language does not necessarily imply that these animals walked about after the fashion of quadrupedal mammals, but his restorations show plainly that such was his conception.

This conception has prevailed in the plaster reproductions of the skeleton of Diplodocus which have been sent abroad by the Carnegie Museum and set up in London, Berlin and Paris; and in the small plaster restorations issued by the American Museum of Natural History. However, the limit of quadrupedal erectness, rigidity, rectangularity, and rectilinearity has quite been reached in the skeleton sent by the last mentioned institution to the Senckenberg Museum, at Frankfort-on-the-Main. In this case the poor beast is made to stand straight-legged and almost on the tips of its digits. On the other band, the American Museum's skeleton of Brontosaurus, a much larger and heavier reptile and one sorely needing the mechanical advantage of straight legs, in case it had to bear its body free from the ground, has been presented to the modern world as having been decidedly bow-legged.

To the present writer it appears that the mammal-like pose attributed to the Sauropoda is one that is not required by their anatomy and one that is improbable. The current conception is one that is easily accounted for. Before exact knowledge of these reptiles had been gained, it was known that the dinosaurs of the other groups, herbivorous and carnivorous, walked erect, after the manner of birds. It was indeed necessary, on account of the length of the fore limbs, to place the Sauropoda on all four feet; but analogy caused it to be supposed that the limbs were disposed, with reference to the vertical plane of the body, similarly to those of the bipedal dinosaurs. The conception of a creeping dinosaur was hardly to be entertained. The straight femora of these reptiles, having the head and the great trochanter moderately developed, lent probability to the idea.

If the straightness of the femora is relied on to support the correctness of the prevailing restorations of the Sauropoda we may call attention to the equally straight femora of sphenodon and of the lizards. Notwithstanding the great size of the carnivorous dinosaur Allosaurus and the fact that the whole weight of its body was commonly borne by the hinder limbs alone, its femora are considerably bent. The prominence and the height of the great trochanter of the Sauropoda do not appear to be such as to have prevented the femora from standing out at right angles with the body. Both the head of the femur and the acetabulum were doubtless invested with much cartilage, so that we can not now be wholly certain about their form and fitting. The same may be said regarding certain other articulations of the limbs. Hatcher has spoken of the character of the articulations and he has expressed the opinion that the habitual support of the body in the air could not have failed to produce closely applied and well-finished articulations, and Osborn had previously expressed the same idea. There is indeed a great difference between the articulations of the limbs of the Sauropoda and those of the Theropoda, such as Allosaurus and Ceratosaurus.

Osborn has found in the large preacetabular process an argument in favor of the ability of Diplodocus to elevate the anterior part of its body. However, Trachodon, which habitually walked on its hind legs has a very insignificant preacetabular process. The crocodiles have a strongly developed process in front of the acetabulum.

It appears to the writer that the structure of the feet of the Sauropoda indicates that the digits were directed somewhat outward, instead of directly forward, as they are placed in the restorations. The strongly developed inner digits would then have come more effectively into contact with the ground than the much reduced outer digits and would have been employed by the animal as a means of pushing itself along. In case the lower end of the radius is placed in front of the ulna, as represented by Hatcher, it appears probable that the foot would be directed more strongly outward than is shown in his restoration.

The writer is not aware that any one has held that the Sauropoda could not, at least while resting, assume a crocodile-like posture, with the abdomen on the ground and the limbs extended outward on each side. If such a position is admitted as possible, the arguments derived from the anatomy in favor of an erect mode of walking are greatly weakened. If such a pose was not assumed, what was the pose? Did Diplodocus and Brontosaurus lie down on their sides, as an ox or a horse does when sleeping? Or did they lie prone, with the limbs drawn up under them, as a dog sometimes does? These positions appear to be improbable. It is worth considering too what disposition Diplodocus made of its elephantine legs while it was swimming with the agility that has been imputed to it.

The weight of Diplodocus and of Brontosaurus furnishes a strong argument against their having had a mammal-like carriage. There will be little dissent from the view that these animals inhabited a country in which marshy lands abounded and that they passed the most of their time in the vicinity of bodies of water. As to weight, Marsh estimated that that of Brontosaurus was more than twenty tons. Each footprint was thought to be about a square yard in extent. The pressure was therefore about 1,100 pounds on each square foot of the ground. What progress could such enormous animals have made through morasses and along mud-depositing rivers, in case they carried themselves as they are represented in the restorations? Without doubt, they would soon have become inextricably mired and would have perished miserably.

Osborn has suggested that Camarasaurus, another sauropod was accustomed to wading about in rivers where the bottoms were sandy and firm. The habits of Diplodocus could have differed little from those of Camarasaurus. It is difficult to understand why an animal whose immediate ancestors must have walked about in a crocodile-like manner, an animal that was stupid and probably slow of movement, an animal which could by means of its long neck reach up from the bottom many feet to the surface and from the surface many feet to the bottom—why such a reptile should need to develop the ability to walk along river bottoms like a mammal. Furthermore, it seems somewhat overgenerous to impute to a reptile so many and so diverse activities as swimming with great facility, walking on river bottoms and on the land with mammal-like gait, and on occasion erecting itself on its hinder legs after the manner of a bird, in order to crop the foliage from the tops of high trees, when this reptile was sixty feet long, weighed many tons, had a brain little larger than one's two thumbs placed side by side, and was provided with a feeble dental apparatus with which to gather food wherewith to support its huge body, and that food of a sort that yielded little energy in proportion to its bulk.

The writer's conception of Diplodocus is that it was eminently amphibious, that it could swim with considerable ease, and that it could creep about on land, with perhaps laborious effort. When feeding it must have swam or crept lazily about, gathering in floating plants and such as were attached loosely to the bottom. If any plants that were relished grew at some depth they could be reached by the long neck; or, if there was foliage twenty feet above the water it could be as easily gathered in. That a Diplodocus ever stood on its hind legs is hardly more probable than that crocodiles may perform the same feat.

The large size of Diplodocus does not preclude the possibility that it could creep about on the land. *Crocodylus robustus*, of Madagascar, is said to attain a length of 10 meters, and yet it doubtless is able to walk as other crocodiles walk. The limb bones of Diplodocus and of Brontosaurus are proportionally as large as those of crocodiles.

It seems to the writer that our museums which are engaged in making mounts and restorations of the great Sauropoda have missed an opportunity to construct some striking presentations of these reptiles that would be truer to nature. The body placed in a crocodile-like attitude would be little, if any, less imposing than when erect; while the long neck, as flexible as that of an ostrich, might be placed in a variety of graceful positions.

A Review of Some Recent Criticisms of the Restorations of Sauropod Dinosaurs . . .

Dr. W. J. Holland

All paleontologists are familiar with the figure of *Brontosaurus excelsus* Marsh, which was originally published in the *American Journal of Science* in August, 1883, and, republished with modifications in the same periodical in 1895. This figure has since been frequently reproduced in text-books. Paleontologists are also familiar with the restoration of the skeleton of *Diplodocus carnegiei* Hatcher, which originally appeared in the *Memoirs of the Carnegie Museum,* and is reproduced in the second volume of the English translation of Zittel's "Text-book of Paleontology," by C. R. Eastman. Since the time when Mr. Hatcher made this restoration the acquisition of new material has thrown much light upon the subject, and certain changes in the pose have been suggested. . . .

In *The Field* (London) of August 26, 1905, Mr. F. W. Frohawk, a well-known English illustrator, published a note, in which he said, among other things:

> The visitor to the Reptile Gallery of the Natural History Museum can not fail to be struck by the extraordinary pose of the gigantic skeleton. . . . It would be interesting to know the reason for mounting the specimen so high on its legs, like some huge pachyderm. As it is a gi-

gantic lizard, why should it not be represented in the attitude usually assumed by such animals? . . . Doubtless there is some good reason for mounting it in such an attitude; if so, information on the subject would be welcome.

No reply was given to this query, except incidentally by Professor (now Sir) L. Ray Lankester, who said in a newspaper interview that "the laterally compressed form of the body, according to the opinion of American students, precludes the idea that the animal could have crawled upon its belly."

Shortly after the restored skeleton of *Brontosaurus excelsus,* which is one of the ornaments of the American Museum of Natural History, had been erected, Messrs. Otto and Charles Falkenbach, assistants in the paleontological laboratory of that museum, made a model, in which they attempted to show the *Brontosaurus* in a crawling attitude. . . . This model was discussed [extensively by paleontologists] . . . and . . . was judged for many reasons to represent the impossible.

In October, 1908, there appeared in Vol. XLII of *The American Naturalist* an article from the pen of Dr. Oliver P. Hay, "On the Habits and Pose of the Sauropod Dinosaurs, especially of Diplodocus." Dr. Hay maintains that in assembling the fossil remains of these animals they should have been given a crocodilian attitude. At the conclusion of his article he sums up his views in the following words:

> It seems to the writer that our museums which are engaged in making mounds and restorations of the great sauropoda have missed an opportunity to construct some striking presentations of these reptiles that would be truer to nature. The body placed in a crocodile-like attitude would be little, if any less imposing than when erect; while the long neck, as flexible as that of an ostrich, might be placed in a variety of graceful positions.

This article of Dr. Hay was followed by a paper from the pen of Dr. Gustav Tornier, who, taking his cue front Dr. Hay, has tried to show that American paleontologists have totally erred in their conception of the structure of the sauropod dinosaurs, and has given his views as to the manner in which the bones of the *Diplodocus* should have been assembled. . . .

In the manner of a man who has made a wonderful discovery, Tornier announces at the outset of his paper that *Diplodocus* is a genuine reptile. . . . No student of the sauropoda has ever doubted this. But having predicated the genuinely reptilian character of the animal, Tornier proceeds thereafter to speak of *Diplodocus* as a lacertilian. . . . There are reptiles and reptiles. Having assured himself of the truly reptilian character of the animal, it was a bold step for him immediately to transfer the creature from the order Dinosauria, and evidently with the skeleton of a *Varanus* and a *Chameleon* before him, to proceed with the help of a pencil, the powerful tool of the closet-naturalist, to reconstruct the skeleton upon the study of which two generations of American paleontologists have expended considerable time and labor, and squeeze the animal into the form which his brilliantly illuminated imagination suggested. The fact that the dinosauria differ radically from existing reptiles in a multitude of important structural points seems not to have greatly impressed itself upon the mind of this astute critic. He intimates that the pelvis of *Diplodocus* is distinctly lacertilian. He states that the great trochanter of the femur, which he does not designate as such, articulated with the ischial peduncle, and takes care to show the point of union. . . . It may be said in passing that Dr. Tornier takes very great liberties with the outlines of the bones. His drawing is very far from accurate. Unfortunately actual experiment shows, *first,* that it is impossible except by smashing the ilium or breaking the femur to jam the head of the latter into the position demanded for it by the learned professor; but, *second,* this is the only time, it is believed, in the history of anatomical science that any one has discovered that the great trochanter of the femur ought to be and is by nature intended to be articulated with the ischial peduncle of the ilium, thus locking the femur into a position utterly precluding all motion whatsoever.

The next step taken by this wonder-working comparative anatomist was to dislocate the knee-joint. This he proceeds to do in a most nonchalant manner, and leaves the articulating end of the femur peering forth into space, while the tibia and the fibula are made to articulate with the posterior edges of the interior and exterior condyles of the femur. Having adopted this change, he succeeds in so lowering the hind quarters of the *Diplodocus* that they must rest upon the anterior extremity of the pubic bones, which, with the fragile ends of the ribs, not much greater in size than those of an ox, have thrown upon them the entire weight of the carcass. To obviate the inconveniences of this pose the lead pencil is again brought into requisition and the anterior vertebrae are hoisted into the air and propped up upon the scapulae, the dorsal ends of which have been glued by a hypothetical superscapula to the lateral processes of the last cervical vertebrae. This transference of the scapula to the Tornerian position is done in order to give, as the author says, an opportunity to so place the scapula that horizontal motion backward and forward may be allowed to the humerus, which he takes pains to inform us is strikingly like that of a *Varanus.* Upon the latter point it is quite possible to differ from the learned critic.

The anterior portion of the trunk having been thus elevated, the fore legs are again dislocated at the juncture of the humerus with the radius and ulna and stuffed underneath the skeleton, while the great neck is thrown upward in the form of a reversed letter "S," the Hogarthian lines of which no doubt suggested themselves to the learned reconstructionist as possessing soulful grace. . . . As a contribution to the literature of caricature the success achieved is remarkable. It reminds us of those creations carved in wood emanating from Nuremberg which were the delight of our childhood, and which came to us stuffed in boxes labeled "Noah's Ark," and stamped "Made in Germany."

I should prefer to end my communication at this point, commending the perusal of the articles by Hay [and] Tornier . . . to the attention of those of you who are familiar with the osteology of the sauropoda as amusing illustrations of the manner in which it is possible for gentlemen possessing entirely inadequate acquaintance with a subject to "darken counsel by words without knowledge."

. . . But let us for the sake of experiment give the femur [of *Diplodocus*] the same relative position which it has in the lacertilia [(as Tornier recommends)], in which the second trochanter plays so great a part. To do this it is necessary to rotate the head of the femur in such a way that the great trochanter will point downward and backward. . . . Of what earthly use the hind limb of the *Diplodocus* could have been to him in such a position I leave for you to determine for yourselves. It has been suggested that kindly nature, to meet the requirements of the case, must have channeled the surface of the earth and provided the *Diplodocus* and its allies with troughs in which they kept their bodies while the feet were employed for the purposes of locomotion along the banks. The *Diplodocus* must have moved in a groove or rut. This might perhaps account for his early extinction. It is physically and mentally bad to "get into a rut."

. . . Dr. Tornier labors long with the scapula and the humerus [of *Diplodocus*]. As I have already stated, he claims that the humerus of *Diplodocus* is startlingly . . . like that of *Varanus*. It is wonderful what a man can see who has determined to see things! If you will simply take the trouble to compare the humerus of the sauropod dinosaurs with that of a *Varanus* I think you will be able without opening your eyes very widely to discover a number of startling differences. In addition to falling into error as to the startling likeness existing between the humerus in the sauropoda and the lacertilia, he makes a multitude of grossly inaccurate and misleading statements in uttering the special plea he makes for his theory. It would be wearisome to recall them. One of the more noticeable misstatements is made when he declares that the coracoid bone belongs on the lower side of the belly—"Bauchunterseite." The coracoid, as we all know, is a sternal element and has nothing whatever to do with the "Bauch," or belly. He ignores the fact that the superior surfaces of the anterior ribs and the lateral processes of the anterior dorsal vertebrae unite to form a surface evidently adapted to the end of providing a field over which the long dorsal blade of the scapula can play. He demands for the scapula a vertical position so as to give the humerus an opportunity, as he says, to move in a horizontal plane backward and forward. He states that a vertical position of the scapula is universal among the recent reptilia, which is not the case. It is true of the lacertilia, but it is not true of the crocodilia. . . .

For the sake of experiment I have placed the scapula in the position demanded for it by Mr. Tornier, and have swung the fore limbs into place as he demands that they shall be put. . . . The result is in every way amusing. It leads in the first place to the entire disarticulation of the humerus from the radius and ulna. But as a secondary consequence it leads to a rather remarkable result, which the Berlin critic did not think of. In the position which Professor Tornier demands for the elements of the fore limb, the foot must fall into a position with the toes turned inwardly, while put into the position which he demands for the hind limbs the toes of the latter necessarily point outwardly. . . . The animal was "pigeon-toed" in front, while its hind feet were planted like those of a grenadier. The animal moved forward with the hind feet [and] backward with its fore feet. If Tornier is right it must necessarily have been somewhat "balled up."

. . . It is in short impossible to articulate the limbs in such a position as to impart to the animal a crawling attitude. We have experimented a score of times and have tried different poses, only to come back again to the position which we have given to the reproduction of *Diplodocus* and which is the position that has generally been accepted by osteologists as the correct position for such animals when standing or moving forward. Our reproductions may be, as they have been contemptuously styled by Hay, "light-legged and straight-legged," but no one who has had the matter practically in hand has yet been able to suggest any way of escaping the conclusion that these creatures were at all events more or less "straight-legged." For their "straight-legged" qualities nature is solely responsible, though I fail, standing before these huge bones, to see why anybody should so describe them. Students of the lacertilia and of the testudinata may sneer, but it is beyond possibility to adopt the suggestions which they from time to time make, that those of us who are engaged in studying the dinosaurs shall squeeze these creatures into the forms with which they are familiar. The critics possibly do not realize that weeks and months and years of study have been spent by those who have been charged with the task of assembling these remains, and that the prescriptions, which they now furnish, have been already tried without their suggestion, and have for good reasons been found wanting. It is easy for a knight of the quill, who has never practically attended the matter, to find fault. The latest attack upon those who have been making a special study of the sauropod dinosaurs has only served in the mind of the speaker to prove the correctness of the careful work which has been done in the past by students on both sides of the Atlantic. Evolution has had something to do since the sauropod dinosaurs walked the earth, and to say simply because the lacertilia of the present day creep and crawl that in Mesozoic times there were no reptiles which walked, is to go further than the facts seem to warrant. . . . [Whales and seals] live in the waters: it does not necessarily follow that the ancestral forms in remote antiquity moved as they move. Because a *Varanus* crawls to-day it does not necessarily follow that a sauropod dinosaur crawled. There is every evidence that they did not crawl, but that the restorations of Marsh, Osborn, and others are substantially correct in many important particulars. . . .

The Tornierian hypothesis may be dismissed, I think, as not within the range of the possible. It has, as one of my learned paleontological friends in Europe jocosely remarked to me, "only this feature to recommend it, that it accounts for the speedy disappearance of the sauropoda because if true, their lives must have been spent in indescribable agony, every joint being dislocated." . . .

Chapter 6

Robert Bakker's 1975 article "Dinosaur Renaissance" is reprinted here from Scientific American. *The opening words throw down the gauntlet against old-fashioned dinosaur "orthodoxy." Bakker proceeds to offer much intriguing evidence in favor of his revisionist interpretations. The article is very well written and copiously illustrated with Bakker's own drawings (unfortunately, the illustrations could not be reproduced here for copyright reasons). Bakker has thrived on controversy and, like him or not, his ideas have unquestionably been a powerful stimulus to paleontology. Talk about the biology of extinct creatures might remind many people of Mark Twain's quip: "There is something fascinating in science. One gets such wholesale returns of conjecture out of such a trifling investment of fact." Actually, though, it is amazing to see how much information really is contained in scraps of fossilized bone.*

The second reading for this chapter is David J. Varricchio's article "Warm or Cold and Green All Over" from National Forum, *published by the Honor Society of Phi Kappa Phi. Varricchio surveys many different lines of evidence relating to dinosaur physiology. He shows how complex the issue is and points to the many gaps in our knowledge. He concludes that the dinosaurs seem to have spanned the gap between ectothermy and endothermy. Since dinosaurs were unique in so many ways, perhaps their physiology was unique also.*

Dinosaur Renaissance

Robert Bakker

The dinosaurs were not obsolescent reptiles but were a novel group of "warm-blooded" animals. And the birds are their descendants.

Recent research is rewriting the dinosaur dossier. It appears that they were more interesting creatures, better adapted to a wide range of environments and immensely more sophisticated in their bioenergetic machinery, than had been thought. In this article I shall be presenting some of the evidence that has led to reevaluation of the dinosaurs' role in animal evolution. The evidence suggests, in fact, that the dinosaurs never died out completely. One group still lives. We call them birds.

The dinosaur is for most people the epitome of extinctness, the prototype of an animal so maladapted to a changing environment that it dies out, leaving fossils but no descendants. Dinosaurs have a bad public image as symbols of obsolescence and hulking inefficiency; in political cartoons they are know-nothing conservatives that plod through miasmic swamps to inevitable extinction. Most contemporary paleontologists have had little interest in dinosaurs; the creatures were an evolutionary novelty, to be sure, and some were very big, but they did not appear to merit much serious study because they did not seem to go anywhere: no modern vertebrate groups were descended from them.

Ectothermy and Endothermy

Dinosaurs are usually portrayed as "cold-blooded" animals, with a physiology like that of living lizards or crocodiles. Modern land ecosystems clearly show that in large animals "cold-bloodedness" (ectothermy) is competitively inferior to "warm-bloodedness" (endothermy), the bioenergetic system of birds and mammals. Small reptiles and amphibians are common and diverse, particularly in the Tropics, but in nearly all habitats the overwhelming majority of land vertebrates with an adult weight of 10 kilograms or more are endothermic birds and mammals. Why?

The term "cold-bloodedness" is a bit misleading: on a sunny day a lizard's body temperature may be higher than a man's. The key distinction between ectothermy and endothermy is the rate of body-heat production and long-term temperature stability. The resting metabolic heat production of living reptiles is too low to affect body temperature significantly in most situations, and reptiles of today must use external heat sources to raise their body temperature above the air temperature—which is why they bask in the sun or on warm rocks. Once big lizards, big crocodiles or turtles in a warm climate achieve a high body temperature they can maintain it for days because large size retards heat loss, but they are still vulnerable to sudden heat drain during cloudy weather or cool nights or after a rainstorm, and so they cannot match the performance of endothermic birds and mammals.

The key to avian and mammalian endothermy is high basal metabolism: the level of heat-producing chemical activity in each cell is about four times higher in an endotherm than in an ectotherm of the same weight at the same body temperature. Additional heat is produced as it is needed, by shivering and some other special forms of thermogenesis. Except for some large tropical endotherms (elephants and ostriches, for example), birds and mammals also have a layer of hair or feathers that cuts the rate of thermal loss. By adopting high heat production and insulation endotherms have purchased the ability to maintain more nearly constant high body temperatures than their ectothermic competitors can. A guarantee of high, constant body temperature is a powerful adaptation because the rate of work output from muscle tissue, heart and lungs is greater at high temperatures than at low temperatures, and the endothermic animal's biochemistry can be finely tuned to operate within a narrow thermal range.

The adaptation carries a large bioenergetic price, however. The total energy budget per year of a population of endothermic birds or mammals is from 10 to 30 times higher than the energy budget of an ectothermic population of the same size and adult body weight. The price is nonetheless justified. Mammals and birds have been the dominant large and medium-sized land vertebrates for 80 million years in nearly all habitats.

In view of the advantage of endothermy the remarkable success of the dinosaurs seems puzzling. The first land vertebrate communities, in the Carboniferous and early Permian periods, were composed of reptiles and amphibians generally considered to be primitive and ectothermic. Replacing this first ectothermic dynasty were the mammal-like reptiles (therapsids), which eventually produced the first true mammals near the end of the next period, the Triassic, about when the dinosaurs were originating. One might expect that mammals would have taken over the land-vertebrate communities immediately, but they did not. From their appearance in the Triassic until the end of the Cretaceous, a span of 140 million years, mammals remained small and inconspicuous while all the ecological roles of large terrestrial herbivores and carnivores were monopolized by dinosaurs; mammals did not begin to radiate and produce large species until after the dinosaurs had already become extinct at the end of the Cretaceous. One is forced to conclude that dinosaurs were competitively superior to mammals as large land vertebrates. And that would be baffling if dinosaurs were "cold-blooded." Perhaps they were not.

Measuring Fossil Metabolism

The third meter of heat production in extinct vertebrates is the predator-prey ratio: the relation of the "standing crop" of a predatory animal to that of its prey.

The ratio is a constant that is a characteristic of the metabolism of the predator, regardless of the body size of the animals of the predator-prey system. The reasoning is as follows: the energy budget of an endothermic population is an order of magnitude larger than that of an ectothermic population of the same size and adult weight, but the productivity—the yield of prey tissue available to predators—is about the same for both an endothermic and an ectothermic population. In a steady-state population the yearly gain in weight and energy value from growth and reproduction equals the weight and energy value of the carcasses of the animals that die during the year; the loss of biomass and energy through death is balanced by additions. The maximum energy value of all the carcasses a steady-state population of lizards can provide its predators is about the same as that provided by a prey population of birds or mammals of about the same numbers and adult body size. Therefore a given prey population, either ectotherms or endotherms, can support an order of magnitude greater biomass of ectothermic predators than of endothermic predators, because of the endotherms' higher energy needs. The term standing crop refers to the biomass, or the energy value contained in the biomass, of a population. In both ectotherms and endotherms the energy value of carcasses produced per unit of standing crop decreases with increasing adult weight of prey animals: a herd of zebra yields from about a fourth to a third of its weight in prey carcasses a year, but a "herd" of mice can produce up to six times its weight because of its rapid turnover, reflected in a short life span.

Now, the energy budget per unit of predator standing crop also decreases with increasing weight: lions require more than 10 times their own weight in meat per year, whereas shrews need 100 times their weight. These two bioenergetic scaling factors cancel each other, so that if the adult size of the predator is roughly the same as that of the prey (and in land-vertebrate ecosystems it usually is), the maximum ratio of predator standing crop to prey standing crop in a steady-state community is a constant independent of the adult body size in the predator-prey system. For example, spiders are ectotherms, and the ratio of a spider population's standing crop to its prey standing crop reaches a maximum of about 40 percent. Mountain boomer lizards, about 100 grams in adult weight, feeding on other lizards would reach a similar maximum ratio. So would the giant Komodo dragon lizards (up to 150 kilograms in body weight). Mammals and birds, on the other hand, reach a maximum predator-prey biomass ratio of only from 1 to 3 percent—whether they are weasel and mouse or lion and zebra.

Some fossil deposits yield hundreds or thousands of individuals representing a single community; their live body weight can be calculated from the reconstruction of complete skeletons, and the total predator-prey biomass ratios are then easily worked out. Predator-prey ratios are powerful tools for paleophysiology because they are the direct result of predator metabolism.

The Age of Ectothermy

The paleobioenergetic methodology I have outlined can be tested by analyzing the first land-vertebrate predator-prey system, the early Permian communities of primitive reptiles and amphibians.

The first predators capable of killing relatively large prey were the finback pelycosaurs of the family Sphenacodontidae, typified by *Dimetrodon,* whose tall-seined fin makes it popular with cartoonists. Although this family included the direct ancestors of mammal-like reptiles and hence of mammals, the sphenacodonts themselves had a very primitive level of organization, with a limb anatomy less advanced than that of living lizards. Finback bone histology was emphatically ectothermic, with a low density of blood vessels, few Haversian canals and the distinct growth rings that are common in specimens from seasonally arid climates.

One might suspect that finbacks and their prey would be confined to warm, equable climates, and early Permian paleogeography offers an excellent opportunity to test this prediction. During the early part of the period ice caps covered the southern tips of the continental land masses, all of which were part of the single southern supercontinent Gondwanaland, and glacial sediment is reported at the extreme northerly tip of the Permian land mass in Siberia by Russian geologists. The Permian Equator crossed what are now the American Southwest, the Maritime Provinces of Canada and western Europe. Here are found sediments produced in very hot climates: thickbedded evaporite salts and fully oxidized, red-stained mudstones. The latitudinal temperature gradient in the Permian must have been at least as steep as it is at present. Three Permian floral zones reflect the strong poleward temperature gradient. The Angaran flora of Siberia displays wood with growth rings from a wet environment, implying a moist climate with cold winters. The Euramerican flora of the equatorial region had two plant associations: wet swamp communities with no growth rings in the wood, implying a continuous warm-moist growing season, and semiarid red-bed evaporite communities with some growth rings, reflecting a tropical dry season. In glaciated Gondwanaland, the peculiar *Glossopteris* flora dominated, with wood from wet environments showing sharp growth rings.

The ectothermy of the finbacks is confirmed by their geographic zonation. Finback communities are known only from near the Permian Equator; no large early Permian land vertebrates of any kind are found in glaciated Gondwanaland. (One peculiar little fish-eating reptile, *Mesosaurus,* is known from southern Gondwanaland, and its bone has sharp growth rings. The animal must have fed and reproduced during the Gondwanaland summer and then burrowed into the mud of lagoon bottoms to hibernate, much as large snapping turtles do today in New England.)

Excellent samples of finback communities are available for predator-prey studies, thanks largely to the lifework of the late Alfred Sherwood Romer of Harvard University. In order to derive a predator-prey ratio from a fossil community one simply calculates the number of individuals, and thus the total live weight, represented by all the predator and prey specimens that are found together in a sediment representing one particular environment. In working with scattered and disarticulated skeletons it is best to count only bones that have about the same robustness, and hence the same preservability, in both predator and prey. The humerus and the femur are good choices for finback communities: they are about the same size with respect to the body in the prey and the predator and should give a ratio that faithfully represents the ratio of the animals in life.

In the earlier early Permian zones the most important finback prey were semi-aquatic fish-eating amphibians and reptiles, particularly the big-headed amphibian *Eryops* and the long-snouted pelycosaur *Ophiacodon*. As the climate became more arid in Europe and America these water-linked forms decreased in numbers, and the fully terrestrial herbivore *Diadectes* became the chief prey genus. In all zones from all environments the calculated biomass ratio of predator to prey in finback communities is very high: from 35 to 60 percent, the same range seen in living ectothermic spiders and lizards.

All three of the paleobioenergetic indicators agree: the finback pelycosaurs and their contemporaries were ectotherms with low heat production and a lizard-like physiology that confined their distribution to the Tropics.

Therapsid Communities

The mammal-like reptiles (order *Therapsida*), descendants of the finbacks, made their debut at the transition from the early to the late Permian and immediately became the dominant large land vertebrates all over the world. The three metabolism-measuring techniques show that they were endotherms.

The earliest therapsids retained many finback characteristics but had acquired limb adaptations that made possible a trotting gait and much higher running speeds. From early late Permian to the middle Triassic one line of therapsids became increasingly like primitive mammals in all details of the skull, the teeth and the limbs, so that some of the very advanced mammal-like therapsids (cynodonts) are difficult to separate from the first true mammals. The change in physiology, however, was not so gradual. Detailed studies of bone histology conducted by Armand Riqles of the University of Paris indicate that the bioenergetic transition was sudden and early: all the finbacks had fully ectothermic bone; all the early therapsids—and there is an extraordinary variety of them—had fully endothermic bone, with no growth rings and with closely packed blood vessels and Haversian canals.

The late Permian world still had a severe latitudinal temperature gradient; some glaciation continued in Tasmania, and the southern end of Gondwanaland retained its cold-adapted Glossopteris flora. If the earliest therapsids were equipped with endothermy, they would presumably have been able to invade southern Africa, South America and the other parts of the southern cold-temperature realm. They did exactly that. A rich diversity of early therapsid families has been found in the southern Cape District of South Africa, in Rhodesia, in Brazil in India—regions reaching to 65 degrees south Permian latitude. Early therapsids as large as rhinoceroses were common there, and many species grew to an adult weight greater than 10 kilograms, too large for true hibernation. These early therapsids must have had physiological adaptations that enabled them to feed in and move through the snows of the cold Gondwanaland winters. There were also some ectothermic holdovers from the early Permian that survived into the late Permian, notably the immense herbivorous caseid pelycosaurs and the bigheaded, seed-eating captorhinids. As one might predict large species of these two ectothermic families were confined to areas near the late Permian Equator; big caseids and captorhinids are not found therapsids in cold Gondwanaland. In the late Permian, there was a "modern" faunal zonation of large vertebrates, with endothermic therapsids and some big ectotherms in the Tropics giving way to an all-endothermic therapsid fauna in the cold south.

In the earliest therapsid communities of southern Africa, superbly represented in collections built up by Lieuwe Boonstra of the South African Museum and by James Kitching of the University of the Witwatersrand, the predator-prey ratios are between 9 and 16 percent. That is much lower than in early Permian finback communities. Equally low ratios are found for tropical therapsids from the U.S.S.R. even though the prey species there were totally different from those of Africa. The sudden decrease in predator-prey ratios from finbacks to early therapsids coincides exactly with the sudden change in bone histology from ectothermic to endothermic reported by Riqles and also with the sudden invasion of the southern cold-temperate zone by a rich therapsid fauna. The conclusion is unavoidable that even early therapsids were endotherms with high heat production.

It seems certain, moreover, that in the cold Gondwanaland winters the therapsids would have required surface insulation. Hair is usually thought of as a late development that first appeared in the advanced therapsids, but it must have been present in the southern African endotherms of the early late Permian. How did hair originate? Possibly the ancestors of therapsids had touch-sensitive hairs scattered over the body as adaptations for night foraging; natural selection could then have favored increased density of hair as the animals' heat production increased and they moved into colder climates.

The therapsid predator-prey ratios, although much lower than those of ectotherms, are still about three times higher than those of advanced mammals today. Such ratios indicate that the therapsids achieved endothermy with a moderately high heat production, far higher than in typical reptiles but still lower than in most modern mammals. Predator-prey ratios of early Cenozoic communities seem to be lower than those of therapsids, and so one might conclude that a further increase in metabolism occurred somewhere between the advanced therapsids of the Triassic and the mammals of the post-Cretaceous era. Therapsids may have operated at a lower body temperature than most living mammals do, and thus they may have saved energy with a lower thermostat setting. This suggestion is reinforced by the low body temperature of the most primitive living mammals: monotremes (such as the spiny anteater) and the insectivorous tenrecs of Madagascar; they maintain a temperature of about 30 degrees Celsius instead of the 36 to 39 degrees of most modern mammals.

Thecodont Transition

The vigorous and successful therapsid dynasty ruled until the middle of the Triassic. Then their fortunes waned and a new group, which was later to include the dinosaurs, began to take over the roles of large predators and herbivores. These were the *Archosauria,* and the first wave of archosaurs were the thecodonts. The earliest thecodonts, small and medium-sized animals found in therapsid communities during the Permian Triassic transition, had an ectothermic bone histology. In modern ecosystems the role played by large freshwater predators seems to be one in which ectothermy is competitively superior to endothermy; the low metabolic rate of ectotherms may be a key advantage because it allows much longer dives. Two groups of thecodonts became large freshwater fish-eaters: the phytosaurs, which were confined to the Triassic, and the crocodilians, which remain successful today. Both groups have ectothermic bone. (The crocodilian endothermy was either inherited directly from the first thecodonts or derived secondarily from endothermic intermediate ancestors.) In most of the later, fully terrestrial advanced thecodonts, on the other hand, Riqles discovered a typical endothermic bone histology; the later thecodonts were apparently endothermic.

The predator-prey evidence for thecodonts is scanty. The ratios are hard to compute because big carnivorous cynodonts and even early dinosaurs usually shared the predatory role with thecodonts. One sample from China that has only one large predator genus, a bigheaded erythrosuchid thecodont, does give a ratio of about 10 percent, which is in the endothermic range. The zonal evidence is clearer. World climate was moderating in the Triassic (the glaciers were gone), but a distinctive flora and some wood growth rings suggest that southern Gondwanaland was not yet warm all year. What is significant in this regard is the distribution of phytosaurs, the big ectothermic fish-eating thecodonts. Their fossils are common in North America and Europe (in the Triassic Tropics) and in India, which was warmed by the equatorial Tethys Ocean, but they have not been found in southern Gondwanaland, in southern Africa or in Argentina, even though a rich endothermic thecodont fauna did exist there.

Did some of the thecodonts have thermal insulation? Direct evidence comes from the discoveries of A. Sharov of the Academy of Sciences of the U.S.S.R. Sharov found a partial skeleton of a small thecodont and named it *Longisquama* for its long scales: strange parachutelike devices along the back that may have served to break the animal's fall when it leaped from trees. More important is the covering of long, overlapping, keeled scales that trapped an insulating layer of air next to its body. These scales lacked the complex anatomy of real feathers, but they are a perfect ancestral stage for the insulation of birds. Feathers are usually assumed to have appeared only late in the Jurassic with the first bird, *Archaeopteryx*. The likelihood that some thecodonts had insulation is supported, however, by another of Sharov's discoveries: a pterosaur, or flying reptile, whose fossils in Jurassic lake beds still show the epidermal covering. This beast (appropriately named *Sordus pilosus,* the "hairy devil") had a dense growth of hair or hairlike feathers all over its body and limbs. Pterosaurs are descendants of Triassic thecodonts or perhaps of very primitive dinosaurs. The insulation in both *Sordus* and *Longisquama,* and the presence of big erythrosuchid thecodonts at the southern limits of Gondwanaland, strongly suggest endothermic thecodonts had acquired insulation by the early Triassic.

The Dinosaurs

Dinosaurs, descendants of early thecodonts, appeared first in the middle Triassic and by the end of the period had replaced thecodonts and the remaining therapsids as the dominant terrestrial vertebrates. Zonal evidence for endothermy in dinosaurs is somewhat equivocal. The Jurassic was a time of climatic optimum, when the poleward temperature gradient was the gentlest that has prevailed from the Permian until the present day. In the succeeding Cretaceous period latitudinal zoning of oceanic plankton and land plants seems, however, to have been a bit sharper. Rhinoceros-sized Cretaceous dinosaurs and big marine lizards are found in the rocks of the Canadian far north, within the Cretaceous Arctic Circle. Dale A. Russell of the National Museums of Canada points out that at these latitudes the sun would have been below the horizon for months at a time. The environment of the dinosaurs would have been far severer than the environment of the marine reptiles because of the lack of a wind-chill factor in the water and because of the ocean's temperature-buffering effect. Moreover, locomotion costs far less energy per kilometer in water than on land, so that the marine reptiles could have migrated away from the arctic winter. These considerations suggest, but do not prove, that arctic dinosaurs must have been able to cope with cold stress.

Dinosaur bone histology is less equivocal. All dinosaur species that have been investigated show fully endothermic bone, some with a blood-vessel density higher than that in living mammals. Since bone histology separates endotherms from ectotherms in the Permian and the Triassic, this evidence alone should be a strong argument for the endothermy of dinosaurs. Yet the predator-prey ratios are even more compelling. Dinosaur carnivore fossils are exceedingly rare. The predator-prey ratios for dinosaur communities in the Triassic, Jurassic and Cretaceous are usually from 1 to 3 percent, far lower even than those of therapsids and fully as low as those in large samples of fossils from advanced mammal communities in the Cenozoic. I am persuaded that all the available quantitative evidence is in favor of high heat production and a large annual energy budget in dinosaurs. Were dinosaurs insulated?

Explicit evidence comes from a surprising source: *Archaeopteryx*. As an undergraduate a decade ago I was a member of a paleontological field party led by John H. Ostrom of Yale University. Near Bridger, Mont., Ostrom found a remarkably preserved little dinosaurian carnivore, *Deinonychus,* that shed a great deal of light on carnivorous dinosaurs in general. A few years later, while looking for pterosaur fossils in European museums, Ostrom came on a specimen of *Archaeopteryx* that had been mislabeled for years as a flying reptile, and he noticed extraordinary points of resemblance between *Archaeopteryx* and carnivorous dinosaurs. After a detailed anatomical analysis Ostrom has now established beyond any reasonable doubt that the immediate ancestor of *Archaeopteryx* must have been a small dinosaur, perhaps one related to *Deinonychus.* Previously it had been thought that the ancestor of *Archaeopteryx,* and thus of birds, was a thecodont rather far removed from dinosaurs themselves.

Archaeopteryx was quite thoroughly feathered, and yet it probably could not fly: the shoulder joints were identical with those of carnivorous dinosaurs and were adapted for grasping prey, not for the peculiar arc of movement needed for wing-flapping. The feathers were probably adaptations not for powered flight or gliding but primarily for insulation. *Archaeopteryx* is so nearly identical in all known features with small carnivorous dinosaurs that it is hard to believe feathers were not present in such dinosaurs. Birds inherited their high metabolic rate and most probably their feathered insulation from dinosaurs; powered flight probably did not evolve until the first birds with flight-adapted joints appeared during the Cretaceous, long after *Archaeopteryx*.

It has been suggested a number of times that dinosaurs could have achieved a fairly constant body temperature in a warm environment by sheer bulk alone; large alligators approach this condition in the swamps of the U.S. Gulf states. This proposed thermal mechanism would not give rise to endothermic bone histology or low predator-prey ratios, however, nor would it explain arctic dinosaurs or the success of many small dinosaur species with an adult weight of between five and 50 kilograms.

Dinosaur Brains and Limbs

Large brain size and endothermy seem to be linked; most birds and mammals have a ratio of brain size to body size much larger than that of living reptiles and amphibians. The acquisition of endothermy is probably a prerequisite for the enlargement of the brain because the proper functioning of a complex central nervous system calls for the guarantee of a constant body temperature. It is not surprising that endothermy appeared before brain enlargement in the evolutionary line leading to mammals. Therapsids had small brains with reptilian organization; not until the Cenozoic did mammals attain the large brain size characteristic of most modern species. A large brain is certainly not necessary for endothermy, since the physiological feedback mechanisms responsible for thermoregulation are deep within the "old" region of the brain, not in the higher learning centers. Most large dinosaurs did have relatively small brains. Russell has shown, however, that some small and medium-sized carnivorous dinosaurs had brains as large as or larger than modern birds of the same body size.

Up to this point I have concentrated on thermoregulatory heat production. Metabolism during exercise can also be read from fossils. Short bursts of intense exercise are powered by anaerobic metabolism within muscles, and the oxygen debt incurred is paid back afterward by the heart-lung system. Most modern birds and mammals have much higher levels of maximum aerobic metabolism than living reptiles and can repay an oxygen debt much faster. Apparently this difference does not keep small ectothermic animals from moving fast: the top running speeds of small lizards equal or exceed those of small mammals. The difficulty of repaying oxygen debt increases with increasing body size, however, and the living large reptiles (crocodilians, giant lizards and turtles) have noticeably shorter limbs, less limb musculature and lower top speeds than many large mammals, such as the big cats and the hoofed herbivores.

The early Permian ectothermic dynasty was also strikingly short-limbed; evidently the physiological capacity for high sprinting speeds in large animals had not yet evolved. Even the late therapsids, including the most advanced cynodonts, had very short limbs compared with the modern-looking running mammals that appeared early in the Cenozoic. Large dinosaurs, on the other hand, resembled modern running mammals, not therapsids, in locomotor anatomy and limb proportions. Modern, fastrunning mammals utilize an anatomical trick that adds an extra limb segment to the forelimb stroke. The scapula, or shoulder blade, which is relatively immobile in most primitive vertebrates, is free to swing backward and forward and thus increase the stride length. Jane A. Peterson of Harvard has shown living chameleonia lizards have also evolved scapular swinging, although its details are different from those in mammals. Quadrupedal dinosaurs evolved a chameleon-type scapula, and they must have had long strides and running speeds comparable to those of big savanna mammals today.

When the dinosaurs fell at the end of the Cretaceous, they were not a senile, moribund group that had played out its evolutionary options. Rather they were vigorous, still diversifying into new orders and producing a variety of big-brained carnivores with the highest grade of intelligence yet present on land. What caused their fall? It was not competition, because mammals did not begin to diversify until after all the dinosaur groups (except birds) had disappeared. Some geochemical and microfossil evidence suggests a moderate drop in ocean temperature at the transition from the Cretaceous to the Cenozoic, and so cold has been suggested as the reason. But the very groups that would have been most sensitive to cold, the large crocodilians, are found as far north as Saskatchewan and as far south as Argentina before and immediately after the end of the Cretaceous. A more likely reason is the draining of shallow seas on the continents and a lull in mountain-building activity in most parts of the world, which would have produced vast stretches of monotonous topography. Such geological events decrease the variety of habitats that are available to land animals, and thus increase competition. They can also cause the collapse of intricate, highly evolved ecosystems; the larger animals seem to be the more affected. At the end of the Permian similar changes had been accompanied by catastrophic extinctions among therapsids and other land groups. Now, at the end of the Cretaceous, it was the dinosaurs that suffered a catastrophe; the mammals and birds, perhaps because they were so much smaller, found places for themselves in the changing landscape and survived.

The success of the dinosaurs, an enigma as long as they were considered "cold-blooded," can be seen as a predictable result of the superiority of their high heat production, high aerobic exercise metabolism and insulation. They were endotherms. Yet the concept of dinosaurs as ectotherms is deeply entrenched in a century of paleontological literature. Being a reptile connotes being an ectotherm, and the dinosaurs have always been classified in the subclass Archosauria of the class Reptilia; the other land-vertebrate classes were the Mammalia and the Aves. Perhaps, then, it is time to reclassify.

Taxonomic Conclusions

What better dividing line than the invention of endothermy? There has been no more far-reaching adaptive breakthrough, and so the transition from ectothermy to endothermy can serve to separate the land vertebrates into higher taxonomic categories. For some time it has been suggested that the therapsids should be removed from the Reptilia and joined with the Mammalia; in the light of the sudden increase in heat production and the probable presence of hair in early therapsids, I fully agree. The term Theropsida has been applied to mammals and their therapsid ancestors. Let us establish a new class Theropsida, with therapsids and true mammals as two subclasses.

How about the class Aves? All the quantitative data from bone histology and predator-prey ratios, as well as the dinosaurian nature of *Archaeopteryx,* show that all the essentials of avian biology—very high heat production, very high aerobic exercise metabolism and feathery insulation—were present in the dinosaur ancestors of birds. I do not believe birds deserve to be put in a taxonomic class separate from dinosaurs. Peter Calton of the University of Bridgeport and I have suggested a more reasonable classification: putting the birds into the Dinosauria. Since bone histology suggests that most thecodonts were endothermic, the thecodonts could then be joined with the Dinosauria in a great endothermic class Archosauria, comparable to the Theropsida. The classification may seem radical at first, but it is actually a good deal neater bioenergetically than the traditional Reptilia, Aves and Mammalia. And for those of us who are fond of dinosaurs the new classification has a particularly happy implication: The dinosaurs are not extinct. The colorful and successful diversity of the living birds is a continuing expression of basic dinosaur biology.

Warm or Cold and Green All Over

David J. Varricchio

Traditional classification of animals has its roots with Linnaeus, an eighteenth-century Swedish scientist. This system became established long before the acceptance of evolutionary concepts. It lacks a unifying theory and simply groups like things. Dinosaur skeletons differ from those of both mammals and birds. Largely on the basis of their primitive features, scientists classified dinosaurs as reptiles together with such living groups as turtles, lizards, snakes, and crocodilians. Over the past twenty years, however, biologists and paleontologists have largely switched from this traditional classification to one based on evolutionary relationships. This system, known as phylogenetic systematics, recognizes as valid only those groups which have a single ancestor and which include all the descendants of that ancestor. Under this system all groups reflect evolutionary relationships that have real biological meaning.

Two important changes occur for dinosaurs under this scheme. First, although dinosaurs remain within the Reptilia, they share a closer ancestry and relationship with crocodilians than they do with other reptilian groups. The second change involves a repositioning of birds. Over the last thirty years, beginning with the work of John Ostrom at Yale, new fossil finds and analyses have demonstrated that birds evolved from the meat-eating line of dinosaurs, the theropods. This theory represents the revival of an old idea, proposed as early as the late 1860s. According to the modern classification, because birds evolved from dinosaurs, they belong to both the Reptilia and Dinosauria.

Under the traditional classification, dinosaurs fall in with the reptiles, a group with strictly cold-blooded, living representatives. Cold-blooded animals, or ectotherms (literally "outside heat"), let their body temperatures fluctuate with the environment or regulate their temperatures by using external heat sources. For example, a frog's temperature will match that of the pond in which it dwells. Alligators and lizards raise their body temperature by basking in the sun or sitting on sun-warmed rocks. In contrast, warm-blooded animals, or endotherms ("inside heat"), namely birds and mammals, regulate their body temperatures largely from within. They generate heat through rapid and constant cellular and muscular activity.

Scientists schooled in the traditional classification system viewed dinosaur warm-bloodedness as unlikely. They reasoned: Dinosaurs were reptiles; living reptiles are cold-blooded; therefore, dinosaurs should have been cold-blooded. In contrast, a modern and more accurate classification shows dinosaurs to sit between their cold-blooded crocodilian cousins and their warm-blooded bird descendants. This intermediate position, borne out by many skeletal features of dinosaurs, does not favor one particular interpretation. Dinosaurs could have been ectothermic, endothermic, or perhaps something in between.

Two complicating problems not likely to disappear include the nature of fossil preservation and the diversity of dinosaurs. A biologist interested in discerning an animal's metabolism would measure body temperature, heart rates, and rates of respiration during both rest and activity. Certainly, a biologist would not sacrifice the animal, remove its flesh, and examine its bones to evaluate whether it maintained its body temperature through external or internal heat sources. For paleontologists, however, generally little remains except bones. Metabolic interpretation of fossil animals thus relies on features such as posture, bone tissue, and braincase size that relate only indirectly to metabolism.

Dinosaurs achieved great diversity through some 160 million years of evolution. Body sizes ranged from the 5-pound, two-legged carnivore, *Compsognathus*, to the 50-ton ponderous herbivore, *Seismosaurus*. The small, medium, and huge include a wide variety of body plans: squat bony tanks, large two-legged herbivores with elaborate grinding teeth, long-limbed toothless forms with unknown diets, giant predators, and many others. This diversity in form possibly represents a corresponding one in metabolism; a herbivore the size of ten elephants and a predator no larger than a chicken undoubtedly possessed different physiologies. The question of dinosaur metabolism is thus a complicated one.

Richard Owen established the Dinosauria back in 1842, based on three partial skeletons. He recognized the unique anatomical features shared by these animals as adaptations to terrestrial life. He even postulated that dinosaurs had a four-chambered heart and a circulatory system approaching that of warm-blooded animals. Nevertheless, although dinosaurs appeared awesome and unique for Owen, they clearly belonged among the reptiles and by default were seen as cold-blooded.

A short time later, and after the discovery of the small theropod *Compsognathus,* both Thomas Huxley, often referred to as "Darwin's Bulldog," and Edward

Drinker Cope, one of America's first dinosaur hunters, recognized a close relationship between birds and dinosaurs. The features uniting the two groups relate to an upright stance, a posture where the legs are held beneath the body and move in a fore-aft plane. Dinosaurs and birds also lengthen their stride by walking on their toes, heel off the ground. These features separate the members of the Dinosauria from other vertebrate groups. Mammals also possess an upright stance, but the construction of pelvis, ankle, and foot suggests that their stance evolved independently.

Recent researchers consider the upright stance of dinosaurs as metabolically significant. All living vertebrates with an upright stance possess warm-blooded physiologies. David Carrier in *Paleobiology* (v. 13, p. 326) has shown that the sideways wiggle found in the sprawling gait of lizards and salamanders prohibits efficient breathing while running. Each step in these ectothermic animals contracts one side of the rib cage and lungs while expanding the other side. This prevents a uniform expansion or contraction of the lungs during movement. Not surprisingly, these animals rely on bursts of activity. Adoption of an upright stance disengages lung from limb function. With the freedom to breathe efficiently during running, dinosaurs may have achieved more sustained activity than their primitive reptilian relatives.

In comparison with cold-blooded animals, endotherms maintain much higher metabolic rates when at rest. Respiration, blood circulation, and digestion generally proceed about ten times faster in warm-blooded animals. If dinosaurs achieved sustained activity, they should show corresponding changes in food processing, motor control, and breathing. Some recent studies have tried to evaluate these processes in dinosaurs. Unfortunately, evaluating metabolic rates in extinct animals remains a speculative endeavor.

Processing Food

Because warm-blooded animals consume about ten times more food than cold-blooded ones, endotherms, particularly plant-eaters, face a difficult task in food processing. Dinosaur herbivores show a range of adaptations, some suggestive of endothermy. The armored ankylosaurs and the long-necked sauropods, two groups of large dinosaurs, possessed relatively unimpressive teeth. Modern mammalian plant eaters, such as cows, horses, rhinos, and elephants, have large grinding teeth, and from our own hairy mammalian perspective, molars seem like a prerequisite for plant eating. Nonetheless, birds, even large herbivorous ones such as ostriches, emus, and rheas, do well despite a lack of teeth. Birds rely on two features: a specially modified region of the stomach, the heavily-muscled gizzard, which crushes food usually with the aid of swallowed stones; and an expanded area of the hindgut that houses plant-digesting bacteria. Because gizzards occur in both crocodilians and birds, dinosaurs' closest living relatives, dinosaurs probably had them too. Several sauropod skeletons, such as the one described by David Gillette in *Seismosaurus,* the Earth Shaker, still contain gizzard stones. Both ankylosaurs and sauropods have large, rotund rib cages, enough room to house a long digestive tract with symbiotic bacteria. Although the lack of large grinding teeth in ankylosaurs and sauropods hardly supports a warm-blooded interpretation, the presence of a gizzard and large gut would not rule it out.

Both the horned dinosaurs (ceratopsians) and the duckbills (iguanodons and hadrosaurs) bear impressive arrays of grinding teeth. In these animals, small teeth fit tightly together to form large crushing platforms. One of these giant pseudo-molars consists of between 50 and 135 functioning teeth. Consequently, these dinosaurs had some 200 to 500 teeth in use at any one time. Greg Erickson, as reported in *Proceedings of the National Academy of Science* (v. 93, p. 14623), examined teeth of these animals microscopically. He found daily growth lines in the teeth, from which he could estimate the rate of tooth formation and replacement. These dinosaurs wore out teeth fast; an inch-long hadrosaur tooth lasted little more than a couple of months. With 500 teeth in use, this rate adds up to some brisk tooth production. Such tooth growth rates rival those of living herbivorous mammals, implying that ceratopsians and duckbills were physiologically capable of producing teeth as fast as warm-blooded animals. Further, it suggests that these dinosaurs processed either endothermic quantities of food or lesser amounts of extra-tough plant material.

Brain Size

Large brains occur almost exclusively in endotherms, and higher activity levels may necessitate more brain matter and nerves. Assuming this idea is roughly accurate, Jim Hopson investigated brain size in dinosaurs. Results were published in an out-of-print but interesting collection of scientific papers, *A Cold Look at the Warm-Blooded Dinosaurs*. He found great variation among dinosaur groups. Large four-legged herbivores such as sauropods, ankylosaurs, stegosaurs, and ceratopsians did not compare particularly well even with living reptiles. Other groups had brains comparable to those of ectotherms today. The largest brains, ones exceeding those of typical reptiles, occur among the carnivorous dinosaurs. Small theropods such as *Troodon* and *Deinonychus,* the dinosaurs with the closest ancestry to birds, possessed brains as large as and similarly shaped to those of their feathered relatives.

Respiration

John Ruben and several colleagues have recently argued (*Science,* v. 273 and 278) that dinosaurs could not breathe like endotherms. They CAT-scanned dinosaur skulls and show dinosaur nasal passages as small and uncomplicated. Living endotherms have respiration rates twenty times those of ectotherms and possess larger nasal passages with elaborate, scroll-like bones or cartilage called turbinates. These structures serve as countercurrent heat exchangers, recapturing heat and water lost to air passing through the lungs. Ruben's team believes that with higher body temperatures and respiration rates, warm-blooded animals could not function without these structures. Further, Ruben interprets the ribs and pelvises of theropod dinosaurs as indicating a breathing style like that of cold-blooded crocodilians.

Ruben's arguments have not convinced everyone. Small, simple nasal passages suggest dinosaurs breathed less efficiently than living endotherms. However, if warm-bloodedness evolved with dinosaurs, we might expect early endothermic physiologies to run differently or less elegantly than ones today. Ruben has also ignored several anatomical features that indicate dinosaurs did not simply breathe like crocodiles. Theropod dinosaurs had stomach ribs with larger muscle scars and more complicated articulations than those found in crocodilians. These probably played a more active role in respiration by assisting in the expansion and contraction of the trunk. Additionally, dinosaurs such as sauropods and theropods bore pockets and openings on their vertebrae and ribs. These structures (first noted by the English anatomist H.G. Seeley in 1870) match ones found in bird bones. Birds employ a unique breathing system involving extra air sacs branching off the lungs, with some additional sacs entering into vertebrae, ribs, and other bones. The pockets and openings in dinosaur bones imply that some were air filled and that dinosaurs might have used a breathing system more akin to that of birds.

Bone Growth

The microscopic features of bone represent the best record of dinosaur physiology. Bone tissue and structures reflect growth and metabolic rates of mineral exchange. In the early 1970s, "Dr. Bob" Bakker recognized that dinosaur bone, like that of mammals and birds, contained large numbers of blood vessels. He used this as one argument for the warm-blooded nature of dinosaurs. In the twenty-five years since his discoveries, paleontologists have continued to examine the microscopic nature of dinosaur bone. Their findings largely show most dinosaurs as metabolically distinct from cold-blooded animals. Nevertheless, some dinosaurs and even primitive birds show tree-ring like growth lines in their bones. These reflect a periodic halt in growth and usually typify bones found in ectotherms, whose growth slows seasonally. Overall, growth rates suggested by bone tissue vary, some being as slow as ectothermic ones, with others being clearly equivalent to those of endotherms.

Skin

One major obstacle to a warm-blooded interpretation of dinosaurs remains dinosaur skin. Among living vertebrates, external insulation such as hair and feathers occurs only in endotherms—birds and mammals. If we consider living animals, warm-bloodedness appears to require some protection from heat loss. Surprisingly, we have a fairly good fossil record of dinosaur skin; skin impressions of ankylosaurs, stegosaurs, iguanodons, hadrosaurs, ceratopsians, sauropods, and even large theropods have turned up with skeletal remains. These all show a pebbly texture suggesting a scaly lizard-like covering.

Until just very recently, skin impressions have been lacking for small theropods. With their air-filled bones, large brains, and close relationship to birds, these small, chicken- to human-sized predators appear to be the best candidates for endotherms among dinosaurs. The last few years have produced several dramatic finds from China. These specimens include at least three chicken-sized theropods and may represent some of the most significant fossil finds ever. Their skeletons preserve internal organs, stomach contents, unlayed eggs, and an outer covering of fibers. The apparently soft and pliable fibers branch, making them more like feathers than hairs. These structures cover much of the body and probably served as insulation.

The past twenty years have seen an explosion in dinosaur discoveries and research that has provided many new insights into dinosaur biology. For example, we know now that some herbivorous dinosaurs traveled in herds, and most of them bore on their heads extravagant horns, crests, or frills. These features served in various social displays: attracting mates, fending off rivals, and establishing pecking orders. Some paleontologists even believe that these herds migrated like modern wildebeest and caribou. A few carnivorous dinosaurs may have hunted in packs, while other dinosaurs nested in colonies and cared for their young.

Although such discoveries and ideas make dinosaurs more interesting, none of these hypothesized behaviors makes dinosaurs warm-blooded. All of their actions can be found among living ectotherms, and none are universal to endotherms.

Geographic Range

The behavioral features do favor warm-bloodedness or something approaching it: the geographic range of many dinosaurs and theropod reproduction. Living cold-blooded animals occur largely within 45 degrees north and south of the equator. In contrast, warm-blooded animals range through almost all latitudes. Several groups of dinosaurs, including ankylosaurs, ceratopsians, hadrosaurs, and theropods, occupied latitudes 70 to possibly 80 degrees away from the equator. Although climates were far warmer then, the drastic seasonal changes in day length associated with these high latitudes would present a severe environmental challenge. Additionally, the ranges of these dinosaurs exceeded the latitudinal distributions of their cold-blooded contemporaries.

Currently, good information exists about reproduction for only two theropod dinosaurs, *Oviraptor* and *Troodon*. Both of these small theropods sit close to the ancestry of birds. Crews of the American Museum of Natural History have discovered exquisitely preserved adult Oviraptors sitting on top of egg clutches. A well-preserved nest with eggs for Troodon suggests that it too sat on its eggs to incubate them. The brooding behavior we associate with living birds apparently evolved first among their theropod dinosaur ancestors. Brooding birds take advantage of their body heat to raise the temperature of the eggs above that of the environment. Oviraptor and Troodon would seem to have had this same capability. Additionally, these two theropods never retained eggs internally, laying them as they were formed. Birds, unlike reptiles, do not retain eggs either. Most primitive living birds such as ostriches, turkeys, and ducks lay large clutches over many days. To ensure that the eggs hatch synchronously, adults do not begin brooding the eggs until the entire clutch has been laid. Initiation of brooding raises the temperature of the embryos and triggers development. If eggs were retained internally, the high body temperature would initiate incubation. Consequently, eggs formed days apart would hatch days apart. Brooding and the lack of egg retention in *Oviraptor* and *Troodon* implies that they too maintained high body temperatures.

The question of cold- or warm-bloodedness in dinosaurs will remain a complex one. Dinosaurs were a varied and diverse group with a fossil record extending over 160 million years. Paleontologists lack any direct evidence for metabolism, and available data are typically incomplete. For example, although there exists good reproductive information for *Troodon,* we lack skin impressions and well-preserved nasal passages for them.

Despite these limitations, a few generalizations persist. Dinosaurs appear to span the gap between cold-and warm-bloodedness. Most dinosaurs lacked some attributes such as insulation, large nasal passages, and elaborate turbinates associated with living endotherms. Further, some species had intermittent growth and ectothermic-like growth rates. Nevertheless, an upright stance and a potential for large size clearly distinguish dinosaurs from their cold-blooded relatives. The freeing of the lungs produced by an upright stance, although not conclusive of warm-bloodedness, certainly allowed for the evolution of sustained activity. Some dinosaurs displayed additional features characteristic of warm-bloodedness. Ceratopsians and hadrosaurs had modest-sized brains, elaborate chewing mechanisms, rapid growth, and geographic distributions extending into polar regions. Small theropods with the same egg-laying and brooding behavior as living birds, large bird-sized brains, and insulation would thus appear to be the near physiological equivalents to their flying descendants.

Chapter 7

Larry Martin's case against the theropod ancestry theory of bird evolution is clearly given in his review "The Big Flap" in The Sciences, *March/April 1998, pp. 39–44. The Sciences was a terrific magazine with top-notch articles and beautiful artwork. Sadly, it is now out of print. Martin's review summarizes many of the recent debates over bird evolution, including Sankar Chatterjee's highly controversial claim to have found a fossil bird from the Triassic— long before* Archaeopteryx *lived. Martin is one of the few specialists who continue to defend Chatterjee's* Protoavis.

Although the next reading, "The Origin of Birds," is from an encyclopedia, the authors, Kevin Padian and Luis Chiappe, are strong proponents of the theropod ancestry hypothesis. They review other models of bird ancestry but make a strong case for their position. The writing is somewhat more technical than the review by Martin, but their anatomical and cladistic arguments should be comprehensible to anyone who has read Chapter 7 in this book.

The Big Flap

Larry D. Martin

A Review of:
The Origin and Evolution of Birds
By Alan Feduccia
Yale University Press, 1996

The Rise of Birds: 225 Million Years of Evolution
By Sankar Chatterjee
The Johns Hopkins University Press, 1997

Taking Wing: Archaeopteryx and the Evolution of Bird Flight
By Pat Shipman
Simon & Schuster, 1998

In 1963, as an undergraduate at the University of Nebraska–Lincoln, I pulled open at dingy drawer in the university's natural history museum. I was looking for the bones of *Hesperornis regalis,* a loonlike, flightless, toothed bird that lived near the end of the Mesozoic era, some 80 million years ago; I wanted to make a model of the bird as a project for an ornithology class. And what lay inside the drawer, cloaked in dust, did indeed bear the yellowing label *"Hesperornis."*

But the bird was not *Hesperornis.* The half skeleton lying there was so small it made me think instead of *Baptornis advenus,* a related but more primitive bird that lived during the same period. The only trace of *Baptornis* I had seen until then was a foot bone—and even that was just a picture in a nineteenth-century book by Othniel C. Marsh, the famous Yale University paleontologist. No remotely complete *Baptornis* specimen had ever been identified. What is more, so few specimens of Mesozoic birds existed in the 1960s that virtually no paleontologists were studying them.

By now the reader has surely guessed the punch line to my little story: The jumble of bones in the drawer that day did indeed turn out to be those of *Baptornis.* But for me, the discovery was much more than a punch line; it was an intellectual turning point, and it led me to make a lifelong study of Mesozoic birds—birds that lived in the age of the dinosaurs.

My story does not really end there: it goes on, with a twist. For ten years I happily studied these animals, making what scientific contributions I could out of profoundly limited evidence. But I never had serious questions about what I (and virtually every other paleontologist of the time) believed to be their basic identity: they were birds—early evolutionary examples of their type, of course, but certainly not dinosaurs.

Then came the sensational conclusions of another Yale paleontologist, John H. Ostrom. In 1973 Ostrom proposed a radical theory that turned the world view of most paleontologists upside down: he claimed that dinosaurs gave rise to birds. Ostrom had studied the bones of the 150-million-year-old *Archaeopteryx lithographica*—considered the oldest known bird specimen since it came to light in Germany in 1861—and he was convinced that *Archaeopteryx* had inherited many of its features directly from a small meat-eating dinosaur.

Ostrom's theory was not new. The English biologist Thomas Henry Huxley had launched the idea of a dinosaurian origin for birds in 1868. But Huxley's idea had long since gone out of favor, replaced by the theory that birds evolved from reptiles that lived before the dinosaurs. Still, Ostrom was certain: a robin twittering outside your window is nothing less than a modern-day dinosaur, a miniature feathered descendant of *Tyrannosaurus rex.*

Now the pendulum has begun to swing back once again. In recent years some investigators—I among them—have become skeptical about the link between birds and dinosaurs. In spite of recent fossil finds that might support a dinosaurian origin for birds, other new evidence contradicting that view is just as strong, if not stronger. Last fall the journal *Science* published two studies, one focusing on lungs and the other on hands, both arguing that dinosaurs are clearly distinct from the birds to which they supposedly gave rise. Each new round of evidence, it seems, begets a louder round from the other side. Eddies of controversy swirl around the beginnings of other vertebrates, to be sure, but the debate about the origins of birds qualifies as something much stormier.

That debate is examined in detail in the three books under review. First, though, full disclosure: I am friends with each author, and my work and opinions are cited in each book. So I will not pretend to impartiality about what has been dubbed the paleontological paternity suit of the century.

I will say, however, that science should be accessible to anyone with an ordinary education, and all three books serve that end. They give the reader the right mix of background and contemporary theory to understand the arguments surrounding the origins of birds and avian flight. The books also serve a wider purpose: they show the reader the reality of science at work, the strengths of science and scientists, along with their weaknesses. And finally they tell a splendid story: how, thirty-five years after I stumbled on *Baptornis* sleeping in the museum drawer, my small and esoteric corner of science has become—unimaginable to me then—the center of an intellectual hurricane.

In the great bird-dinosaur debate, the participants huddle in two camps, which paleontologists have nicknamed "ground up" and "trees down." The ground-up camp sides with Ostrom: it holds that birds evolved from theropods, an order of dinosaurs that includes *T. rex* and other predators that walked upright. More specifically, according to the theory, birds belong to a specialized group of theropods called maniraptors. That group includes *Velociraptor mongoliensis,* one of the more memorable stars of the films *Jurassic Park* and *The Lost World.*

The ground-up partisans' view *Archaeopteryx*—with its toothed jaws, long lizardlike tail, feathered body and modern-looking wings—as a missing link between dinosaurs and birds. And they have recently claimed more support for their theory. In October 1996, at a meeting of the Society of Vertebrate Paleontology in New York City, paleontologists revealed the discovery of *Sinosauropteryx prima,* a chicken-size dinosaur unearthed in northeast China. Some supporters of the ground-up theory maintain that the strange, bristlelike fibers on the four extant specimens' backs are the precursors of feathers. And in May 1997 another alleged missing link came to light: the Argentine paleontologists Fernando E. Novas and Pablo F. Puerta reported the discovery of a Great Dane–size theropod dinosaur, *Unenlagia comahuensis,* with arms that might have folded up against its body like the wings of a bird.

But the ground-up theory has a fatal flaw: it has to assume that a theropod dinosaur, a decidedly earthbound creature, somehow managed to lift off and take wing. In *The Origin and Evolution of Birds* Alan Feduccia argues persuasively that such an evolution of flight would have been biophysically impossible. Theropods have long vertical hind legs that, like the centerpiece of a teeter-totter, balance the body and long tail: their front legs, in contrast, are short. That is just the wrong build for flying or even for climbing trees. Moreover, a bird's body is flattened from top to bottom, which makes it aerodynamically ideal for flight. Theropods were flattened from side to side.

Feduccia goes far beyond those criticisms. An ornithologist and evolutionary biologist at the University of North Carolina at Chapel Hill, he makes the best case available to the educated public for a reasonable alternative to the ground-up theory: the trees-down theory. Unlike the proponents of the ground-up theory, trees-down advocates think that birds are not closely related to dinosaurs—that they are more like cousins than sons or daughters. Instead of evolving directly from dinosaurs, birds evolved from dinosaur ancestors known as thecodonts.

Thecodonts include small reptiles that lived during the Triassic period, some 245 million years ago. Given their foot structure and body proportions, it seems that some lived in trees and may have leaped between branches, eventually evolving elongated, featherlike scales on the back edges of their arms, for gliding. According to the theory, those scales made possible the evolution of wings and, eventually, powered flight. Feduccia discusses several small Triassic thecodonts that were probably arboreal, including *Megalancosaurus preonensis* and *Longisquama insignis*.

But birds do more than fly: they also walk on two legs, and so the trees-down theory must explain bipedalism as well as flight. To that end, Feduccia devotes part of his book to extending an idea I published in the 1983 volume of *Current Ornithology*. Birds, I suggested, might have developed their two-footedness through vertical clinging and leaping. Feduccia shows how, in two species of lemur that use such locomotion, the hair on the back of the arms has become long and featherlike. And in 1994 at the International Ornithological Congress in Vienna, I showed that *Archaeopteryx* had a vertical walking posture. Theropods held their backs horizontal when they walked; they could not have climbed tree trunks, as *Archaeopteryx* did.

A truly original scenario for the genesis of avian flight comes from Sankar Chatterjee, a paleontologist at Texas Tech University in Lubbock and a strong supporter of the bird-dinosaur link. In *The Rise of Birds* he proposes that flight evolved in small, unknown Triassic theropods that were already spending part of their time in trees. I find it hard to argue the details of the life of an imaginary animal, but I enjoyed the whimsical suggestion that such a creature lived near water and often leaped from trees, breaking its falls with a splash.

The Rise of Birds is the most radical of the three books under review. Chatterjee maintains that two partial skeletons and some isolated bones he discovered in 1983 in the Texas desert belong to a Triassic bird, which he calls *Protoavis* (Latin for "first bird") *texensis*. The fossils are only a few million years younger than those of the oldest known dinosaurs, and some 75 million years older than those of *Archaeopteryx*. If *Protoavis* is accepted as a bird, therefore, it will knock *Archaeopteryx* off its perch as the earliest known avian species. But Chatterjee's anatomical interpretations of the *Protoavis* bones would demand that *Archaeopteryx* belong to a line that branched off *before Protoavis*. That logic would almost ensure that birds antedate the oldest fossil record for dinosaurs—a viewpoint opposite the one Chatterjee claims to hold. At the very least, Chatterjee's interpretations put him in the awkward spot of promoting the extravagant idea that birds evolved from an elusive group of dinosaurs that lived earlier than any dinosaur specimens yet unearthed.

Chatterjee's find exacerbates a discomfiting paradox for the ground-up theory: the dinosaurs that most closely resemble birds come mainly from the very late Cretaceous period. They are much younger fossils than those of *Archaeopteryx*. Because related animals usually become less alike with the passage of time, one would expect that the most birdlike dinosaurs should be even older than *Protoavis*.

But they are not. In fact, the resemblance between dinosaurs and birds *increases* with the passage of time. Such a historical pattern is a hallmark of convergence, the development of superficial similarities in species that have evolved independently of one another.

Another difficulty for Chatterjee's account is that it casts doubt on a cladistic analysis that has been the chief prop for the dinosaurs-to-birds theory. That analysis was done in 1982 by the paleontologist Kevin Padian of the University of California at Berkeley, and was expanded in 1986 by Padian's former student, the paleontologist Jacques A. Gauthier, now of Yale. (Cladistic analysis is a system of biological classification used by most paleontologists to define evolutionary relations uniquely by shared characteristics not found in ancestral groups.) Padian and Gauthier's studies describe birds as the ultimate innovation of the dinosaurs. But if their view is correct, and if *Protoavis* turns out to be a bird, then every branch of the dinosaurian evolutionary tree would already have appeared by the late Triassic period, some 210 million years ago. Even more implausibly, dinosaur evolution would have stood still for the subsequent 145 million years of its known fossil record.

With all those difficulties, is it any wonder that supporters of the bird-dinosaur link have heaped more criticism on their mate Chatterjee than on anyone who opposes the dinosaurian origin of birds?

Several paleontologists have gone so far as to dismiss the bones Chatterjee calls *Protoavis* as a hodgepodge of unrelated fragments that shed no new light on the bird-dinosaur question. But I have examined the fossil and I can say, as the paleontologist Lawrence M. Witmer of Ohio University in Athens does in the foreword to *The Rise of Birds,* that some of the bones are well preserved and fit together reasonably well as parts of a single skeleton. The fossil is hardly less informative than the original *Deinonychus antirrhopus* fossils, which figured heavily in Ostrom's first paper resuscitating the dinosaurs-to-birds theory. (*Deinonychus* was a man-size, birdlike theropod dinosaur from the early Cretaceous period, equipped with a vicious killing claw.) In his paper, Ostrom misidentified a bone from the shoulder of *Deinonychus* as one from the hip.

Chatterjee's anatomical descriptions of *Protoavis* are not unreasonable. A key feature of the fossil may be the special structures for jaw movement he describes that are a unique feature of modern birds but are not known in *Archaeopteryx*. Most of Chatterjee's descriptions, however, lack the rigor needed to settle such contentious issues, and his restorations tend toward the fanciful. His restoration of the *Archaeopteryx* skull, for instance, makes several errors of position and proportion. Such mistakes cast doubt on his restoration of the *Protoavis* skull, which is based on fossil remains that are much more fragmentary.

Skepticism is a reasonable stance in any scientific investigation. But it is virtually indispensable when confronting a juggernaut, a theory you want so much to believe that it casts a powerful spell over your own good sense. In the 1970s I had accepted the dinosaurian theory of bird origins, dutifully comparing the upper arm bones of *Baptornis* with Ostrom's avian avatar *Deinonychus*. Much of the early success of the bird-dinosaur theory as expressed by Ostrom came from another idea developed in the early 1970s by Ostrom and Robert T. Bakker, who was then Ostrom's student at Yale and is now a paleontologist at the Tate Museum of Casper College in Wyoming. Ostrom and Bakker suggested that dinosaurs were warm-blooded.

Warm-bloodedness seized the imagination of the public and paleontologists alike. The possibility that dinosaurs shared with mammals and birds certain advanced traits of intelligence, activity and complexity of behavior was hugely appealing. The burning question then became how to study physiology and behavior—attributes that do not fossilize well—in animals so long dead. The suggestion that birds were living dinosaurs answered that question.

In retrospect, it is probably telling that most of the scientific support for the dinosaurian origin of birds came from the people studying dinosaurs, who were delighted to learn that their subjects were still alive. As Feduccia points out, most ornithologists did not like the theory then, and they do not like it now. I began to grow disenchanted with the bird-dinosaur link when I compared the eighty-five or so anatomical features seriously proposed as being shared by birds and dinosaurs. To my shock, virtually none of the comparisons held up. For example, the characteristic upward-projecting bone on the inner ankle in dinosaurs lies on the outer ankle in birds. In some cases I even discovered that the supposedly shared features occurred on entirely different bones. That is a bit like saying that you and I are related because my nose resembles your big toe.

The confusion over anatomy stems in part from spotty ornithological literature. Although many ornithologists study the songs, brilliant plumage and behavior of birds, few choose to scrutinize the smelly bones and muscles. By the same token, dinosaur specialists who advocate a bird-dinosaur link have been largely content to leave avian anatomy to the ornithologists. So it is not surprising that the literature is vague about many aspects of the avian skeleton.

The consequence is that the supporters of a bird-dinosaur relation often learn, to their horror, that a certain aspect of dinosaur anatomy is not in fact "just the way it is in birds," as they had previously announced. Damage control then usually takes one of three tactical forms. The investigators may simply ignore the inconsistency, because so much other support exists for their hypothesis. They may change the interpretation of dinosaurian anatomy to match the avian model. Or they may agree that modern birds have a certain anatomical construction, but assert that *Archaeopteryx* is different and more like a dinosaur. (Indeed, the existing anatomical knowledge about both dinosaurs and *Archaeopteryx* is just blurry enough to leave a broad middle ground where all the bird-dinosaur comparisons that do not precisely match can be justified nonetheless.)

Such Band-Aids have been applied to almost every anatomical feature that supposedly links dinosaurs and birds. When the burden of ad hoc repairs became too heavy for me, I had to abandon the theory altogether. It was a disappointment. How wonderful it would have been if dinosaurs had escaped extinction! But I would also like to believe that Anastasia survived the murder of the Czar's family. The hard facts—Hollywood movies notwithstanding—have pretty much laid both hopes to rest.

A few simple principles could ensure at least a dash of rigor in evaluating whether some proposed anatomical feature supports a link between birds and dinosaurs. The proposed feature ought to be visible in both birds and dinosaurs. Its constriction ought to be similar enough across both groups that it likely occurred in the common ancestor. I would discourage the use of features that are widely distributed among unrelated groups. They represent easy solutions to widely shared morphological problems. For example, the elongation of the finger bones just behind the claws, a trait shared by dinosaurs and *Archaeopteryx,* occurs in many flying and climbing animals, including bats and flying reptiles.

In *Taking Wing,* a wonderful exploration of the way bird flight may have begun, Pat Shipman, an anthropologist at Pennsylvania State University in State College, makes a determined effort to sort out the differences between my anatomical interpretations and those of Ostrom. In part, those differences arise from cladistic analysis, which often cites such a large number of anatomical traits that only the most superficial effort may be made to determine whether those traits are homologous. As the reader may have surmised by now, I do not necessarily believe that cladistics reveals truth, though I have used cladistic analysis. I prefer to analyze traits individually, without regard to any theory of avian origins. And I would commend to any student Feduccia's well-documented argument in *The Origin and Evolution of Birds* that investigators who focus on superficial similarities get entangled in evolutionary convergences.

A classic example of a cladistic trap is the trait that convinced Ostrom of his theory: the half-moon bone in the wrists of dinosaurs and birds. A half-moon shape is a sign that the animal rotates its hand extensively in a single plane. In birds the bone enables the flight feathers to be tucked against the body when the wing is folded. In theropod dinosaurs the bone enables the hand to be tucked against the body while the animal is running. Both birds and dinosaurs have fewer wrist bones than do their more primitive relatives. Some of the wrist bones of birds and dinosaurs, along with part of the hand and two fingers, were lost during evolutionary development. Thus it is crucial in the bird-dinosaur debate to make certain that the bones being compared are homologous, and that the observed similarities actually came from a common ancestor.

So here is where the rubber meets the road. Ostrom says *Deinonychus* had a wrist "almost identical" to *Archaeopteryx.* But that statement cannot be true. *Deinonychus* has only two bones in the wrist, though its half-moon bone may be the result of a fusion with another bone. *Archaeopteryx* resembles all other known birds in having a total of four bones arranged in two rows. One row of two bones is connected to the arm, the other row to the hand.

Ostrom saw only half as many bones in the wrist of *Deinonychus.* He reported that both of them, including the half-moon bone, lay in the row connected to the arm. (The row connected to the hand, he believed, did not exist.) Hence by Ostrom's own logic the half-moon bone in *Deinonychus* occurs in a different part of the wrist from where it occurs in birds. Therefore, though the half-moon bones in the wrists of birds and dinosaurs look alike, they develop from completely different wrist bones during the animals' growth. Birds and dinosaurs must have evolved the trait separately. The half-moon bone should have been dismissed immediately from consideration as a homologous bone, a shared characteristic. Instead it has continued to be cited for decades.

How did such a blunder take place? One can begin to appreciate the disarray into which the data can fall by comparing Chatterjee's descriptions of wrists with those of Shipman. In his book Chatterjee has a drawing of a *Deinonychus* wrist with *three* bones arranged in two rows. He also shows *Archaeopteryx* with two rows of wrist bones, as in modern birds. Shipman, by contrast, uses Ostrom's *Deinonychus* wrist with a single row of bones and indicates that the same arrangement occurs in *Archaeopteryx*.

Who is right? They agree only that somehow the wrists of the two animals support the theory that birds evolved from dinosaurs. It turns out that Chatterjee is more correct when it comes to the wrist of *Archaeopteryx and* Shipman is more correct about *Deinonychus.* But the moral of the story is that such poor attention to detail has been repeated with almost every feature cited to support a bird-dinosaur relation. No wonder Feduccia's book has an undercurrent of righteous outrage, or that it has been bitterly attacked by the practitioners of the faulty logic it exposes.

In spite of such technical shortcomings, the books under review all make fine introductions to a fascinating scientific story. Taken together, they paint a rich picture of what Shipman calls the "brilliant deductions, wild speculations, penetrating analyses, and amazing insights" that feed into the bird-dinosaur controversy. Chatterjee's *The Rise of Birds* strongly supports the theropod hypothesis. Shipman's *Taking Wing* offers a more-or-less balanced approach, though she concludes by supporting the dinosaur theory. Feduccia's *The Origin and Evolution of Birds* opposes any dinosaur connection. Of the three, Feduccia's is the most scholarly—indeed, it is a landmark publication in ornithology. No other work has presented so much of the original data or discussed it so clearly. The book has won high praise from distinguished ornithologists. It has also been harshly criticized by a subset of the bird-dinosaur school for its less-than-glowing treatment of cladistic theory, though the book presents and discusses almost all the pertinent cladograms.

So the controversy remains. As with most paleontological battles, the basic problem for the bird-dinosaur debate is a lack of evidence on either side. The theropod hypothesis suffers from the utter unbelievability of flight evolution from the ground up; its partisans also lack fossils from the relevant time period. The trees-down group, too, lacks fossil evidence: no partly feathered thecodont has ever been found.

If the recently discovered *Sinosauropteryx* were a feathered dinosaur, as some paleontologists have claimed, that would certainly go a long way toward convincing me that birds are dinosaurs. Unfortunately, a team of paleontologists, including me, sent by the Academy of Natural Sciences in Philadelphia last spring to examine the specimens, found no evidence for feathers or for structures that would give rise to them. And the zoologist John A. Ruben of Oregon State University in Corvallis recently dissected a sea snake's tail to show that frayed collagen fibers under the skin can look feathery.

That state of limbo has quickened the pace of studies bringing new evidence to light; in that sense, the controversy has led to a great deal of productive science. Recently the paleontologists Mark A. Norell and Peter J. Makovicky of the American Museum of Natural History in New York City and the paleontologist James M. Clark of George Washington University in Washington, D.C., reported finding a supposedly diagnostic avian feature, the wishbone, in a *Velociraptor* specimen. But the wishbone (two clavicles fused together) also occurs in the small nondinosaurian Triassic reptile *Longisquama*. As Feduccia points out in *The Origin and Evolution of Birds, Longisquama* also has long, featherlike scales. If it were a dinosaur it might well be hailed as the centerpiece of the bird-dinosaur argument.

A study reported last October by Feduccia and the developmental biologist Ann C. Burke, also of the University of North Carolina at Chapel Hill, has dealt a particularly hard blow to the dinosaur theory. Feduccia and Burke showed that although both birds and dinosaurs have lost two digits of the hand, those digits are not the same. Whereas birds lost digits one and five, theropod dinosaurs lost digits four and five. Another study, led by Ruben and reported last November, has suggested that dinosaurs, like crocodiles, had piston-driven breathing, unlike the lung system found in birds. And so the battle rages.

As I weigh those recent finds, it looks to me as if the dinosaur connection is in trouble. Yet old desires die hard. A colleague of mine recently told me that the dinosaur hypothesis should be maintained because no clear counterhypothesis exists to replace it. I found that suggestion dangerously similar to arguing that I should follow some kind of religious belief because if I do not, I will be without faith.

As scientists, though, we must remind ourselves that unjustified belief is no better than a person's opinion—in fact, the two are synonymous. We do not share our faith with one another—only evidence and logical argument. If our knowledge has gaps, the questions remain open, and we must be content with the search.

Bird Origins

Kevin Padian

University of California
Berkeley, California, USA

Luis M. Chiappe

American Museum of Natural History
New York, New York, USA

Since the 1970s it has come to be nearly universally accepted that birds evolved from small carnivorous dinosaurs most closely related to Dromaeosaurids, probably sometime in the Middle to early Late Jurassic. *Archaeopteryx* is the first known bird, now represented by seven skeletons and a feather from the Late Jurassic Solnhofen limestones of Germany. Other records of Late Jurassic birds so far have been questionable or apocryphal, although research in the past decade continues to unearth new Early Cretaceous birds that are only slightly more derived than *Archaeopteryx* (Chiappe, 1995; Padian and Chiappe, 1997). These new finds help to fill the stratigraphic and morphological gaps between *Archaeopteryx* and more derived, Late Cretaceous birds such as *Hesperornis* and *Ichthyornis,* which have been known for well over a century.

Hypotheses of Bird Origins

As reviewed by Gauthier (1986) and Witmer (1991), there are three major hypotheses of bird origins. One is that they evolved from an unspecified group of basal archosaurs characterized by the disused waste-basket term "thecodonts." A second is that they share an immediate common ancestor with crocodylomorphs. A third is that they evolved from small theropod dinosaurs. Other suggestions have been made, including common ancestry with lizards, pterosaurs, or mammals, but these ideas were based on only superficial resemblances in a few features and were discredited long ago (Gauthier, 1986).

"Thecodont"Hypothesis
This can be traced to the early 1900s but reached its most detailed statement in 1926 with the English-language publication of Gerhard Heilmann's classic *The Origin of Birds* (an earlier version, *Fuglenes Afstamning,* was published in Danish in 1916). Heilmann's book was an exceptionally thorough consideration of avian biology, including skeletal anatomy, embryology, musculature, pterylosis, paleontology, and many other subjects. Heilmann found that theropod dinosaurs were most similar of all fossil groups to *Archaeopteryx* and other birds, but he rejected a theropod ancestry because theropods lacked clavicles; hence, under his interpretation of Dollo's law of evolutionary irreversibility, the clavicles (furcula) of birds could not have reevolved from a theropod precursor. He concluded that the origin of birds must have been among more archaic Archosaurs . . . , perhaps forms related to *Ornithosuchus* or *Euparkeria,* which had clavicles. Clavicles have since been found in the basal ceratosaurian theropods *Coelophysis* and *Segisaurus,* and a fully formed furcula has been recently discovered in tetanuran theropods ranging from *Allosaurus* and tyrannosaurs to *Velociraptor, Oviraptor,* and *Ingenia.* Some critics contend that the avian furcula is a neomorph not homologous to the reptilian clavicles, partly because the latter are apparently lost in ornithodirans (they are absent in pterosaurs and not known in any nontheropodan dinosauromorph) and partly because in some recent birds the furcula seems to be composed of both dermal and endochondral bone. Regardless of these facts, there is no doubt about theropod monophyly, so the homology of the ceratosaurian clavicles and the tetanuran furcula would not seem to be in question; also, there is no mistaking the identity in shape and position of the boomerang-shaped furcula in nonavian tetanurans, *Archaeopteryx,* and other birds.

Heilmann's retreat to a thecodont hypothesis was a default argument; as he recognized, no features that linked any particular basal archosaur to birds are also not found in theropods, usually with greater similarity. Since Heilmann's work, other authors have advocated a thecodont hypothesis, and the approach is very much the same. No specific candidate among basal archosaurs has been presented as the direct ancestor or the closest known animal to birds; rather, a range of forms with one or two supposedly bird-like characters is advanced, even though most of their character states are far more primitive than those in theropods (see Witmer, 1991, for a thoughtful review). Tarsitano (1991) has advanced most forcefully the idea that the origin of birds is to be found among such "avimorph thecodonts," but cladistic analyses have found that all these animals are more closely related to other forms quite distant from birds. For example, *Cosesaurus* and *Megalancosaurus* are aquatic prolacertiform archosauromorphs, *Scleromochlus* is the closest known sister group to pterosaurs, and *Lagosuchus* and *Lagerpeton* are closest to basal dinosaurs (Gauthier, 1986; Sereno, 1991; Benton, 1988; Padian and Chiappe, 1997). *Protoavis* (Chatterjee, 1991,

1995, 1997) has been advanced as a Triassic bird but has been met with substantial skepticism (reviewed in Padian and Chiappe, 1997); there are apparently even more differences of opinion about the interpretation of its morphology than there are about *Archaeopteryx*. The question has been made more difficult by the circumstance that, in the more than two decades since this controversy was renewed, no advocate of a thecodont ancestry has produced a cladogram incorporating all, or even any, of the available evidence that would support such a case. Cladistic analyses are not infallible but at least they are explicit: If the weight of evidence supports a different interpretation than the several independent analyses that have placed birds within the theropods, then this result would be very interesting to see expressed in cladistic terms.

Crocodilian Hypothesis

This should more properly be termed the "crocodylomorph" hypothesis because its advocates regard the ancestry of birds as complete before true crocodiles evolved. Indeed, it is within the sphenosuchian crocodylomorphs, outside Crocodylia, that bird-like characters appear to have been most pronounced. A. D. Walker (e.g., 1977) has been the chief advocate of this view, based on his detailed studies of the braincase, quadrate, ear region, and other features of *Sphenosuchus,* an Early Jurassic crocodylomorph from South Africa. His view has been generally supported by L. D. Martin and his students (e.g., Martin, 1983), although they have tended to draw similarities to birds from true crocodilians as much as from crocodylomorphs. Many of these similarities are valid but have been shown to apply either to a more general level among archosaurs or to have evolved convergently in certain crocodilians and early birds (but not present in the hypothesized common ancestor of both groups) (Gauthier, 1986). Again, no cladogram incorporating all the available evidence has to date supported a crocodylomorph origin of birds.

Theropod Hypothesis

This had its roots in the 1860s with T. H. Huxley, who noted in a series of papers (e.g., 1870) a suite of 35 characters shared uniquely by birds and theropod dinosaurs (reviewed by Gauthier, 1986, pp. 4–6). Many of these are still considered valid today, whereas others have turned out to be more general to dinosaurian or other archosaurian groups. Desmond (1982) and Gauthier (1986) have both noted that Huxley's hypothesis, presented to the Geological Society of London in 1870, was contested by Harry Govier Seeley, who "thought it possible that the peculiar structure of the hinder limbs of the Dinosauria was due to the functions they performed, rather than to any actual affinity with birds." This shadow of potential convergence, though not explicitly tested either by Seeley or anyone since, nonetheless not only frustrated the acceptance of Huxley's views at the time but also continues to be contended by opponents of the theropod hypothesis (e.g., Feduccia, 1996).

In the 1970s John Ostrom, in a series of papers (e.g., 1975a,b, 1976b), demonstrated the detailed similarities of *Archaeopteryx* to theropod dinosaurs. Although he did not specify a taxon within Theropoda to which birds might be directly connected, his comparisons tended to run to the dromaeosaurid *Deinony-*

chus, which he had recently described in a monograph (Ostrom, 1969). As it turned out, dromaeosaurids were found to be the closest sister group to birds in several independent cladistic analyses beginning in the early 1980s by Padian, Gauthier, Benton, Sereno, Holtz, Perle *et al.,* and others (see Padian and Chiappe, 1997). Synapomorphies that link dromaeosaurids and *Archaeopteryx* include the presence of dorsal, caudal, and rostral tympanic recesses, the semilunate carpal, thin metacarpal III, longer pubic peduncle, posteroventrally directed pubis with only a posteriorly projecting foot, shortened ischium, and other features of the skull, pectoral girdle, and hindlimb (Gauthier, 1986; Padian and Chiappe, 1997).

A great many characters classically considered "avian" apply to more general levels within Theropoda. Basal theropods have lightly built bones and a foot reduced to three main toes, with the first usually held off the ground and the fifth lost. Closer to birds, the fifth and fourth digits of the hand are progressively reduced and lost, the skeleton (especially the vertebrae) becomes lighter, and the tail becomes shorter as its vertebrae partially interlock through the elongation of zygapophyses to reinforce its stiffness. In Coelurosaurs (*sensu* Gauthier, 1986, the sister taxon to Carnosaurs), contrary to the picture suggested by opponents of the theropod hypothesis, the forelimbs become progressively longer until they are nearly as long as the hindlimbs in some dromaeosaurs; the first toe (hallux) begins to rotate behind the metatarsus, although it does not descend to the point seen in perching birds; the metatarsals become longer; and the scapular blade becomes longer and more strap-like. The presence of the furcula may turn out to be a general tetanuran character, and it is not yet clear how general the calcified sternum in adults may be (it is known, for example, in some oviraptorids, dromaeosaurs, tyrannosaurs, and sinraptorids).

Holtz (1994, 1996) reevaluated the phylogenetic relationships of Theropoda, and his conclusions uphold Gauthier's (1986) comprehensive analysis, with some adjustments that do not affect the position of birds within Theropoda.

Clavicles, Digits, Feathers, and Stratigraphy

Individual characters are sometimes advanced as conclusive evidence that birds could not have descended from theropod dinosaurs. For example, it is claimed that the digits of the bird hand are II-III-IV, whereas they are I-II-III in theropods; that the semilunate carpal bones of maniraptorans and other theropods are different from those of *Archaeopteryx* and the birds; and that the ascending process of the astragalus is not the same in theropods and birds. [Feduccia (1996) reviews these claims favorably, and to obviate extensive literature citations readers are referred to his work for historical background and strong advocacy.] These statements amount to hypotheses that the characters are not homologs but homoplasies. Since Darwin's day, homologies have been recognized as features inherited from common ancestors; even to non-Darwinians, such as Richard Owen, homologies were established by similarity of morphology, position, development, and histological structure. Roth (1988) proposed that, like monophyletic taxa, homologies are defined by ancestry but diagnosed by features such as the criteria just listed. In comparative biology, the recognition of homologous structures in two or more organisms can be tested using the phylogenetic distributions of other, presumably independent characters and by including more taxa in the analysis.

Proceeding according to this method, the hypothesis that the furcula of birds is not homologous to the clavicles of theropod dinosaurs is weak both because undoubted nonavian tetanuran theropods have boomerang-shaped clavicles in the same position as those of *Archaeopteryx* and other basal birds and because basal theropods have structures that are manifestly similar in morphology and position to clavicles in other tetrapods.

Digits of the Bird Hand

The three remaining digits of the bird hand are sometimes regarded as I, II, and III and sometimes as II, III, and IV. The theropod hand is unquestionably I, II, and III, and opponents of the theropod hypothesis of avian descent are unanimous in their contention that the bird hand is II, III, and IV (therefore, the digits could not be homologous between the two groups). This is often inaccurately portrayed as a difference between paleontologists and ornithologists: Feduccia (1996, p. 2), for example, reproduced a figure of the avian skeleton by Lucas and Stettenheim with the digits numbered II, III, and IV, but reproduced on page 7 of the same work two figures, one by Van Tyne and Berger and another by Burton and Milne, in which the digits are numbered I, II, and III. These illustrations are all by ornithologists. Classical ornithologists from W. K. Parker to Proctor and Lynch (1993) have agreed, and Heilmann (1926) examined the problem at length and reached the same conclusion: I, II, and III.

The evidence for II, III, and IV comes entirely from some interpretations of the ontogeny of the hand in some living birds. As Heilmann (1926) noted, interpretations on this basis have varied. Some have been based on an assumption of Morse's "law" that digits must be lost from the both sides inward (as in the bird foot: 5, then 1; and in horses' feet: 5, then 1, 4, and 2), although this pattern must surely be related to the weight-bearing function of the locomotory structures. Some authors have even claimed to find *Anlagen* of four digits in the bird hand. Hinchliffe and Hecht (see Hecht and Hecht, 1994) have strongly advocated the modern developmental case for II, III, and IV based on the presence of an ephemeral "element X" medial and palmar to the wrist in early ontogeny that is taken for a remnant of digit I. However, as Shubin (1994) and Padian and Chiappe (1997) note, *Anlagen* do not appear with labels on them but have to be interpreted. To accept the II-III-IV view, the first digit, including its carpal and metacarpal, has to be lost, and digits II, III, and IV have to assume the precise forms, articulations, and proportions of the original digits I, II, and III. If this is possible, developmental biology has so far not provided examples or mechanisms from living tetrapods that support this process. Unfortunately, we do not have similar embryological stages for *Archaeopteryx* or any other Mesozoic birds or dinosaurs, so we cannot examine phylogenetically the hypothesis of element X outside living birds.

In favor of the I-II-III hypothesis, the theropod hand follows a consistent pattern of reduction and loss of elements from the lateral side medially (Shubin, 1994; Padian and Chiappe, 1997). Crocodilians and other ornithodiran outgroups have a five-fingered mantis in which the third digit is longest. In Dinosauria the fourth and fifth digits are reduced in size, and the phalangeal formula is reduced to 2-3-4-3-2. In Saurischia the second digit becomes the longest, and the fourth digit's phalanges are reduced. In Theropoda the fifth digit is reduced to a nubbin of the metacarpal or lost altogether, and the fourth digit is quite small. In tetanurans all trace of the outer two digits is lost. In tyrannosaurs [which von Huene (1914, 1920, 1926) and Novas (1992) showed are not carnosaurs but actually gigantic coelurosaurs], the third digit is lost, thus proving the invalidity of Morse's law at least in the case of dinosaurs. Given the suite of dozens of synapomorphies from other parts of the skeleton that also support the placement of birds within theropods, it is difficult not to accept the manifest similarities of other features of the hand of *Archaeopteryx* as homologous to those of theropods, including 14 characters related to the form and proportion of hand and wrist elements in theropods (Gauthier, 1986: characters 21–26, 43–46, 61, 62, 75, and 76).

Semilunate Carpal

In maniraptorans, and perhaps at a more general level within tetanurans, a half-moon-shaped wrist element overlaps the bases of metacarpals I and II (Ostrom, 1969, 1975a,b, 1976b; Gauthier, 1986). The rounded proximal surface allows the hand to swivel sideways, a feature that appears to have been related to predation in basal maniraptorans and exapted for the flight stroke in birds (Gauthier and Padian, 1985). There appears to be little question about the morphological similarity of this bone in *Archaeopteryx* and troodontids, oviraptors, and dromaeosaurs such as *Deinonychus* and *Velociraptor;* the question surrounds its homology. Carpal elements are often incomplete or unknown for Mesozoic theropods: They may not have been preserved because they were not ossified, perhaps because the specimen in question was not adult; or they may have been removed or destroyed by taphonomic processes, preserved out of position and so unrecognized and not collected, or collected and not correctly identified. Furthermore, as in other amniotes, during ontogeny many of these elements attained better definition and frequently fused, and phylogenetically fusion appears to have increased (Gauthier, 1986). There is no simple answer to the identity of some elements of the wrist in theropods (Feduccia, 1996), but some points are clear.

Contrary to some inferences (Feduccia, 1996), theropods have not "lost" a row of carpals, but sometimes they are incompletely ossified or preserved. Ostrom (1969 *et passim*) identified the semilunate carpal in *Deinonychus* and *Archaeopteryx* as the radiale and the element next to it as the ulnare, but this cannot be because the radiale contacts the radius and does not contact the metacarpals in any tetrapod. Comparison to basal theropods, such as *Coelophysis* and *Syntarsus,* in which the manus is preserved, demonstrates a row of carpals between those contacting the radius and ulna (*de facto* the radiale and ulnare) and the metacarpals. A single distal carpal, identified by both Colbert and Raath as a fusion of distal carpals 1 and 2, overlaps metacarpal II, and this element is exactly in the position of the semilunate carpal of tetanurans (Gauthier, 1986). In birds, the radiale and ulnare have become more tightly associated with the radius and ulna as the sideways flexion of the wrist has evolved between the proximal and distal carpal rows (Padian and Chiappe, 1997); the distal carpals have become associated immovably with the metacarpals (Ostrom, 1976a).

Hence, the semilunate carpal of birds and other tetanurans is the result of the fusion of distal carpals 1 and 2; the radiate and ulnare are not lost but are not always preserved and should be sought in association with the radius and ulna; and a third distal carpal is usually present, if ossified, in tetanurans, but the fourth distal carpal, like the fifth, has been lost in tetanurans along with these digits.

Pubic Foot

The end of the pubis is unexpanded in basal dinosaur groups, including all ornithischians, sauropodomorphs, and basal theropods (ceratosaurs, including *Coelophysis, Syntarsus, Ceratosaurus,* etc.). In tetanurans it is expanded fore and aft: This can be seen both in carnosaurs such as *Allosaurus* and in coelurosaurs such as ornithomimids and tyrannosaurs. In maniraptorans, such as the dromaeosaurs *Deinonychus, Velociraptor,* and *Adasaurus,* the anterior projection of this foot is severely reduced or lost, as it is in *Archaeopteryx* and the birds. Moreover, in dromaeosaurids, as in birds, the pubis itself is retroverted (and convergently in therizinosaurids), although apparently not to the extent seen in living birds. Herrerasaurids, seen variously as basal theropods, basal saurischians, or a sister taxon to dinosaurs, also have a development of the distal pubis similar in some respects to the theropod pubic foot, but this is expected to be a convergence because they are not otherwise closer to birds than are dromaeosaurids. Likewise, the Triassic archosaur *Postosuchus* (like other poposaurids) appears to have a similar expansion of the distal pubis, but this is clearly a convergence because *Postosuchus* is closely allied not to birds or dinosaurs but rather to crocodylomorphs.

Ascending Process of the Astragalus

In ornithodiran ornithosuchians, posture and gait have changed from the general sprawled or semierect reptilian condition to a more upright stance and a parasagittal gait. The ankle flexes mesotarsally so that the proximal ankle elements (astragalus and calcaneum) are associated with the lower leg and the distal tarsals with the metatarsals and phalanges. Because parasagittal gait virtually eliminates rotation at the knee, and skeletal mass is concentrated medially, the tibia and fibula have the same action. The tibia becomes the dominant element and the fibula is reduced, and the same happens to their corresponding proximal tarsals. Astragalus and calcaneum frequently fuse in adults, and in no cases do they rotate against each other, as in crurotarsal archosaurs. The astragalus expands transversely in basal ornithodirans such as pterosaurs (in which the proximal tarsals are always fused to each other and to the tibia and the fibula is reduced to a splint), *Marasuchus/Lagosuchus,* and *Lagerpeton.* Hence, for the first time, the astragalus articulates with both the tibia and fibula. A dorsal process of the astragalus separating these two articulations begins in basal ornithodirans and is known in all taxa in which the elements are preserved. It is continuous with an ascending process that is posterior to the lower leg bones in *Lagerpeton,* medial in *Marasuchus/Lagosuchus,* but anterior in dinosaurs and much more expanded in theropods (especially tetanurans) than in any other taxa. Based on congruence with other characters, it would appear parsimonious to conclude that birds carry on this basal ornithodiran feature because it is seen to trend through Theropoda.

The identity of this ossification, however, is at issue; it is held by some to be different from the avian "pretibial" bone (summarized in Feduccia, 1996), which is a separate ossification. Its shape and position on the tibia also varies, but generally it is centered anterolaterally, as is the pretibial bone. Feduccia (1996, p. 75) misinterpreted the variation in the ascending process described by Welles and Long (1974): They identified five morphological types, but these are not of independent phylogenetic origin, and in fact the types they identified are mostly not characteristic of any natural taxa within theropods; they are simply morphological types that vary for reasons apparently connected with relative size or functional features. As Gauthier (1986, p. 29) noted, the ascending process may be a separate ossification in *Dilophosaurus* (as S. P. Welles first discovered) and other theropods, as it is in birds. Moreover, although the pretibial bone fuses with the precociously developed calcaneum in neognath birds, this is not the case in ratites and tinami in which, as in other theropods, it fuses with the astragalus, and in all cases it is located on the anterolateral side of the tibia. Hence, the pretibial bone of birds appears to be homologous to the ascending process of the astragalus in theropods.

Stratigraphic Disjunction

A difficulty regarded as insurmountable by opponents of the theropod origin of birds is the presumption that the taxa identified as closest to *Archaeopteryx* among theropods—the dromaeosaurids—do not appear in the fossil record until Albian-Aptian times (perhaps 110 mya: *Deinonychus,* Cloverly Formation, Wyoming), whereas *Archaeopteryx* comes from Late Jurassic (Tithonian) times (approximately 150 mya). The apparent absence of earlier records of dromaeosaurids, although puzzling, is not unusual in the Mesozoic fossil record: For example, although stegosaurs and ankylosaurs are regarded as sister taxa that must have diverged by the late Early Jurassic, stegosaurs are not known before the Bathonian-Callovian (approximately 170 mya), whereas before the 1980s, ankylosaurs were not known before the Aptian-Albian (approximately 110 mya; Weishampel *et al.,* 1990). The situation is not unique to dinosaurs. No one doubts today that marsupials and placentals are sister taxa within mammals, and monotremes are their sister taxon. Hence, the split between therians (marsupials + placentals) and monotremes must have taken place before the first recognizable marsupials and placentals evolved. However, the first marsupials and placentals are known from Early Cretaceous times (approximately 100 mya), whereas until recently, monotremes were not known until the Oligocene (approximately 20 mya), a disjunction of 80 million years—over twice that between *Archaeopteryx* and *Deinonychus* (Carroll, 1988)! Moreover, small maniraptorans are not at all absent from Late Jurassic sediments: Jensen and Padian (1989) described a collection of bones pertaining to small maniraptorans from the Dry Mesa Quarry (Late Jurassic: ?Tithonian; Morrison Formation, Colorado). These bones unfortunately could not be identified to the generic level but nonetheless indicated that if they are not bones of birds, then they are certainly those of their sister taxon, the dromaeosaurids. These arguments would appear to dispose of the fatality of the stratigraphic argument to the theropod hypothesis.

In summary, birds, as Gauthier (1986) pointed out, must be considered dinosaurs because phylogenetic analysis clearly indicates that they evolved from dinosaurs. They are not only dinosaurs but also saurischian, theropodan, tetanuran, and maniraptoran dinosaurs. Arguments to the contrary have been proposed for 20 years since the theropod hypothesis was advanced by Ostrom, but these can no longer be considered matters of evidence. Rather, it is a question of whether one uses the methods of modern comparative biology. Issues related to the origin of flight in birds and other topics with starkly contrasting viewpoints are discussed at length in Hecht *et al.* (1985), Schultze and Trueb (1991), Feduccia (1996), and Padian and Chiappe (1997).

References

Benton, M. J. (Ed.) (1988). *The Phylogeny and Classification of the Tetrapods, Volume 1.* Clarendon, Oxford.

Carroll, R. L. (1988). *Vertebrate Paleontology and Evolution.* Freeman, New York.

Chatterjee, S. (1991). Cranial anatomy and relationships of a new Triassic bird from Texas. *Philos. Trans. R. Soc. London B* 332, 277–346.

Chatterjee, S. (1995). The Triassic bird *Protoavis. Archaeopteryx* 13, 15–31.

Chatterjee, S. (1997). *Protoavis* and the early evolution of birds. *Palaeontographica,* in press. [Abstract A].

Chiappe, L. M. (1995). The first 85 million years of avian evolution. *Nature* 378, 349–355.

Desmond, A. J. (1982). *Archetypes and Ancestors: Palaeontology in Victorian London, 1850–1875.* Muller, London.

Feduccia, A. (1996). *The Origin and Evolution of Birds.* Yale Univ. Press, New Haven, CT.

Gauthier, J. (1986). Saurischian monophyly and the origin of birds. *Mem. California Acad. Sci.* 8, 1–55.

Gauthier, J., and Padian, K. (1985). Phylogenetic, functional, and aerodynamic analyses of the origin of birds and their flight. In *The Beginnings of Birds* (M. K. Hecht, J. H. Ostrom, G. Viohl, and P. Wellnhofer, Eds.), pp. 185–197. Freunde des Jura-Museums, Eichstatt.

Hecht, M. K., and Hecht, B. M. (1994). Conflicting developmental and paleontological data: The case of the bird manus. *Acta Paleontol. Polonica* 38(3/4), 329–338.

Hecht, M. K., Ostrom, J. H., Viohl, G., and Wellnhofer, P. (1985). *The Beginnings of Birds.* Freunde des Jura-Museums, Eichstatt.

Heilmann, G. (1926). *The Origin of Birds.* pp. 210. Appleton, New York.

Holtz, T. R., Jr. (1994). The phylogenetic position of the Tyrannosauridae: Implications for theropod systematics. *J. Paleontol.* 68(5), 1100–1117.

Holtz, T. R., Jr. (1996). Phylogenetic taxonomy of the Coelurosauria (Dinosauria: Theropoda). *J. Paleontol.* 70, 536–538.

Huxley, T. H. (1870). Further evidence of the affinities between the dinosaurian reptiles and birds. *Q. J. Geol. Soc. London* 26, 12–31.

Jensen, J. A., and Padian, K. (1989). Small pterosaurs and dinosaurs from the Uncompahgre Fauna (Brushy Basin Member, Morrison Formation: ?Tithonian), Late Jurassic, Western Colorado. *J. Paleontol.* 63, 364–373.

Martin, L. D. (1983). The origin of birds and of avian flight. In *Current Ornithology* (R. F. Johnston, Ed.), Vol. 1, pp. 106–129. Plenum, New York.

Novas, F. E. (1992). La evolucion de los dinosaurios carnivoros. In *Los Dinosaurios y su entorno biotico* (J. L. Sanz and A. D. Buscalione, Eds.), pp. 125–163. Instituto "Juan Valdes," Cuenca, Spain.

Ostrom, J. H. (1969). Osteology of *Deinonychus antirrhopus,* an unusual theropod from the Lower Cretaceous of Montana. *Bull. Peabody Museum Nat. History Yale Univ.* 30, 1–165.

Ostrom, J. H. (1974). *Archaeopteryx* and the origin of flight. *Q. Rev. Biol.* 49, 27–47.

Ostrom, J. H. (1975a). The origin of birds. *Annu. Rev. Earth Planet. Sei.* 3, 35–57.

Ostrom, J. H. (1975b). On the origin of *Archaeopteryx* and the ancestry of birds. *Proc. CNRS Colloq. Int. Prob. Act. Paleontol.-Evol. Vertebr.* 218, 519–532.

Ostrom, J. H. (1976a). Some hypothetical anatomical stages in the evolution of avian flight. *Smithsonian Contrib. Paleobiol.* 27, 1–27.

Ostrom, J. H. (1976b). *Archaeopteryx* and the origin of birds. *Biol. J. Linnean Soc.* 8, 91–182.

Padian, K., and Chiappe, L. M. (1997). The early evolution of birds. *Biol. Rev.,* in press.

Proctor, N. S., and Lynch, P. J. (1993). *Manual of Ornithology: Avian Structure and Function.* Yale Univ. Press, New Haven, CT.

Roth, V. L. (1988). The biological basis of homology. In *Ontogeny and Systematics* (C. J. Humphries, Ed.), pp. 1–26. Columbia Univ. Press, New York.

Schultze, H.-P., and Trueb, L. (Eds.) (1991). *Origins of the Higher Groups of Tetrapods.* Cornell Univ. Press, Ithaca, NY.

Sereno, P. (1991). Basal archosaurs: phylogenetic relationships and functional implications. *J. Vertebr. Paleontol.* 11(Suppl. 4), 1–53.

Shubin, N. (1994). History, ontogeny, and evolution of the archetype. In *Homology: The Hierarchical Basis of Comparative Biology* (B. K. Hall, Ed.), pp. 248–271. Academic Press, New York.

Tarsitano, S. (1991). *Archaeopteryx:* Quo Vadis? In *Origins of the Higher Groups of Tetrapods* (H.-P. Schultze and L. Trueb, Eds.), pp. 541–576. Cornell Univ. Press, Ithaca, NY.

von Huene, F. (1914). Das naturliche System der Saurischia. *Zentralblatt Mineral. Geol. Paleontol. B* 1914, 154–158.

von Huene, F. (1920). Bemerkungen zur Systematik and Stammesgeschichte einiger Reptilien. *Zeitschrift Indukt. Abstammungslehre Vererbungslehre* 24, 162–166.

von Huene, F. (1926). The carnivorous Saurischia in the Jura and Cretaceous formations, principally in Europe. *Revista Museo de La Plata* 29, 35–167.

Walker, A. D. (1977). Evolution of the pelvis in birds and dinosaurs. In *Problems in Vertebrate Evolution* (S. M. Andrews, R. S. Miles, and A. D. Walker, Eds.), pp. 319–357. Academic Press, New York.

Walker, A. D. (1985). The braincase of *Archaeopteryx*. In *The Beginnings of Birds* (M. K. Hecht, J. H. Ostrom, G. Viohl, and P. Wellnhofer, Eds.), pp. 123–134. Freunde des Jura-Museums, Eichstatt.

Weishampel, D. B., Dodson, P., and Osmolska, H. (Eds.) (1990). *The Dinosauria*. Univ. of California Press, Berkeley.

Welles, S. P., and Long, R. A. (1974). The tarsus of theropod dinosaurs. *Ann. South African Museum* 64, 191–218.

Witmer, I. (1991). Perspectives on avian origins. In *Origins of the Higher Groups of Tetrapods* (H.P. Schultze and L. Trueb, Eds.), pp. 427–466. Cornell Univ. Press, Ithaca, NY.

Chapter 8

The readings for this chapter consist of two debates, an earlier one (1990) between impact theorists Walter Alvarez and Frank Asaro and defender of the volcanic theory Vincent Courtillot, and a later one (1997) between catastrophist Dale Russell and gradualist Peter Dodson. In the earlier debate, the battle lines are clearly drawn, and the way forward is not very clear. Though the debate here is well mannered and does not reflect the extreme acrimony of some of the encounters (e.g., between Charles Officer and Luis Alvarez), the two sides are quite far apart. The second debate, from just a few years later, shows that considerable progress had been made. Russell and Dodson still disagree about many points, but they have discovered rather large areas of agreement. This is an encouraging sign. Within just a few years advocates of theories that had been sharply polarized came to find evidence that moved them toward consensus. Of course, the debate may never be settled to everyone's satisfaction, but complete agreement is not necessary for science to progress.

An Extraterrestrial Impact

*Accumulating evidence suggests an asteroid
or comet caused the Cretaceous extinction*

Walter Alvarez and Frank Asaro

About 65 million years ago something killed half of all the life on the earth. This sensational crime wiped out the dinosaurs, until then undisputed masters of the animal kingdom, and left the humble mammals to inherit their estate. Human beings, descended from those survivors, cannot avoid asking who or what committed the mass murder and what permitted our distant ancestors to survive. For the past dozen years researchers from around the world, in disciplines ranging from paleontology to astrophysics, have mustered their observational skills, experimental ingenuity and theoretical imagination in an effort to answer these questions. Those of us involved in it have lived through long months of painstaking measurement, periods of bewilderment, flashes of insight and episodes of great excitement when parts of the puzzle finally fell into place.

We now believe that we have solved the mystery. Some 65 million years ago a giant asteroid or comet plunged out of the sky, striking the earth at a velocity of more than 10 kilometers per second. The enormous energy liberated by that impact touched off a nightmare of environmental disasters, including storms, tsunamis, cold and darkness, greenhouse warming, acid rains and global fires. When quiet returned at last, half the flora and fauna had become extinct. The history of the earth had taken a new and unexpected path.

Other suspects in the dinosaur murder mystery, such as sea level changes, climatic shifts and volcanic eruptions, have alibis that appear to rule them out. Some issues, however, are still unclear: Where was the impact site? Was it a single or multiple impact? Have such impacts occurred on a regular, periodic timetable? What is the role of such catastrophes in evolution?

The puzzle presented by a mass extinction is both like and unlike that of a more recent murder. There is evidence—chemical anomalies, mineral grains and isotopic ratios instead of blood or fingerprints or torn matchbooks—scattered throughout the world. No witnesses remain, however, and no chance exists of obtaining a confession. The passage of millions of years has destroyed or degraded most of the evidence in the case, leaving only the subtlest clues.

Indeed, it is difficult even to be sure which of the individual fossils that survive are those of victims killed by the impact. But paleontologists know there must have been victims because fossil-bearing sedimentary rocks show a great discontinuity 65 million years ago. Creatures such as dinosaurs and ammonites, abundant for tens of millions of years, suddenly disappeared forever. Many other groups of animals and plants were decimated.

This discontinuity defines the boundary between the Cretaceous period, during which dinosaurs reigned supreme, and the Tertiary, which saw the rise of the mammals. (It is known as the KT boundary after *Kreide,* the German word for "Cretaceous.")

When we began to study the KT boundary, we wanted to find out just how long the extinction had taken to occur. Was it sudden—a few years or centuries—or was it a gradual event that took place over millions of years? Most geologists and paleontologists had always assumed that the extinction had been slow. (These fields have a long tradition of gradualism and are uncomfortable with invoking catastrophes.) Because dinosaur fossils are relatively rare, their age provides little detailed information on the duration of the extinction. It was possible to view the extinction of dinosaurs as gradual.

When paleontologists looked at the fossils of pollen or single-celled marine animals called foraminifera, however, they found the extinction to be very abrupt. In general, smaller organisms produce more abundant fossils and so yield a sharper temporal picture.

The extinction also appears more sudden as paleontologists study closely the fossil record for medium-size animals such as marine invertebrates. Among these are the ammonites (relatives of the modern chambered nautilus), which died out at the end of the Cretaceous period. The best record of their extinction is found in the coastal outcrops of the Bay of Biscay on the border between Spain and France.

In 1986 Peter L. Ward and his colleagues at the University of Washington made detailed studies of these outcrops at Zumaya in Spain. Ward found that the ammonites appeared to die out gradually—one species disappearing after another over an interval of about 170 meters, representing about five million years. But in 1988 Ward studied two nearby sections in France and found evidence that these ammonite species actually survived right up to the KT boundary. The apparent gradual extinction at Zumaya was merely the artifact of an incomplete fossil record. If organisms whose fossils are well preserved died out abruptly, then it is likely that others that perished about the same time, such as dinosaurs, whose remains are more sparsely preserved, did so as well.

This establishes that the extinction was abrupt in geologic terms, but it does not establish how many years this extinction took, because it is a major accomplishment to date a rock to an accuracy of a million years. Intervals in the geologic records can be determined with precision only to within 10,000 years (.01 Myr), a period longer than the entire span of human civilization.

The duration of the mass extinction that marks the KT boundary can be estimated more precisely than this. In the deep-water limestones at Gubbio in Italy, a thin layer of clay separates Cretaceous and Tertiary sediments. The layer, discovered by Isabella Premoli Silva of the University of Milan, is typically about one centimeter thick. In the 1970s one of us (Alvarez) was part of a group that found the clay falls within a six-meter thickness of limestone deposited during the .5-Myr period of reversed geomagnetic polarity designated 29R. On the face of it, this suggests that the clay layer, and the mass extinction it marks, represents a span of no more than .001 Myr, about 1,000 years.

Jan Smit of the University of Amsterdam did a similar study of sediments at Caravaca in southern Spain, where the stratigraphic record is even more precise, and estimated the extinction lasted no more than 50 years. By geologic standards this is blindingly fast!

Our work on the KT boundary began in the late 1970s when we and our Berkeley colleagues Luis W. Alvarez and Helen V. Michel tried to develop a more accurate way to determine how long the Gubbio KT clay layer took to be deposited. Our efforts failed, but they did provide a crucial first clue to the identity of the mass killer. (That is what detectives and scientists need: a lot of hard work and an occasional lucky break.)

The method depended on the rarity of iridium in the earth's crust—about .03 part per billion as compared with 500 parts per billion, for example, in the primitive stony meteorites known as carbonaceous chondrites. Iridium is rare in the earth's crust because most of the planet's allotment is alloyed with iron in the core.

We suspected that iridium would enter deep-sea sediments, such as those at Gubbio, predominantly through the continual rain of micrometeorites, sometimes called cosmic dust. This constant infall would provide a clock: the more iridium in a sedimentary layer, the longer it must have taken to lay down. Moreover, iridium could be measured at very low concentrations by means of neutron-activation analysis, a technique in which neutron bombardment converts the metal into a radioactive and hence detectable form.

One scenario we considered was that the KT boundary clay layer formed over a period of about 10,000 years when organisms that secrete calcareous shells died out, and so no calcium carbonate (which makes up most of the limestone) was deposited. Most layers at Gubbio contain about 95 percent calcium carbonate and 5 percent clay; the boundary layer contains 50 percent clay. If this scenario was correct, the ratio of iridium to clay would be the same in the boundary clay as in higher and lower layers. If clay deposition had slowed at the same time as calcium carbonate deposition, the ratio would be higher than that in adjacent rocks.

In June of 1978 our first Gubbio iridium analyses were ready. Imagine our astonishment and confusion when we saw that the boundary clay and the immediately adjacent limestone contained far more iridium than any of our scenarios predicted—an amount comparable to that in all the rest of the rock deposited during the 500,000 years of interval 29R.

Clearly, this concentration could not have come from the usual sprinkling of cosmic dust. For a year we debated possible sources, testing and rejecting one idea after another. Then in 1979 we proposed the one solution that had survived our testing: a large comet or asteroid about 10 kilometers in diameter had struck the earth and dumped an enormous quantity of iridium into the atmosphere.

Since we first proposed the impact hypothesis, so much confirming evidence has come to light that most scientists working in the field are persuaded that a great impact occurred. More than 100 scientists in 21 laboratories in 13 countries have found anomalously high levels of iridium at the KT boundary at about 95 sites throughout the world. The anomaly has been found in marine and nonmarine sediments, at outcrops on land and in oceanic sediment cores. Further, we have analyzed enough other sediments to know that iridium anomalies are very rare. As far as we know, the one at the KT boundary is unique.

The iridium anomaly is well explained by impact because the ratio of iridium to elements with similar chemical behavior, such as platinum, osmium, ruthenium, rhodium and gold, is the same in the boundary layer as it is in meteorites. Miriam Kastner of the Scripps Institution of Oceanography, working with our group, has determined that the gold-iridium ratio in the carefully studied KT boundary at Stevns Klint in Denmark agrees to within 5 percent with the ratio in the most primitive meteorites (type I carbonaceous chondrites).

Indeed, the ratios of all the platinum-group elements found in the KT boundary give evidence of extraterrestrial origin. George Bekov of the Institute of Spectroscopy in Moscow and one of us (Asaro) have found that the relative abundances of ruthenium, rhodium and iridium can distinguish stony meteorites from terrestrial samples. Analysis of KT boundary samples from Stevns Klint, Turkmenia in the Soviet Union and elsewhere support the impact hypothesis.

So do ratios of isotopes. Jean-Marc Luck, then at the Institute of Physics of the Earth in Paris, and Karl K. Turekian of Yale University found that most of the osmium in KT boundary samples from Denmark and New Mexico could not have come from a continental source, because the abundance of osmium 187 is too low. The ratio of osmium 187 to osmium 186 is higher in continental rocks than in meteorites or in the earth's mantle because those rocks are relatively enriched in rhenium, whose radioactive isotope, rhenium 187, decays to osmium 187. The osmium in KT samples must be extraterrestrial or from the earth's mantle.

Not only does the composition of rocks at the KT boundary suggest impact, but so does their mineralogy. In 1981 Smit discovered another telltale clue: mineral spherules as large as a millimeter in diameter in the Caravaca KT clay. (Alessandro Montanari of Berkeley confirmed their presence in Italy as well.) The spherules originated as droplets of basaltic rock, shock-melted by impact and rapidly cooled during ballistic flight outside the atmosphere, then chemically altered in the boundary clay. They are the basaltic equivalent of the more silica-rich glassy tektites and microtektites that are the known result of smaller impacts. The basaltic chemistry suggests that the impact took place on oceanic crust.

In addition to the spherules, shocked grains of quartz have been discovered by Bruce F. Bohor of the U.S. Geological Survey in Denver and Donald M. Triplehorn of the University of Alaska. Painstaking studies by E. E. Foord, Peter J. Modreski and Glen A. Izett of the USGS show that the grains carry the multiple intersecting planar "lamellae"—bands of deformation—symptomatic of hypervelocity shock. Such grains are found only in known impact craters, at nuclear test sites, in materials subjected to extreme shock in the laboratory—and in the KT boundary.

There is in fact a candidate crater beneath the glacial drift at Manson, Iowa; it lies in a quartz-rich bedrock, and its location is suitable to explain the size and abundance distribution of the shocked quartz grains. At 32 kilometers in diameter, the crater is too small to have been formed by the single body posited as having caused the extinction. Nevertheless, detailed studies of the crater show it to have an age indistinguishable from that of the KT boundary, and so it probably played a part in the mystery.

How would an impact disperse shocked and molten materials around the globe? A 10-kilometer asteroid moving at more than 10 kilometers per second would ram a huge hole in the atmosphere. When it hit the ground, its kinetic energy would be converted to heat in a nonnuclear explosion 10,000 times as strong as the total world arsenal of nuclear weapons. Some vaporized remains of the asteroid and rock from the ground near the impact point would then be ejected through the hole before the air had time to rush back in.

The fireball of incandescent gas created by the explosion would also propel material out of the atmosphere. The fireball of an atmospheric nuclear explosion expands until it reaches the same pressure as the surrounding atmosphere, then rises to an altitude where its density matches that of the surrounding air. At that point, usually around 10 kilometers high, the gas spreads laterally to form the head of the familiar mushroom cloud.

Computer models of explosions with energies of 1,000 megatons—about 20 times the energy of the largest nuclear bombs but only 1/100,000 the energy of the KT impact—have shown that the fireball never reaches pressure equilibrium with the surrounding atmosphere. Instead, as the fireball expands to altitudes where the density of the atmosphere declines significantly, its rise accelerates and the gas leaves the atmosphere at velocities fast enough to escape the earth's gravitational field. The fireball from an even greater asteroid impact would simply burst out the top of the atmosphere, carrying any entrained ejecta with it, sending the material into orbits that could carry it anywhere on the earth.

The impact of a comet-size body on the earth, creating a crater 150 kilometers in diameter, would clearly kill everything within sight of the fireball. Researchers are refining their understanding of the means by which an impact would also trigger extinction worldwide. Mechanisms proposed include darkness, cold, fire, acid rain and greenhouse heat.

In our original paper, we proposed that impact-generated dust caused global darkness that resulted in extinctions. According to computer simulations made in 1980 by Richard P. Turco of R&D Associates, O. Brian Toon of the National Aeronautics and Space Administration and their colleagues, dust lofted into the atmosphere by the impact of a 10-kilometer object would block so much light that for months you would literally be unable to see your hand in front of your face.

Without sunlight, plant photosynthesis would stop. Food chains everywhere would collapse. The darkness would also produce extremely cold temperatures, a condition termed impact winter. (After considering the effects of the impact, Turco, Toon and their colleagues went on to study nuclear winter, a related phenomenon as capable of producing mass extinctions today as impact winter was 65 million years ago.)

In 1981 Cesare Emilliani of the University of Miami, Eric Krause of the University of Colorado and Eugene M. Shoemaker of the USGS pointed out that an oceanic impact would loft not only rock dust but also water vapor into the atmosphere. The vapor, trapping the earth's heat, would stay aloft much longer than the dust, and so the impact winter would be followed by greenhouse warming. More recently John D. O'Keefe and Thomas J. Ahrens of the California Institute of Technology have suggested that the impact might have occurred in a limestone area, releasing large volumes of carbon dioxide, another greenhouse gas. Many plants and animals that survived the extreme cold of impact winter could well have been killed by a subsequent period of extreme heat.

Meanwhile John S. Lewis, G. Hampton Watkins, Hyman Hartman and Ronald G. Prinn of the Massachusetts Institute of Technology have calculated that shock heating of the atmosphere during impact would raise temperatures high enough for the oxygen and nitrogen in the air to combine. The resulting nitrous oxide would eventually rain out of the air as nitric acid—an acid rain with a vengeance. This mechanism may well explain the widespread extinction of marine invertebrate plants and animals, whose calcium carbonate shells are soluble in acidic water.

Another killing mechanism came to light when Wendy Wolbach, Ian Gilmore and Edward Anders of the University of Chicago discovered large amounts of soot in the KT boundary clay. If the clay had been laid down in a few years or less, the amount of soot in the boundary would indicate a sudden burning of vegetation equivalent to half of the world's current forests. Jay Meos of the University of Arizona and his colleagues have calculated that infrared radiation from ejecta heated to incandescence while reentering the atmosphere could have ignited fires around the globe.

Detailed studies of the KT boundary sediments may eventually provide evidence supporting a particular killing mechanism. For example, dissolution patterns in the Italian limestone show that bottom waters were acidic immediately after the extinction. And work we have done with William Lowrie of ETH-Zurich shows that those waters also changed briefly from their normal oxidizing state to a reducing condition, possibly because of the massive death of marine organisms.

It has always been a major disappointment that no one has found the 150-kilometer crater a 10-kilometer impacting object should have produced. The crater might be hidden under the Antarctic ice sheet, or it might have been on the 20 percent of the earth's surface that has subsequently been consumed in subduction zones at the edges of oceanic plates. The evidence regarding the location is contradictory: the basaltic spherules in the boundary clay point to an impact on the ocean floor, but the shocked quartz grains argue for a continental hit.

A newly emerging point of view suggests, unlikely as it may seem, that the KT extinction may have been caused by two or more nearly simultaneous impacts. Shoemaker and Piet Hut of the Institute for Advanced Study in Princeton, N. J., have identified a number of mechanisms that could yield multiple impacts, either on the same day or over the course of many years. Double or multiple craters have been found on the earth, the moon and other planets, suggesting that some asteroids may consist of two or more objects mutually orbiting one another. Alternatively, the earth may have been struck by two or more large fragments of a comet nucleus in the process of breaking up.

Multiple impacts over longer periods could have occurred if a dispersing comet nucleus left several large fragments in an earth-crossing orbit. Such impacts could also occur randomly if some other factor increased the average number of comets in the inner solar system. Although not one scenario has won out, collectively they indicate that multiple impacts are not as improbable as might be thought.

The comet theory gains credibility from the discovery of apparently extraterrestrial materials near the KT boundary. Meixun Zhao and Jeffrey L. Bada of the University of California at San Diego analyzed chalk layers just above and below the KT boundary in Denmark. They found amino acids that are not used by life on the earth but do occur in carbonaceous chondrite meteorites. It seems unlikely that amino acids could survive the heat of a large impact, and they in fact do not appear in the KT boundary itself.

Kevin Zahnle and David Grinspoon of NASA have proposed that dust from a disintegrating comet entered the earth's atmosphere over an extended period and carried these extraterrestrial amino acids with it. During that interval the impact of a large fragment of the comet would have caused the KT extinction.

An apparently unrelated line of inquiry, based on statistical rather than chemical analyses, has yielded a hypothesis explaining how comets could hit the earth periodically. In 1984 David M. Raup and John J. Sepkoski, Jr., of the University of Chicago published an analysis of the fossil record, which seemed to indicate that mass extinctions have occurred at 32-million-year intervals. Like most scientists working with the KT boundary, we were very skeptical of their results. But astrophysicist Richard A. Muller of the University of California at Berkeley reexamined Raup and Sepkoski's data and convinced himself that the periodicity was real.

Muller, Marc Davis of Berkeley and Hut hypothesized that a dim, unrecognized companion star orbiting the sun every 32 million years (which they provisionally dubbed Nemesis) might regularly disturb the orbits of comets on the outer fringe of the solar system. The disturbance would send a million-year storm of comets into the inner solar system, greatly increasing the chance of a large impact (or multiple impacts) on the earth. Daniel Whitmire of the University of

Southwestern Louisiana and Albert Jackson of Computer Sciences Corporation independently proposed the same hypothesis.

When Muller showed one of us (Alvarez) the paper proposing Nemesis, I was very skeptical. I remember telling him that I thought it was "an ingenious solution to a nonproblem" because I was not convinced of Raup and Sepkoski's evidence for periodic mass extinctions. If the hypothesis was correct, I pointed out, terrestrial impact craters should show the same periodicity in their ages. Muller and I found, to his delight and to my surprise, that crater ages do show essentially the same periodicity as mass extinctions. Since then I have felt that the hypothesis must at least be taken seriously.

It turns out, however, that it is very difficult to find a dim red star close to the sun when one has no idea where to look. Muller and Saul Perlmutter of Berkeley are now about halfway through a computerized telescopic search for a star with the characteristics of Nemesis; they expect to finish in a couple of years. Meanwhile new analysis of crater ages and extinction dates has raised questions about whether they actually are periodic. The small numbers of events and the sketchy information available make the question difficult to answer unequivocally.

Murder suspects typically must have means, motive and opportunity. An impact certainly had the means to cause the Cretaceous extinction, and the evidence that an impact occurred at exactly the right time points to opportunity. The impact hypothesis provides, if not motive, then at least a mechanism behind the crime. How do other suspects in the killing of the dinosaurs fare?

Some have an air-tight alibi: they could not have killed all the different organisms that died at the KT boundary. The venerable notion that mammals ate the dinosaurs' eggs, for example, does not explain the simultaneous extinction of marine foraminifera and ammonites.

Stefan Gartner of Texas A&M University once suggested that marine life was killed by a sudden huge flood of fresh water from the Arctic Ocean, which apparently was isolated from other oceans during the late Cretaceous and filled with fresh water. Yet this ingenious mechanism cannot account for the extinction of the dinosaurs or the loss of many species of land plants.

Other suspects might have had the ability to kill, but they have alibis based on timing. Some scientific detectives have tried to pin the blame for mass extinction on changes in climate or sea level, for example. Such changes, however, take much longer to occur than did the extinction; moreover, they do not seem to have coincided with the extinction, and they have occurred repeatedly throughout the earth's history without accompanying extinctions.

Others consider volcanism a prime suspect. The strongest evidence implicating volcanoes is the Deccan Traps, an enormous outpouring of basaltic lava in India that occurred approximately 65 million years ago. Recent paleomagnetic work by Vincent E. Courtillot [see "A Volcanic Eruption"] and his colleagues in Paris confirms previous studies. They show that most of the Deccan Traps erupted during a single period of reversed geomagnetic polarity, with slight overlaps into the preceding and succeeding periods of normal polarity. The Paris team has found that the interval in question is probably 29R, during which the KT extinction occurred, although it might be the reversed-polarity interval immediately before or after 29R as well.

Because the outpouring of the Deccan Traps began in one normal interval and ended in the next, the eruptions that gave rise to them must have taken place over at least .5 Myr. Most workers interested in mass extinction therefore have not considered volcanism a serious suspect in a killing that evidently took place over .001 Myr or less.

Some researchers have argued that, contrary to the fossil record, the KT extinctions took place over many thousands of years and that volcanism can account for quartz grains, spherules and the iridium anomaly.

In 1983 William H. Zoller and his colleagues at the University of Maryland at College Park discovered high concentrations of iridium in aerosols from Kilauea volcano in Hawaii collected on filters 50 kilometers away; however, the ratio between iridium and other rare elements in the volcanic aerosols does not match the ratio found at the KT boundary. The ratio of gold to iridium in the Kilauea aerosols is more than 35 times that in the KT boundary at Stevns Klint.

There has also been debate as to whether an explosive volcanic eruption might produce shocked quartz. It now seems agreed, however, that volcanic explosions can produce some deformation but that the distinctive multiple lamellae seen in the KT boundary quartz can only be formed by impact shocks. In addition, John McHone of Arizona State University has found that they contain stishovite, a form of quartz produced only at pressures far greater than those of volcanic eruptions. And Mark H. Anders of Columbia University and Michael R. Owen of St. Lawrence University have used a technique known as cathode luminescence, in which an electric field causes quartz to glow, to determine the origin of the KT grains. The colors produced by the grains are not volcanic; they argue instead for impact on an ordinary sedimentary sandstone.

Moreover, basaltic spherules in the KT boundary argue against explosive volcanism in any case; spherules might be generated by quieter forms of volcanism, but then they could not be transported worldwide.

The apparent global distribution of the iridium anomaly, shocked quartz and basaltic spherules is strong evidence exonerating volcanism and pointing to impact. Eruptions take place at the bottom of the atmosphere; they send material into the high stratosphere at best. Spherules and quartz grains, if they came from an eruption, would quickly be slowed by atmospheric drag and fall to the ground.

Nevertheless, the enormous eruptions that created the Deccan Traps did occur during a period spanning the KT extinction. Further, they represent the greatest outpouring of lava on land in the past quarter of a billion years (although greater volumes flow continually out of mid-ocean ridges). No investigator can afford to ignore that kind of coincidence.

It seems possible that impact triggered the Deccan Traps volcanism. A few minutes after a large body hit the earth the initial crater would be 40 kilometers deep, and the release of pressure might cause the hot rock of the underlying mantle to melt. Authorities on the origin of volcanic provinces, however, find it very difficult to explain in detail how an impact could trigger large-scale basaltic volcanism.

In the past few years the debate between supporters of each scenario has become polarized: impact proponents have tended to ignore the Deccan Traps as irrelevant, while volcano backers have tried to explain away evidence for impact by suggesting that it is also compatible with volcanism. Our sense is that the argument is a Hegelian one, with an impact thesis and a volcanic antithesis in search of a synthesis whose outlines are as yet unclear.

Even in its present incompletely solved form, the mystery of the KT mass killing carries a number of lessons. The late 18th and early 19th centuries, when the study of the earth was first becoming a science, was a period marked by a long battle between catastrophists—who thought that sudden great events were crucial to the evolution of the planet—and uniformitarians—who explained all history in terms of gradual change.

Steven J. Gould of Harvard University has shown how the uniformitarians so thoroughly won this battle that generations of geology students have been taught catastrophism is unscientific. The universe, however, is a violent place, as astronomy has shown, and it is now becoming clear that the earth has also had its violent episodes.

Evidence that a giant impact was responsible for the extinctions at the end of the Cretaceous has finally rendered the catastrophic viewpoint respectable. Future geologists, with the intellectual freedom to think in both uniformitarian and catastrophic terms, have a better chance of truly understanding the processes and history of the planet than did their predecessors.

Catastrophes have an important role to play in evolutionary thinking as well. If a chance impact 65 million years ago wiped out half the life on the earth, then survival of the fittest is not the only factor that drives evolution. Species must not only be well adapted, they must also be lucky.

If chance disaster occasionally wipes out whole arrays of well-adapted organisms, then the history of life is not preordained. There is no inevitable progress leading inexorably to intelligent life—to human beings. Indeed, Norman Sleep of Stanford University and his colleagues have suggested that in the very early history of the earth, when impacts were more frequent, incipient life may have been extinguished more than once.

Impact catastrophes may also prevent evolution from bogging down. The fossil record indicates that in normal times each species becomes increasingly well adapted to its particular ecological niche. Thus, it becomes ever more difficult for another species to evolve into that niche.

As a result, the rate of evolution slows. Wholesale removal of species by impact, however, provides a great opportunity for the survivors to evolve into newly vacant niches. (We have heard graduate students compare this situation with the excellent job prospects they would face if half of all tenured professors were suddenly fired.) Indeed, the fossil record shows that the rate of evolution accelerated immediately after the end of the Cretaceous.

Among the happy survivors of the KT extinction were the early Tertiary mammals, our ancestors. When dinosaurs dominated the earth, mammals seem always to have been small and insignificant. Warm-blooded metabolism, small size, large number or other traits may have suited them to endure the harsh conditions imposed by impact—or they may just have been lucky. And with the removal of the huge reptiles from the scene, mammals began an explosive phase of evolution that eventually produced human intelligence. As detectives attempting to unravel this 65-million-year-old mystery, we find ourselves pausing from time to time and reflecting that we owe our very existence as thinking beings to the impact that destroyed the dinosaurs.

A Volcanic Eruption

What dramatic event 65 million years ago killed most species of life on the earth? The author argues it was a massive volcanic eruption

Vincent E. Courtillot

The mysterious mass extinction that took place 65 million years ago has been attributed to either the impact of a large asteroid or a massive volcanic eruption. Both hypotheses presume that clouds of dust and chemical changes in the atmosphere and oceans created an ecological domino effect that eradicated large numbers of animal and plant families. The geologic record generally is consistent with either scenario; the central issue has been how rapid the event was. New evidence implies that the mass extinction occurred over tens or even hundreds of thousands of years. Such a duration closely corresponds to an episode of violent volcanic eruptions in India that occurred at the time of the mass extinction. Moreover, other extinction events also appear to be roughly simultaneous with periods of major volcanic activity.

The conventional divisions of geologic history reflect times of significant geologic and biological change. The mass extinction 65 million years ago defined the end of the Mesozoic era, when reptiles enjoyed great evolutionary success, and the beginning of the Cenozoic era, when mammals became extremely prevalent. Because the last period of the Mesozoic is the Cretaceous and the first period of the Cenozoic the Tertiary, the time of the most recent mass extinction is called the Cretaceous-Tertiary, or KT, boundary.

At this boundary the dinosaurs met their demise and, even more remarkable, 90 percent of all genera of protozoans and algae disappeared. John J. Sepkoski, Jr., and David M. Raup of the University of Chicago conclude that from 60 to 75 percent of all species vanished then. Equally important, many species, among them the ancestors of human beings, survived.

In 1980 Luis W. and Walter Alvarez (father and son) of the University of California at Berkeley, along with their colleagues Frank Asaro and Helen V. Michel, discovered unusually high concentrations of the metal iridium—from 10 to 100 times the normal levels—in rocks dating from the KT boundary in Italy, Denmark and New Zealand. Iridium is rare in the earth's crust but can be relatively abundant in other parts of the solar system. The Berkeley group therefore concluded that the iridium came from outer space, and thus the asteroid hypothesis was born.

A large asteroid impact would have cloaked the earth with a cloud of dust, resulting in darkness, suppression of photosynthesis, the collapse of food chains and, ultimately, mass extinction. The iridium is contained in a thin layer of clay whose chemical composition differs from that of the layers both above and below the boundary. Alvarez's group interpreted the clay as being the altered remains of the dust thrown up by the impact. In this view the boundary layer was laid down in less than one year, a flickering instant in geologic time. Other unusual findings at the KT boundary, most notably quartz crystals that appear to have been subjected to extremely powerful physical shocks, also could be explained by an asteroid impact.

An alternative to the asteroid hypothesis had already been brewing for some time. As early as 1972 Peter R. Vogt of the Naval Research Laboratory in Washington, D.C., pointed out that extensive volcanism had taken place at roughly the time of the KT boundary, principally in India. The volcanism produced extensive lava flows, known as the Deccan Traps (*deccan* means "southern" in Sanskrit, and trap means "staircase" in Dutch). Vogt suggested that the traps might be connected to the many changes that took place at the end of the Cretaceous period.

In the mid-1970s Dewey M. McLean of the Virginia Polytechnic Institute proposed that volcanoes could produce mass extinctions by injecting vast amounts of carbon dioxide into the atmosphere that would trigger abrupt climate changes and alter ocean chemistry. Charles B. Officer and Charles L. Drake of Dartmouth College analyzed sediments from KT boundary sections and concluded that the iridium enrichment and other chemical anomalies at the boundary were not deposited instantaneously but rather over a period of 10,000 to 100,000 years. They also argued that the anomalies were more consistent with a volcanic rather than meteoritic origin.

The amount of time represented by the clay layer at the KT boundary emerged as a major point of contention. Dating a 100-million-year-old rock with a precision of one part in 1,000 (that is, to within 100,000 years) is not yet possible. Yet much of the debate focuses on whether the boundary clay was deposited in less than one year (as would be expected from an impact) or in 10,000 (from an extended period of volcanism).

The sheer size of the Deccan Traps suggests that their formation must have been an important event in the earth's history. Individual lava flows extend well over 10,000 square kilometers and have a volume exceeding 10,000 cubic kilometers. The thickness of the flows averages from 10 to 50 meters and sometimes reaches 150 meters. In western India the accumulation of lava flows is 2,400 meters thick (more than a quarter the height of Mount Everest). The flows may have originally covered more than two million square kilometers, and the total volume may have exceeded two million cubic kilometers.

An important, unresolved question was whether the date and duration of Deccan volcanism are compatible with the age and thickness of the KT boundary. Until recently the lava samples from the Deccan Traps were thought to range in age from 80 to 30 million years (estimated by measuring the decay of the radioactive isotope potassium 40 in rocks). Whether this range was real or just reflected an error in measurement was unknown. So in 1985 I joined forces with a number of colleagues to try to clarify the picture.

One important clue emerged from the fact that the Deccan rocks are basalts, volcanic rocks rich in magnesium, titanium and iron that are rather strongly magnetic. When basaltic lava cools, the magnetization of tiny crystals of iron-titanium oxides in the rock becomes frozen, aligned with the earth's magnetic field. The polarity of the field occasionally reverses, so that the magnetic north pole becomes south and vice versa. These brief reversals—about 10,000 years long—occur in random fashion at a rate that has varied from about one reversal every million years at the end of the Cretaceous to roughly four every million years in recent times.

Jean Besse and Dither Vandamme at the Institute of Physics of the Earth in Paris and I found that more than 80 percent of the rock samples from the Deccan Traps had the same, reversed polarity. Had the volcanism truly continued from 80 to 30 million years ago, we would have expected to find approximately equal numbers of normal- and reverse-magnetized samples, because tens of reversals took place during that 50-million-year stretch.

In fact, the thickest (1,000-meter-thick) exposed sections of the traps record only one or two reversals. We therefore concluded in 1986 that Deccan volcanism began during an interval of normal magnetic activity, climaxed in the next, reversed interval, then waned in a final, normal interval. Judging from the usual frequency of reversals, our results implied that the volcanism could not have lasted much more than one million years.

If so, the spread of ages found by potassium 40 dating must have been wrong. My colleagues Henri Maluski of the University of Montpellier and Gilbert Féraud of the University of Nice and other researchers used a newer, more reliable technique—argon-argon dating—to determine how much potassium 40 had decayed during the lifetime of the rock samples. Their results confirmed that the Deccan flows were laid down over a relatively brief period. Age estimates for the Deccan lavas now cluster between 64 and 68 million years, and much of the remaining scatter in ages may result from alteration of the samples or differing laboratory standards.

Although accurate dating of sedimentary rock is difficult, recent findings by Ashok Sahni of the University of Chandigarh, J. J. Jaeger of the University of Montpellier and their colleagues further narrow estimates of the age of the Deccan Traps. Sediments immediately below the Deccan flows contain dinosaur fossil fragments that seem to date from the Maastrichtian stage, the last eight million years of the Cretaceous. Dinosaur and mammalian teeth and dinosaur egg fragments that appear to be of Maastrichtian age have also been found in layers of sediment between the flows. This implies that Deccan volcanism began during the very last stage of the Cretaceous.

More precise data come from oil-exploration wells on the east coast of India, which crossed three thin trap flows, each separated by a layer of sedimentary rock. The lowest level of lava rests on sedimentary layers that contain fossils of a plankton called *Abatomphalus mayaroensis,* which thrived during the last one million years of the Cretaceous and became extinct shortly thereafter. The sedimentary rock layers between the lava flows also contain fossils from the exact same time, but the layers above the flows do not.

A. mayaroensis fossils appear in strata with normal magnetic polarity that lie below (before) the KT boundary and disappear at the boundary itself, which is located in the next, magnetically reversed set of strata.

The most reasonable conclusion from the various evidence is that Deccan volcanism began during the last normal magnetic interval of the Cretaceous, climaxed during the following reversed interval (at or very near the Cretaceous-

Tertiary boundary) and ended in the first normal magnetic interval of the Ceno-zoic era.

Magnetic and fossil studies together reduce the estimated duration of Dec-can volcanism to about 500,000 years, the best time resolution that can be ob-tained using present techniques. The fact that Deccan volcanism—one of the largest and fastest episodes of lava flow of the past 250 million years—coincided with the KT boundary to within the best time accuracy now attainable made it hard for us to escape the conclusion that a link existed between the Deccan Traps and the mass extinction.

Having established that the Deccan Traps erupted roughly simultaneously with the extinction at the end of the Cretaceous period, we next sought to deter-mine whether a volcanic eruption could explain the observed features of the KT boundary layers. In general, either a huge volcanic eruption or an asteroid impact could plausibly have produced these features.

The unusual iridium-rich deposit that appears to have been laid down simul-taneously around the earth need not have come from outer space. William H. Zoller, Ilhan Olmez and their colleagues at the University of Maryland at College Park discovered unusual iridium enhancements in particles emitted by the Kilauea volcano in Hawaii. J. P. Toutain and G. Meyer of the Institute of Physics of the Earth found iridium in particles emitted by another volcano, the Piton de la Four-naise on the island of Réunion, which (as discussed below) is related to the Deccan volcanism. Iridium-rich volcanic dust has been found embedded in the Antarctic ice sheet, thousands of kilometers from the source volcanoes.

The composition of the clay at the boundary layer differs from that of the clays above and below the layer. The usual mineral in clay, illite, is replaced by smectite, which can be created when basaltic rock is altered. Recent studies of the mineralogy of the KT boundary clay at Stevns Klint in Denmark led W. Crawford Elliott of Case Western Reserve University and his co-workers and Birger Schmitz of the University of Göteborg to conclude that the clay consists of a distinctive kind of smectite that in fact is altered volcanic ash.

The KT boundary clay can be simulated by mixing 10 parts of material from the earth's crust with one part of material from common stony meteorites. The earth's mantle (the layer below the crust), however, has a composition similar to that of stony meteorites and so could generate the same chemical anomalies. Karl K. Turekian of Yale University and Jean-Marc Luck, then at the Institute of Physics of the Earth, found that the relative abundance of the elements rhenium and os-mium in the clay resembles the ratio in both meteorites and in the earth's mantle.

Peculiar physical features in material from the KT boundary also can be ex-plained by either hypothesis. Boundary layers contain large numbers of tiny spherules. Some spherules consist of clay minerals that appear to be altered re-mains of molten basaltic droplets, but it is impossible to say whether they origi-nated as volcanic ejecta or from oceanic crust melted by an asteroid impact. Mat-ters are somewhat confused by the fact that at least some of the spheres turned out to be round fossil algae or even recent insect eggs that contaminated the material.

The discovery of shocked, deformed grains of quartz crystal in KT boundary layers, first made by Bruce F. Bohor and Glen A. Izett of the U.S. Ge-ological Survey in Denver, is often considered the strongest evidence in favor of the impact hypothesis. Such shocked grains had been found previously only from known impact craters (such as Meteor Crater in Arizona) or from sites of underground nuclear explosions. They are produced by dynamic shock stress at more than 100,000 times atmospheric pressure, but shocked structures can be produced at much lower pressures if the rock is heated before the shock occurs, as would be the case in a volcanic eruption.

As magma rises to the earth's surface, it decompresses and releases dissolved gases. At the same time, the magma often cools and thickens. If it cools particularly quickly, it becomes so stiff that the gases cannot escape. Pressure therefore builds up, possibly leading to an explosion and powerful shock waves. Such stresses might be sufficient to shock quartz crystals if the temperatures and duration were great enough.

Magma that is rich in silicate material is viscous and especially prone to provoke explosive eruptions; examples of silicic volcanism include Vesuvius and Mount St. Helens. In 1986 Neville L. Carter of Texas A&M University and his associates discovered evidence of shock features similar to those at the KT boundary in rocks from some geologically recent silicic volcanic explosions, such as the large Toba, Sumatra, eruption of 75,000 years ago. Using transmission electron microscopy, Jean-Claude Doukhan of the University of Lille recently found that shock features produced by laboratory impact, meteorite impact and those observed in samples from the KT boundary are all different from one another in some respects and that the similarity between laboratory and meteorite features has been overstated. Shock features from KT samples are decorated with microscopic bubbles that are not observed in samples from meteorite impacts and that seem to indicate a higher formation temperature, compatible with a volcanic origin.

Explosive silicic volcanism commonly precedes periods of relatively quiet, Deccan-type (flood basaltic) volcanism, during which basaltic lava flows freely and copiously. Ten to 15 percent of the volume of lava from known Deccan-type flows erupts in episodes of explosive silicic volcanism. A rising plume of hot magma would melt its way through the continental crust, producing the viscous silicic (acidic) magmas that lead to explosive volcanism.

The unusual chemical and physical features in the KT boundary layers are present worldwide. An asteroid impact could have propelled material into the stratosphere, where it would have been transported around the globe. On the other hand, Richard B. Stothers and his co-workers at the National Aeronautics and Space Administration's Goddard Space Flight Center in Greenbelt, Md., modeled the manner in which fountains of lava, such as those from Kilauea in Hawaii, expel dust and ejecta. When scaled up to the dimensions of the Deccan volcanism, their models predict that large amounts of material should also be lofted into the stratosphere. Atmospheric circulation would distribute material rather evenly between the two hemispheres, no matter where it was originally emitted.

The appalling consequences of an asteroid impact and a massive volcanism would be quite similar. The first effect would have been darkness resulting from large amounts of dust (either impact ejecta or volcanic ash) into the atmosphere. The darkness would have halted photosynthesis, causing food chains to collapse. Such environmental trauma appears to be reflected in the fossil record. Freshwater creatures were much less affected than land- or sea-based ones, perhaps because freshwater animals did not feed on vascular plants (as do many land-dwelling animals) or on photosynthetic plankton (an important food source for marine vertebrates that was devastated at the end of the Cretaceous).

Life would also have been confronted by large-scale toxic acid rain. The heat of a large impact would have triggered chemical reactions in the atmosphere that would in turn produce nitric acid. Alternatively, volcanic eruptions would have emitted sulfur that would form sulfuric acid in the air. The environmental effects of sulfur-rich volcanism can be significant even in the case of fairly moderate eruptions. The 1783 eruption at Laki, Iceland, killed 75 percent of all livestock and eventually 24 percent of the country's population, even though it released only 12 cubic kilometers of basaltic lava. The event was followed by strange dry fogs and an unusually cold winter in the Northern Hemisphere.

Using the Kilauea eruption as a model, Terrence M. Gerlach of Sandia National Laboratory in Albuquerque estimated that the Deccan Traps injected up to 30 trillion tons of carbon dioxide, six trillion tons of sulfur and 60 billion tons of halogens (reactive elements such as chlorine and fluorine) into the lower atmosphere over a few hundred years. The emissions from the Laki eruption seem to have been far greater than would be expected from simply scaling up the figures for Kilauea, so the estimates may represent a lower limit. Airborne sulfur and dust from a 1,000-cubic-kilometer lava flow could decrease average global temperatures by three to five degrees Celsius (five to nine degrees Fahrenheit).

Other factors could contribute to an opposite effect, however. Marc Javoy and Gil Michard, both of the Institute of Physics of the Earth and the University of Paris, propose that sulfur dioxide from Deccan volcanoes turned the ocean surface acidic, killing the algae that normally extract carbon dioxide from the atmosphere and then carry it to the ocean bottom when they die. Acidic ocean waters also would have dissolved carbonate sediments at the bottom, releasing trapped carbon dioxide. Altogether atmospheric carbon dioxide levels would shoot up to about eight times the present concentration, producing a rise in temperature of five degrees C (nine degrees F). The interaction between cooling from dust and warming from carbon dioxide (which may occur on widely different time scales) is unclear, but the resulting climate gyrations probably would have been especially traumatic for the global ecosystem. Both the asteroid and volcanic hypotheses predict overlapping cooling and warming effects.

So far the evidence discussed has been equally consistent with both hypotheses. But many details suggest that the mass extinction and odd physical processes that occurred at the end of the Cretaceous took place over hundreds of thousands of years. This period is comparable to the duration of Deccan volcanism but incompatible with a sudden asteroid impact.

A number of paleontologists have pointed out that the extinction at the end of the Cretaceous was not a single, instantaneous event. Extinction rates appear to have started to increase up to a million years before the KT boundary. Even near the boundary, the pattern is not uniform: for instance, planktonic foraminifera and nanoplankton (microscopic calcareous algae) species exhibit different patterns of extinction and recovery. This ragged sequence is known as stepwise mass extinction.

One of the most thorough recent studies of the pattern of extinctions was conducted by Gerta Keller of Princeton University. When she analyzed the well-preserved sections of the KT boundary in Tunisia and Texas, Keller found evidence for a first phase of extinction (also seen in the macrofossil record) that began 300,000 years before the KT iridium event and for another extinction event that took place 50,000 years after the boundary. Keller attributes the first event to falling sea levels and global cooling.

Other evidence confirms that the earth experienced not one but many disruptions at the end of the Cretaceous. Abrupt change occurred, for example, in the abundance of carbon 13 and oxygen 18 (respectively, light and heavy versions of these elements, whose concentrations vary according to the ocean temperature and acidity and to the number of living creatures present). Extinctions and carbon 13 fluctuations observed in strata in Spain occur in magnetic intervals that fit the same normal-reversed-normal polarity pattern found in the Deccan Traps.

Even the iridium appears to display a number of fine fluctuations near the KT boundary. Robert Rocchia and his colleagues at the Atomic Energy Commission and National Center for Scientific Research in Gif-sur-Yvette and Saclay, France, found secondary iridium peaks above and below the primary iridium layer (corresponding to time intervals of about 10,000 years) in KT boundary clay in Spain and Denmark. Rocchia, I and our colleagues found that the layer of iridium enrichment in Gubbio, Italy, seems spread over about 500,000 years. The much discussed shocked quartz crystals exhibit a similar pattern of distribution. Officer and Carter discovered that shocked minerals extend through four meters of the Gubbio section, again corresponding to a time span of about 500,000 years.

James C. Zachos of the University of Rhode island and his co-workers measured the chemical composition of microscopic fossils from the North Pacific seafloor and found that the productivity of open-sea marine life was suppressed at the time of the KT boundary and for about 500,000 years thereafter. They also concluded that significant environmental changes, including cooling, began at least 200,000 years before the boundary.

Some proponents of the impact theory, most prominently Piet Hut of the Institute for Advanced Study in Princeton, N. J., and his colleagues, quickly substituted a series of comet impacts for the single asteroid impact to explain these findings. The search for an all-encompassing answer also led to the suggestion that the Deccan Traps might mark the site of the asteroid impact, but there are many difficulties with that idea. No traces of an impact have been found in India. Robert S. White of the University of Cambridge has shown that large impacts cannot trigger massive volcanism, because the section of the mantle just below the lithosphere (the relatively rigid crust and upper mantle) does not normally contain large reserves of molten rock. Moreover, Deccan volcanism started during a normal geomagnetic interval, a few hundred thousand years before the reversed magnetic interval containing the KT iridium anomaly and the clay layer.

During the Cretaceous period, volcanism increased, the sea level rose and fell drastically and the global mantle shifted significantly. The Cretaceous period and the one that preceded it, the Jurassic, were also times of major continental breakups. Between 120 and 85 million years ago, the earth's magnetic field did not undergo a single magnetic reversal, but 15 to 20 million years before the KT boundary, the field started reversing again. Reversal frequency, which indicates activity in the earth's core and at the core-mantle boundary, has increased regularly since then to about once every 250,000 years at present.

All these features can be related to an episode of energetic mantle convection that began tens of millions of years before the KT boundary. To me, the existence of overlapping short- and long-term geodynamic, geologic and paleontological anomalies points to a common internal cause.

What might that cause be? A likely answer comes from the theory of mantle hot spots, developed most prominently by W. Jason Morgan of Princeton University and others. Peter L. Olson and Harvey Singer of Johns Hopkins University developed a model that may explain these regions of persistent volcanic activity. A plume of hot, low-density and low-viscosity material rises from the lowermost parts of the mantle, forming a quasi-spherical head as it pushes its way through cooler, thicker mantle. The head keeps growing as long as it is fed by a conduit of molten rock rising from below.

White and Dan I. McKenzie, also of Cambridge, along with Mark Richards and Robert A. Duncan of Oregon State University and myself, think that as a hot mantle plume rises, the crust above the plume lifts and stretches, leading to continental rifting [see "Volcanism at Rifts," by Robert S. White and Dan P. McKenzie; *Scientific American,* July, 1989]. The plume material decompresses as it reaches the surface and so melts rapidly (in less than one million years). The head of the plume would elevate a large area of crust, so that when the magma finally breaks through to the surface, it runs rapidly downhill, producing extensive flows.

The Deccan eruptions could have followed the arrival of such a head at the base of the lithosphere. Volcanism from a hot plume would be rapid and highly episodic. Individual flows would be extruded in days or weeks; the next flow would follow years to thousands of years later. The far-reaching ecological consequences of each flow could explain the stepwise mass extinctions.

The giant mantle plume that produced the Deccan Traps should have left structural and dynamic relics. In 1987 the Ocean Drilling Program, led by Duncan, explored and dated an undersea chain of volcanoes that extends from southwest India, near the Deccan Traps, to Réunion, the active volcano east of Madagascar. Réunion is a hot spot volcano—one powered by a deep, rising flow of hot magma from the mantle—that burned its way through the Indian and African continents as they drifted over it. The ages of the Réunion seamounts increase steadily from zero to two million years around Réunion itself to 55 to 60 million years just south of the Deccan Traps.

Richards, Duncan and I believe that the Réunion hot spot may represent the tail of hot magma that would be expected to follow in the wake of the plume that produced the traps. Besse, Vandamme and I verified that the mantle hot spot now beneath Réunion was located precisely under the Deccan Traps at the end of the Cretaceous. There is no trace of the hot spot from before the KT boundary; the episode of violent Deccan volcanism appears to mark the appearance of the hot spot at the surface of the earth.

The internal geologic activity associated with a rising mantle plume fits the behavior of the earth's magnetic field at the time of the KT boundary. Slow convection of the molten iron in the earth's outer core—10 kilometers per year—is thought to produce the earth's magnetic field. Instabilities at the boundary between the core and the mantle above it may cause magnetic reversals.

Heat escaping from the core raises the temperature and so lowers the density of material in the deepest layer of the mantle (called the D"), which grows thicker until it becomes unstable and forms rising plumes of magma. Long durations with few or no magnetic reversals, such as the span from 120 to 85 million years ago, indicate a lack of outer core activity and the growth of the D" layer.

About 80 million years ago the layer broke up, sending enormous hot magma plumes upward. At this point, flow of heat from the core to the mantle would have increased, and magnetic reversals would have resumed. At typical mantle velocities of about one meter a year, the plumes would have traveled a few million years before reaching the surface, where their sudden decompression of the plumes would have led to explosive volcanism followed by large lava flows. Smaller, secondary plumes would not have reached the surface but could have accelerated mantle convection, seafloor spreading, sea level changes and other geologic disruptions that took place during the Cretaceous.

This kind of geologic upheaval may be a natural consequence of the fact that the earth is an active, complex heat engine composed of layers that have vastly different physical and chemical properties. Smooth, well-regulated mantle convection and brutal, plumelike instabilities are perhaps just two extremes of the ways in which the earth's internal heat escapes to the outside.

If this is indeed the way the earth functions, similar catastrophes should have taken place. In fact, most major, relatively recent extinction events (those since the Mesozoic era began 250 million years ago) seem to correlate in time with a large flood basalt eruption. Interestingly, the longest known period during which the earth's magnetic field did not reverse also ended with the largest mass extinction, the one that marked the dawn of the Mesozoic era. More than 95 percent of marine species disappeared at that time. The 250-million-year-old Siberian Traps are a prime candidate for having caused this extinction.

Both the asteroid impact and volcanic hypotheses imply that short-term catastrophes are of great importance in shaping the evolution of life. This view would seem to contradict the concept of uniformitarianism, a guiding principle of geology that holds that the present state of the world can be explained by invoking currently occurring geologic processes over long intervals. On a qualitative level, volcanic eruptions and meteorite impacts happen all the time and are not unusual. On a quantitative level, however, the event witnessed by the dinosaurs is unlike any other of at least the past 250 million years.

Magnetic reversals in the earth's core and eruptions of large plumes in the mantle may be manifestations of the fact that the earth is a chaotic system. Variations in the frequency of magnetic reversals and breakup of continents over the past few hundred million years hint that the system may be quasi-periodic: catastrophic volcanic episodes seem to occur at intervals of 200 million years, with lesser events spaced some 30 million years apart.

It is tempting to speculate that the dawn of the Paleozoic era 570 million years ago, when multicellular life first appeared, might have coincided with one such episode. Large extinctions abruptly open broad swaths of ecological space that permit new organisms to develop. Events that at first seem to have been disasters may in fact have been agents essential in the evolution of complex life.

The Extinction of the Dinosaurs

A Dialogue between a Catastrophist and a Gradualist

Dale A. Russell and Peter Dodson

The concept of extinction as a significant process in the history of life emerged early in the nineteenth century with the realization that many organisms in the fossil record were no longer living (Rudwick 1985). Georges Cuvier, studying invertebrate and vertebrate fossils of the Paris Basin, observed abrupt changes between organic remains preserved in succeeding sedimentary series. He postulated the intervention of violent events or catastrophes that destroyed one biota and cleared the way for another. Conversely, Charles Lyell saw extensive sedimentary sequences containing slightly differing molluscan assemblages as demonstrating gradual change on a time scale of millions of years. Charles Darwin was deeply impressed with Lyell's uniformitarianism, and formulated his theory of evolution by natural selection in terms of gradual change. The gradualist paradigm was reinforced by numerous subsequent paleontological studies, such as the classic interpretations of the history of horses by Othniel Charles Marsh and William Diller Matthew.

Until recently, the extinction of the dinosaurs was the object of unconstrained speculation rather than systematic study. Many theories were advanced (see review in Dodson and Tatarinov 1990), most of which were from a gradualistic point of view. An early attempt to invoke an extraterrestrial cause for dinosaurian extinction (a nearby supernova; Russell and Tucker 1971) did not generate wide support, but did prepare the conceptual stage for the asteroid/comet impact hypothesis of Alvarez et al. (1980). The latter was based on the discovery of anomalously high concentrations of iridium at the Cretaceous-Tertiary boundary in marine strata, initially near Gubbio, Italy, and later at many localities around the world. The impact extinction hypothesis was immediately and vigorously challenged by Clemens (e.g., Clemens et al. 1981; Archibald and Clemens 1982; Clemens 1982), who was arguing from the fossil vertebrate record. Although it is now widely accepted that an important impact occurred at the end of the Cretaceous (Ward 1995), disagreement continues as to its biological effects (Archibald 1996). The long-enduring debate between gradualism and catastrophism in the history of life has become focused on the demise of the dinosaurs.

Two Scenarios

The dinosaur extinction debate thus evokes two contrasting scenarios. A gradualist scenario envisages a changing-world hypothesis, and draws heavily on the observation that terrestrial vertebrate communities, absent the dinosaurs, show substantial continuity across the Cretaceous-Tertiary boundary. Healthy, diverse dinosaur populations existed some millions of years before the end of the Cretaceous. In response to a variety of biological, physical, climatological, oceanographic, volcanologic, and tectonic factors, no one of which was necessarily decisive, dinosaurs went into decline. At first the decline was barely noticeable. Later, populations were markedly reduced in diversity compared to their earlier states, with reduced populations. They were now more susceptible to extinction by physical causes. The efficient cause of extinction thus may be irrelevant. What is of interest is why they had become so susceptible to extinction after 160 million years of success.

In contrast, the impact extinction scenario postulates that healthy populations of dinosaurs were exterminated by the effects of the collision of a large bolide with the earth. The impact probably produced a variety of severe stresses, the more important of which may have included an interval of planet-wide darkness and acid rain. When terrestrial communities re-formed following the catastrophe, dinosaurs were no longer present.

The two scenarios contrast greatly in nearly every way. It would seem that definitive resolution is to be expected. Was the tempo of dinosaurian extinction restricted to the geologic instant of a bolide impact, or did it span several hundred thousand to several million years? Can the primary data of the dinosaurian record provide insight into the causes of their extinction, or is the extinction of dinosaurs better revealed from other sources of data? The present authors are proponents of opposing interpretations of the extinction of the dinosaurs (for example, compare the catastrophism of Russell [1982, 1984, and 1989] with the gradualism favored by Dodson and Tatarinov [1990]). Our disparate viewpoints nevertheless contain a broad consensus. It is our hope that others will find our points of agreement and disagreement to be as stimulating as we have.

Points of Agreement

1. For more than 160 million years, dinosaurs were the largest land dwelling animals on all of the major land areas of the world. During this time they underwent a major adaptive radiation, and the functional beauty of their skeletal structures is an enduring marvel. The ultimate disappearance of the dinosaurs in no way negates their signal importance in the history of life on land, nor should they serve as cultural icons for failure.

2. In a phylogenetic sense, dinosaurs are not extinct, for birds are theropodan descendants (but see Feduccia 1996 for a dissenting view). For the purposes of this review, however, the term *dinosaur* connotes what cladists might term "non-avian dinosauromorph." We thus (unrepentantly) use a paraphyletic rather than a monophyletic (holophyletic) "Dinosauria." Whatever the scientific merits of the latter, the former is widely understood, and avoids such circumlocutions as "non-avian dinosaur."

3. The terminal-Cretaceous extinctions collectively rate among the five greatest extinctions of all time (Sepkoski 1992). The biotic turnover was planetary in scope, and varyingly affected terrestrial, aquatic and marine ecosystems.

4. Mass extinction is a complex phenomenon. It should not be presumed that all organisms on land and in the seas that became extinct during the Maastrichtian did so for the same reason, or at the same tempo. Demonstration of a cat-

astrophic extinction for one group (e.g., planktonic foraminiferans) does not constitute evidence for catastrophic extinction of another (e.g., salamanders).

5. The dinosaurian record, as presently known, shows a peak in global diversity during early Maastrichtian [the latest stage of the Cretaceous Period—ed.] time (Dodson 1990), suggesting that the causes of the decline of the dinosaurs occurred within the final 3 million years of the Cretaceous.

6. One can seldom make statistically significant statements on trends in dinosaurian diversity because of the limited nature of the skeletal record. Fewer than 1,000 articulated skeletons or partial skeletons of dinosaurs are available worldwide to document the final 10 million years of their existence.

7. Beyond North America, dinosaurs of Maastrichtian age are known from all of the continents of the world except Australia (although a New Zealand occurrence has been documented). Maastrichtian records of uncertain substage correlation include Alaska (Clemens and Nelms 1993), Antarctica (Campanian-Maastrichtian *fidé* Hooker et al. 1991;, Argentina and Bolivia (Gayer et al. 1992), China (Mateer and Chen 19921, Mongolia (Jerzykiewicz and Russell 1991), New Zealand (Wiffen and Molnar 1988, 1989), and Siberia (Nessov and Starkov 1992). Late Maastrichtian records include Belgium and the Crimea (Russell 1982), eastern North America (Russell 1982), Egypt (Barthel and Herrmann-Degen 1981), France and adjacent Spain (Weishampel 1990; Buffetaut and Le Loeuff 1991; Feist 1991; Galbrun et al. 1993), India (Jaeger et al. 1989), Romania (Weishampel et al. 1991), and Siberia (Nessov and Starkov 1992). Many of these records refer to occurrences of fragmentary or disassociated bones; only the Mongolian (Nemegt) badlands have yielded relatively complete skeletons and an adequate sample of the assemblage to which the dinosaurs belonged. However, the terminal-Cretaceous-basal Paleocene (Maastrichtian-Danian) record in Mongolia is interrupted by a sedimentary hiatus. Taken together, these records demonstrate the worldwide survival of dinosaurs only into Maastrichtian time. In no case do they document a temporal series of two or more Maastrichtian assemblages, or a Maastrichtian-Danian succession.

8. By far the most complete record of biotic changes across the Cretaceous-Paleocene boundary in terrestrial environments is confined to the Western Interior of North America. Sedimentation was relatively continuous across the northern portion of this region, and well-sampled fossil assemblages are available in relative abundance.

9. Mesozoic chronofaunas characterized by the dominance of dinosaurs ended with the Cretaceous Period. Dinosaurs, which included all land vertebrates exceeding 25 kilograms in weight, were not a trivial component of the Hell Creek assemblage. There is no evidence that dinosaurian herbivores were in the process of being replaced by large-bodied mammalian herbivores, such as the 500-kilogram *Coryphodon* of the early Cenozoic, during Hell Creek time.

10. Eustatic sea level changes near the end of the Cretaceous did not exceed those which occurred earlier during the dinosaurian era (cf. Haq et al. 1988), and thus were probably not a primary cause of dinosaurian extinctions.

11. It is currently tenable to postulate that the final disappearance of the dinosaurs coincided with a bolide strike (see Hildebrand 1993).

12. There is at present no compelling evidence that any dinosaur survived into Paleocene time (e.g., Rigby et al. 1987; Van Valen 1988), although neither of

us would rule out the possibility. Indeed, one of the most amazing features of the end-Cretaceous extinctions is the absence of a credible record of any dinosaurian taxon from strata of early Tertiary age anywhere.

13. A signal aspect of the extinctions is the survival of many different varieties of organisms in terrestrial ecosystems. Survival patterns place important constraints on the magnitude of any physical stresses involved (e.g., Buffetaut 1990). For example, widespread freezing of soils is thereby precluded. Moreover, the ecology of organisms that survived and of those that became extinct may provide insight into the nature of the environmental stresses that led to the observed extinctions.

Points Supporting Alternative Models of Extinction:
A Catastrophic Decline of Dinosaurs

Dale A. Russell

1. Within the Western Interior of North America (see above), the diversity of large dinosaurs in sediments of Hell Creek and equivalent strata of late Maastrichtian age may not have been as great as in the underlying Horseshoe Canyon Formation and equivalent strata of early Maastrichtian age. However, disarticulated teeth and bones indicate that the diversity of smaller, more derived dinosaurs was comparable (cf. Late Maastrichtian occurrences of *Ornithomimus* and *Struthiomimus* [Russell 1972]; Dromaeosauridae [Carpenter 1982]; *Aublysodon, Richardoestesia,* and Troodontidae [Currie et al. 1990]; cf. *Chirostenotes* [D.A.R., personal observation]; *Stygimoloch* and *Stegoceras* [Goodwin 1989]).

2. Few articulated specimens of large dinosaurs have been collected from relatively limited exposures of the Scollard Formation, a Hell Creek equivalent which outcrops in the Red Deer Valley of Alberta. However, specimens are evidently as abundantly preserved there (4.7 per sq km) as in the highly productive badlands of Campanian age in Dinosaur Provincial Park, Alberta (3.9 per sq km; Béland and Russell 1978).

3. The broad expansion of Hell Creek–age deltas toward the east (Gill and Cobban 1973), warmer climates (Johnson and Hickey 1990), and change to dicot-dominated canopy forests (Wing et al. 1993; K. R. Johnson, personal communication 1993) constitute environmental changes of a magnitude sufficient to produce important changes between the older Campano-Maastrichtian and younger Hell Creek dinosaurian communities. The latter assemblages were possibly representative of more densely forested environments (Russell 1989).

4. It has not been feasible to resolve increments of time of less than about 1 million years using relatively rare dinosaurian fossils (Sheehan et al. 1991). Large changes in leaf, pollen, and spore assemblages (Johnson 1992) are associated with trace-element anomalies linked to a bolide impact and time scales of days to tens to thousands of years. The bolide trace element signature is global in distribution (Hildebrand 1993).

5. Apparently in both marine and terrestrial environments, the Cretaceous ended with an abrupt collapse in green plant productivity associated with the bolide trace-element signature. Marine and terrestrial animals belonging to food chains based on organic detritus tended to dominate post-extinction assemblages.

However, those dependent directly or indirectly on living plant tissues (e.g., dinosaurs on land, and planktonic foraminifera and mosasaurs in the sea) are postulated to have died on time scales consistent with starvation (Arthur et al. 1987; Sheehan and Fastovsky 1992; Olsson and Liu 1993).

A relatively parsimonious interpretation of the foregoing points is that dinosaur-dominated assemblages prospered in the Western Interior of North America until they were altered by regional topographic and climatic changes which began in middle Maastrichtian time. Several million years after they had achieved a new balance regionally, these assemblages were decimated by a catastrophic environmental deterioration resulting from the impact of a comet. Like the bolide trace-element signature, dinosaurian extermination was global in extent.

According to this interpretation, the late Maastrichtian dinosaur record is not special. It was preceded by an interval (Campanian through early Maastrichtian time) for which the record was more complete, generating the illusion of subsequent decline. This interpretation predicts that the extinction of the dinosaurs is unrelated to gradual changes in dinosaurian diversity, or to environmental stresses that would have precluded the survival of many Hell Creek microvertebrates. It also predicts that, should a relatively complete record be documented at other terrestrial sites around the world, no dinosaur-dominated assemblages will ever be found stratigraphically above the bolide signature.

Points Supporting Alternative Models of Extinction:
A Gradual Decline of Dinosaurs

Peter Dodson
1. The record of non-dinosaurian terrestrial and freshwater aquatic vertebrates, including fishes, amphibians, turtles, lizards, champsosaurs, crocodiles, multituberculates, and placental mammals, shows substantial continuity across the Cretaceous-Tertiary boundary (Hutchison and Archibald 1986; Sloan et al. 1986; Sullivan 1987; Archibald and Bryant 1990; Archibald 1996; MacLeod et al. 1997). Plant communities also show continuity (McIver 1991), although significant disruptions have been noted (Johnson et al. 1989; Johnson 1992). These observations suggest that terrestrial communities did not suffer a devastating catastrophe, but responded to changing environmental conditions as noted by various authors (e.g., Sloan et al. 1986; Johnson 1992).

2. Significant environmental changes occurred in the marine realm *during* the Maastrichtian (Barrera 1994; Ward 1995; MacLeod et al. 1997) up to 6 million years prior to the Cretaceous-Tertiary boundary. While some important Maastrichtian extinctions appear to have been abrupt or even catastrophic (particularly those of planktonic foraminifera), others, including those of reef-forming rudistid clams, inoceramid clams, and belemnites, took place up to 3 million years prior to the end of Maastrichtian time (Kauffman 1988; Ward 1990, 1995). Until recently (e.g., Ward et al. 1986), it was thought that ammonites also disappeared at least several hundred thousand years before the end of the Cretaceous. A number of species were then recorded at the Cretaceous-Tertiary boundary, and it can now be claimed that ammonites were victims of a catastrophe (e.g., Marshall 1995). However, it is clear that by the end of the Maastrichtian, ammonites were

already greatly reduced in comparison with their diversity during Campanian or early Maastrichtian time.

The observation that dinosaurs became extinct during the Maastrichtian does not imply that they perished in a global terminal Maastrichtian catastrophe. Many discussions of dinosaur extinction (e.g., Alvarez and Asaro 1990; Courtillot 1990; Glen 1990) have *assumed* that because geological, geochemical, geophysical, and astrophysical evidence has been documented for a terminal-Cretaceous catastrophe, the case for a catastrophic dinosaur extinction is also documented. Does dinosaur extinction best correspond to the planktonic foraminiferan model of catastrophic extinction or to the rudistid/inoceramid model of gradual disappearance?

3. The fossil record of dinosaurs, however incomplete (Dodson 1990; Dodson and Dawson 1991), can provide some insight into the nature of the extinction of dinosaurs. Articulated specimens provide the best taxonomic resolution, but do so at the expense of statistically significant sample sizes. Articulated specimens provide striking evidence of an apparently worldwide decline in dinosaur diversity during the late Maastrichtian. The number of recorded dinosaur genera during this interval is only about 30 percent of that during early Maastrichtian time. In the late Maastrichtian there are about eighteen well-characterized genera of dinosaurs, fourteen of which are from western North America.

Only in western North America can the presence of apparently healthy dinosaurian communities be demonstrated during the late Maastrichtian. Was this region an oasis in a changing world? Sheehan et al. (1991) and Sheehan and Fastovsky (1992) have examined the stratigraphic distribution of isolated dinosaur bones, which are diagnostic only to family level. It has been argued on the basis of this low-resolution component of the fossil record that dinosaur diversity was undiminished during the 2.25-million-year interval represented by the terminal-Cretaceous Hell Creek Formation of Montana. Hurlbert and Archibald (1995) have argued vigorously that the statistics used by Sheehan et al. (1991) are insufficient to support the claimed decline in diversity. The record of articulated dinosaurs suggests that the large dinosaur fauna of Hell Creek was dominated by a few common species *(Edmontosaurus, Triceratops, Tyrannosaurus)*, unlike the more evenly diversified assemblages of the Judith River (Oldman) and Horseshoe Canyon formations of late Campanian and early Maastrichtian age from Alberta (Russell 1984; Dodson 1990; Weishampel 1990).

The worldwide terminal-Cretaceous decline in dinosaur diversity is mirrored in the stratigraphic sequence exposed along the Red Deer River of Alberta. This is the only region in the world where three successive dinosaur-bearing formations occur. A tabulation of articulated skeletons from the Judith River to the Horseshoe Canyon to the Scollard Formation suggests a diversity decline from thirty to eighteen to nine genera, respectively (Dodson 1990). Moreover, three important groups of previously successful dinosaurs had disappeared by the late Maastrichtian: the Centrosaurinae (short-frilled ceratopsids), the Lambeosaurinae (crested hadrosaurids), and the Nodosauridae (armored dinosaurs). Were these dinosaurs, like rudistids, inoceramids, and belemnites in the seas, bellwethers of increasing environmental stress?

4. In the Pyrenees, dinosaurs are reported to have disappeared from the fossil record between 1 million and 350,000 years before the end of the Cretaceous (Hansen 1990, 1991; Galbrun et al. 1993). Interestingly, reports of diversity

trends in dinosaur egg assemblages show a parallel decline in diversity. In the Aix Basin in southeastern France, five egg types in the early Maastrichtian (Rognacian) are succeeded by two egg types in the middle Maastrichtian, but only one egg type survives into the late Maastrichtian (Vianey-Liaud et a1. 1994). Erben et al. (1979) failed to note the taxonomic change in egg types through the Maastrichtian, mistakenly attributing all to *Hypselosaurus.* They thereby falsely concluded that eggshell thinning contributed to dinosaur extinction. In the Nanxiong Basin, Guandong Province, in southwestern China, twelve egg types are documented in the early Maastrichtian, but only a single type survives to the Cretaceous-Tertiary boundary, where it becomes extinct (Zhao 1994).

5. Gradual extinction resulted from a variety of environmental stresses, no one of which was in itself sufficient to cause the extinction of the dinosaurs. In western North America, such stresses included a general trend toward cooling temperatures (notwithstanding a short-term peak in temperature at the boundary itself), decreasing equability (Axelrod and Bailey 1968), the draining of the epicontinental seaway, orogeny, and volcanism (Courtillot 1990; Hansen 1990, 1991). Even biological factors such as temperature-dependent sex determination (Paladino et a1. 1989) may have been a contributing factor, if ecological segregation of upland dinosaur breeding grounds (Horner 1984; Horner and Gorman 1988) from those of lowland rattles and crocodiles was maintained.

6. A logical sequel of the above points is the prediction that dinosaur dominated assemblages of post-Cretaceous age may be found at low latitude. If the bolide impact scenario is correct, it is conceivable that dinosaurs survived at high latitude, far removed from the hypothesized Yucatan impact site.

The fossil record is consistent with a gradual worldwide disappearance of the dinosaurs. Yet most of the discussion of dinosaur extinction is centered on western North America, because here alone do vertebrate bearing sediments of Paleocene age overlie latest Cretaceous dinosaur bearing strata. It can be argued that if sections were better sampled, or if new boundary sections were found (perhaps in Argentina, China, Thailand, Antarctica, or elsewhere), a different picture of dinosaur extinction would emerge. On the basis of what is currently known, however, the overwhelming pattern of continuity of the terrestrial biota is too great to be compatible with fully apocalyptic accounts of a bolide impact scenario. Global wildfire (Wohlback et al. 1988) seems too far-fetched to be credible, and the survival of pH-sensitive aquatic vertebrates argues against acid rain as a major factor in extinction (D'Hondt et al. 1994). The possibility remains that a bolide impact constituted a coup de grâce for the last surviving dinosaur populations.

We have been pleasantly surprised to discover a broad area of common agreement. We concur that the latest Cretaceous dinosaurian record is far too incomplete to support either the catastrophic or the gradualistic model in a statistically meaningful manner. Indeed, Raup and Jablonski (1993) find that even the record of late Maastrichtian marine bivalve assemblages is too small to support statistical analyses. We differ in our assessment of which data are of greater significance. As we have described them, the two extinction models surely exist only in our imaginations. The truth lies in nature, which through the scientific method continually reveals ever fascinating constellations of data which render the pursuit of scientific knowledge so enjoyable. We are confident that nature has much yet to teach us about the extinction of the dinosaurs.

References

Alvarez, L. W; W. Alvarez, F. Asaro; and H. V. Michel. 1980. Extraterrestrial cause for the Cretaceous-Tertiary extinction. *Science* 208: 1095–1 108.

Alvarez, W, and F. Asaro. 1990. An extraterrestrial impact. *Scientific American* 263 (4): 78–84.

Archibald, J. D. 1996. *Dinosaur Extinction and the End of an Era.* New York: Columbia University Press.

Archibald, J. D., and L. J. Bryant. 1990. Differential Cretaceous/Tertiary extinctions of nonmarine vertebrates: Evidence from northeastern Montana. Geological Society of America Special Paper 247: 549–562.

Archibald, J. D., and W. A. Clemens. 1982. Late Cretaceous extinctions, *American Scientist* 70: 377–385.

Arthur, VI. A.; J. C. Zachos; and D. S. Jones. 1987. Primary productivity and the Cretaceous/Tertiary boundary event in the oceans. *Cretaceous Research* 8: 43–54.

Axelrod, D. 1., and H. P. Bailey. 1968. Cretaceous dinosaur extinction. *Evolution* 22:595–611.

Barrera, E. 1994. Global environmental changes preceding the Cretaceous-Tertiary boundary: Early-late Maastrichtian transition. *Geology* 22: 877–880.

Barthel, K. W, and W Herrmann-Degen. 1981. Late Cretaceous and Early Tertiary stratigraphy in the Great Sand Sea and its SE margins (Farafra and Dakhla Oases), SW Desert. Egypt. *Mitteilungen den Bayerischen Staatssammlung für Paleontologie and Historische Geologie* 21: 141–182.

Beland, P., and D. A. Russell. 1978. Paleoecology of Dinosaur Provincial Park (Cretaceous), Alberta, interpreted from the distribution of articulated remains. *Canadian Journal of Earth Sciences* 15: 1012–1024.

Buffetaut, E. 1990. Vertebrate extinctions and survival across the Cretaceous-Tertiary boundary. *Tectonophysics* 171: 337–345.

Buffetaut, E., and J. Le Loeuff. 1991. Late Cretaceous dinosaur faunas of Europe: Some correlation problems. *Cretaceous Research* 12: 159–176.

Carpenter, K. 1982. Baby dinosaurs from the Late Cretaceous Lance and Hell Creek formations and a description of a new species of theropod. *University of Wyoming Contributions to Geology* 20: 123–134.

Clemens, W. A. 1982. Patterns of extinction and survival of the terrestrial biota during the Cretaceous/Tertiary transition. Geological Society of America Special Paper 190: 407–4t3.

Clemens, W. A.; J. D. Archibald, and L. J. Hickey. 1981. Out with a whimper not a bang. *Paleobiology* 7: 293–298.

Clemens, W A., and L. G. Nelms. 1993. Paleoecological implications of Alaskan terrestrial vertebrate fauna in latest Cretaceous time at high paleolaritudes. *Geology* 21: 503–506.

Courtillot, V. E. 1990. A volcanic eruption. *Scientific American* 263 (4): 85–92.

Currie, P. J.; J. K. Rigby; and R. E. Sloan. 1990. Theropod teeth from the Judith River Formation of southern Alberta, Canada. In K. Carpenter and P. J. Currie (eds.), *Dinosaur Systematics: Perspectives and Approaches,* pp. 107–125. Cambridge: Cambridge University Press.

D'Hondt, S.; M. E. Q. Pilson; H. Sigurdsson; and S. Carey. 1994. Surface-water acidification and extinction at the Cretaceous-Tertiary boundary. *Geology* 22: 983–986.

Dodson, P. 1990. Counting dinosaurs: How many kinds were there? *Proceedings of the National Academy of Science* U.S.A. 87: 7608–7612.

Dodson, P., and S. D. Dawson. 1991. Making the fossil record of dinosaurs. *Modern Geology* 16: 3–15.

Dodson, P., and L. P. Tatarinov. 1990. Dinosaur extinction. In D. B. Weishampel, P. Dodson, and H. Osmolska (eds.), *The Dinosauria,* pp. 55–62. Berkeley: University of California Press.

Erben, H. K.; J. Hoefs; and K. H. Wedepohl. 1979. Paleobiological and isotopic studies of eggshells from a declining dinosaur species. *Paleobiology* 4: 380–414.

Feduccia, A. 1996. *The Origin and Evolution of Birds.* New Haven, Conn.: Yale University Press.

Feist, M. 1991. Charophytes at the Cretaceous-Tertiary boundary. Geology Society of America Abstracts with Programs 23 (5): A358.

Galbrun, B.; M. Feist; F. Columbo; R. Rocchia; and Y. Tambareau. 1993. Magnetostratigraphy and biostratigraphy of Cretaceous-Tertiary continental deposits, Ager Basin, Province of Lerida, Spain. *Palaeogeography, Palaeoclimatology, Palaeoecology* 102: 41–52.

Gayer, M.; L. G. Marshall; and T. Sempere. 1992. The Mesozoic and Paleocene vertebrates of Bolivia and their stratigraphic context: A review. *Revista Techica de Yacimientos Petroliferos Fiscales de Bolivia* 12 (3–4): 393–433.

Gill, J. R., and W. A. Cobban. 1973. Stratigraphy and geologic history of the Montana Group and equivalent rocks, Montana, Wyoming and North and South Dakota. U.S. Geological Survey Professional Paper 776: 1–37.

Glen, W. 1990. What killed the dinosaurs? *American Scientist* 78: 354–369.

Goodwin, M. B. 1989. New occurrences of pachycephalosaurid dinosaurs from the Hell Creek Formation, Garfield County, Montana. *Journal of Vertebrate Paleontology* 9 (Supplement to no. 3): 23A.

Hansen, H. J. 1990. Diachronous extinctions at the KIT boundary: A scenario. Geological Society of America Special Paper 247: 417–424.

Hansen, H. J. 1991. Diachronous disappearance of marine and terrestrial biota at the Cretaceous-Tertiary boundary. *Contributions from the Paleontological Museum University of Oslo* 364: 31–32.

Haq, B. U.; J. Hardenbol; and P. R. Vail. 1988. Mesozoic and Cenozoic chronostratigraphy and cycles of sea-level change. Society of Economic Paleontologists and Mineralogists Special Publication 42: 71–108.

Hildebrand, A. R. 1993. The Cretaceous/Tertiary boundary impact (or the dinosaurs didn't have a chance). *Journal of the Royal Society of Canada* 87: 77–118.

Hooker, J. J.; A. C. Milner; and S. E. K. Sequeira. 1991. An ornithopod dinosaur trom the Late Cretaceous of west Antarctica. *Antarctic Research* 3: 331–332.

Horner, J. R. 1984. Three ecologically distinct vertebrate faunal communities from the Late Cretaceous Two Medicine Formation of Montana, with discussion of evolutionary pressures induced by interior seaway fluctuations. Montana Geological Society 1984 Field Conference, Northwestern Montana, pp. 299–303.

Horner, J. R., and J. Gorman. 1988. *Digging Dinosaurs.* New York: Workman.

Hurlbert, S. H., and J. D. Archibald. 1995. No statistical evidence for sudden (or gradual) extinction of dinosaurs. *Geology* 23: 881–884.

Hutchison, J. H., and J. D. Archibald. 1986. Diversity of turtles across the Cretaceous/Tertiary boundary in northeastern Montana. *Palaeogeography, Palaeoclimatology, Palaeoecology* 55: 1–22.

Jablonski, D. 1991. Extinctions: A paleontological perspective. *Science* 253: 754–757.

Jaeger, J. J.; V. Courtillot; and P. Tapponier. 1989. Paleontological view of the ages of the Deccan Traps, the Cretaceous/Tertiary boundary, and the India-Asia collision. *Geology* 17: 316–319.

Jerzykiewicz, T., and D. A. Russell. 1991. Late Mesozoic stratigraphy and vertebrates of the Gobi Basin. *Cretaceous Research* 12: 345–377.

Johnson, K. R. 1992. Leaf-fossil evidence for extensive floral extinction at the Cretaceous-Tertiary boundary, North Dakota, USA. *Cretaceous Research* 13:91–117.

Johnson, K. R., and L. J. Hickey. 1990. Megafloral change across the Cretaceous/Tertiary boundary in the northern Great Plains and Rocky Mountains, U.S.A. Geological Society of America Special Paper 247: 433–444.

Johnson, K. R.; D. L. Nichols; M. Attrep; and C. J. Orth. 1989. High-resolution leaf-fossil record spanning the Cretaceous/Tertiary boundary. *Nature* 340:708–711.

Kauffman, E. G. 1988. The dynamics of marine stepwise mass extinction. *Revista Espanola de Paleontologia extraordinario:* 54–71.

MacLeod, N.; P. F. Rawson; P. L. Forey; F. T. Banner; M. K. Boudagher-Fadel; P. R. Brown; J. A. Burnett; P. Chambers; S. Culver; S. E. Evans; C. Jeffery; M. A. Kaminski; A. R. Lord; A. C. Milner; A. R. Milner; N. Morris; E. Owen; B. R. Rosen; A. B. Smith; P. D. Taylor; E. Urquhart; and J. R. Young. 1997. The Cretaceous-Tertiary biotic transition. *Journal of the Geological Society, London* 154: 265–292.

Marshall, C. R. 1995. Distinguishing between sudden and gradual extinctions in the fossil record: Predicting the position of the Cretaceous-Tertiary iridium anomaly using the ammonite fossil record on Seymour Island. Antarctica. *Geology* 23:731–734.

Mateer, N. J., and P. J. Chen. 1992. A review of the nonmarine Cretaceous-Tertiary transition in China. *Cretaceous Research* 13: 81–90.

McIver, E. E. 1991. Floristic change in the northern deciduous forests of western Canada during the Maastrichtian and Paleocene. Geological Society of America Abstracts with Program 23 (5): A358.

Nessov, L. A., and A. I. Starkov. 1992. Cretaceous vertebrates from the Gusino-ozerskaia Basin of Transbaikalia and their value for determining the age and depositional environment of the sediments. *Geologiia i geofizika,* no. 6: 10–19. (In Russian.)

Olsson, R. K., and C. J. Liu. 1993. Controversies on the placement of the Cretaceous-Paleocene boundary and the K/P mass extinction of planktonic formanifera. *Palaios* 8: 127–139.

Paladino, E V.; P. Dodson; J. K. Hammond, and J. R. Spotila. 1989. Temperature-dependent sex determination in dinosaurs: Implications for population dynamics and extinction. Geological Society of America Special Publication 238:63–70.

Raup, D. M., and D. Jablonski. 1993. Geography of end-Cretaceous marine bivalve extinctions. *Science 260:* 971–973.

Rigby, J. K. Jr.; K. R. Newman; J. Smit; S. Van der Kars; R. E. Sloan; and J. K. Rigby. 1987. Dinosaurs from the Paleocene part of the Hell Creek Formation, McCone County, Montana. *Palaios* 2: 296–302.

Rudwick, M. J. S. 1985. *The Meaning of Fossils.* Chicago: University of Chicago Press.

Russell, D. A. 1972. Ostrich dinosaurs from the Late Cretaceous of North America. *Canadian Journal of Earth Sciences* 9: 375–402.

Russell, D. A. 1982. A paleontological consensus on the extinction of the dinosaurs. Geological Society of America Special Paper 190: 401–405.

Russell, D. A. 1984. The gradual decline of the dinosaurs: Fact or fallacy? *Nature* 307: 360–361.

Russell, D. A. 1989. *An Odyssey in Time: the Dinosaurs of North America.* Toronto: University of Toronto Press.

Russell, D. A., and W. Tucker. 1971. Supernovae and the extinction of the dinosaurs. *Nature* 229: 553–554.

Sepkoski, J. J. 1992. Phylogenetic and ecologic patterns in the Phanerozoic history of marine biodiversity. In N. Eldredge (ed.), *Systematics, Ecology and the Biodiversity Crisis,* pp. 77–100. New York: Columbia University Press.

Sheehan, P. M., and D. E. Fastovsky. 1992. Major extinctions of land-dwelling vertebrates at the Cretaceous-Tertiary boundary, eastern Montana. Geology 20:556–560.

Sheehan, P. M.; D. E. Fastovsky; R. G. Hoffmann; C. B. Berghaus; and D. L. Gabriel. 1991. Sudden extinction of the dinosaurs: Latest Cretaceous, upper Great Plains, U.S.A. *Science* 254: 835–839.

Sloan, R. E.; J. K. Rigby Jr.; L. M. Van Valen; and D. Gabriel. 1986. Gradual dinosaur extinction and simultaneous ungulate radiation in the Hell Creek Formation. Science 232: 629–633.

Sullivan, R. M. 1987. A reassessment of reptilian diversity across the Cretaceous-Tertiary boundary. *Natural History Museum of Los Angeles County Contributions to Science* 391: 1–26.

Van Valen, L. M. 1988. Paleocene dinosaurs or Cretaceous ungulates in South America. *Evolutionary Monographs 10:* 1–79.

Vianey-Liaud, M.; P. Mallan; O. Buscail; and C. Montgelard. 1994. Review of French dinosaur eggshells: Morphology, structure, mineral and organic composition. In K. Carpenter, K. F. Hirsch, and J. R. Horner (eds.),

Dinosaur Eggs and Babies, pp. 151–183. Cambridge: Cambridge University Press.

Ward, P. D. 1990. The Cretaceous/Tertiary extinctions in the marine realm: A 1990 perspective. Geological Society of America Special Paper 247: 425–432.

Ward, P. D. 1995. After the fall: Lessons and directions from the K/T debate. Palaios 10: 530–538.

Ward, P. D.; J. Wiedmann; and J. F. Mount. 1986. Maastrichtian molluscan biostratigraphy and extinction patterns in a Cretaceous/Tertiary boundary section exposed at Zumaya, Spain. *Geology* 14: 899–903.

Weishampel, D. B. 1990. Dinosaurian distribution. In D. B. Weishampel, P. Dodson, and H. Osmólska (eds.), *The Dinosauria,* pp. 63–139. Berkeley: University of California Press.

Weishampel, D. B.: D. Grigorescu; and D. B. Norman. 1991. The dinosaurs of Transylvania. *National Geographic Research and Exploration* 7: 196–215.

Wiffen, J., and R. E. Molnar. 1988. First pterosaur from New Zealand. *Alcheringa* 12: 53–59.

Wiffen, J., and R. E. Molnar. 1989. An Upper Cretaceous ornithopod from New Zealand. *Geobios 22:* 531–536.

Wing, S. L.; L. J. Hickey; and C. C. Swisher. 1993. Implications of an exceptional fossil flora for Late Cretaceous vegetation. *Nature* 363: 342–344.

Wohlback, W. S.; I. Gilmour; E. Anders; C. J. Orth; and R. R. Brooks. 1988. Global fire at the Cretaceous-Tertiary boundary. *Nature* 334: 665–669.

Zhao, Z. 1994. Dinosaur eggs in China: On the structure and evolution of eggshells. In K. Carpenter, K. F. Hirsch, and J. R. Horner (eds.), *Dinosaur Eggs and Babies,* pp. 184–203. Cambridge: Cambridge University Press.

Document Credits and Acknowledgments

Documents listed with no credits or acknowledgments are public domain.

Chapter Four

Url Lanham, "Cope's Revenge." From *The Bone Hunters,* Url Lanham. Reprinted with
permission. Copyright © 1991 by Dover Publications. All rights reserved.

Letter from Earl Douglass to W. J. Holland. Reprinted with permission of The
Carnegie Museum of Natural History, Pittsburgh. From the archives of the
Department of Vertebrate Paleontology.

Chapter Six

Robert T. Bakker, "Dinosaur Renaissance." Reprinted with permission. Copyright ©
1975 by Scientific American, Inc. All rights reserved.

David J. Varricchio, "Warm or Cold and Green All Over." Republished with permission of the Honor Society of Phi Kappa Phi. From *National Forum,* Vol. 78, No.
3. Copyright © 1998. Permission conveyed through Copyright Clearance
Center, Inc.

Chapter Seven

Larry D. Martin, "The Big Flap." Republished with the permission of The New York
Academy of Sciences. From *The Sciences,* Vol. 38, No. 2. Copyright © 1998.
Permission conveyed through Copyright Clearance Center, Inc.

Kevin Padian and Luis M. Chiappe, "Bird Origins." Republished with permission of
Academic Press. From *Encyclopedia of Dinosaurs,* edited by Philip J. Currie and
Kevin Padian. Copyright © 1997. Permission conveyed through Copyright
Clearance Center, Inc.

Chapter Eight

Walter Alvarez and Frank Asaro, "An Extraterrestrial Impact." Reprinted with permission. Copyright © 1990 by Scientific American, Inc. All rights reserved.

Vincent E. Courtillot, "A Volcanic Eruption." Reprinted with permission. Copyright
© 1990 by Scientific American, Inc. All rights reserved.

Dale A. Russell and Peter Dodson, "The Extinction of the Dinosaurs: A Dialogue between a Catastrophist and a Gradualist." Republished with the permission of

Index